Contents

Delmar Publishers

an International Thomson Publishing company ITP®

Albany • Bonn • Boston • Cincinnati • Detroit • London • Madrid
Melbourne • Mexico City • New York • Pacific Grove • Paris • San Francisco
Singapore • Tokyo • Toronto • Washington

NOTICE TO THE READER

Printed in the United States of America

For more information, contact:

Delmar Publishers Inc.
3 Columbia Circle, Box 15015
Albany, New York 12212-5015

International Thomson Publishing Europe
Berkshire House
168-173 High Holborn
London, WC1V7AA
England

Thomas Nelson Australia
102 Dodds Street
South Melbourne, 3205
Victoria, Australia

Nelson Canada
1120 Birchmont Road
Scarborough, Ontario
Canada, M1K 5G4

Delmar Staff:
Publisher: Robert D. Lynch
Editor: Paul Shepherdson
Production Manager: Larry Main
Art & Design Coordinator: Mary Beth Vought

International Thomson Editores
Campos Eliseos 385, Piso 7
Col Polanco
11560 Mexico D F Mexico

International Thomson Publishing GmbH
Königswinterer Str. 418
53227 Bonn
Germany

International Thomson Publishing Asia
221 Henderson Road
#05-10 Henderson Bldg.
Singapore 0315

International Thomson Publishing Japan
Hirakawacho Kyowa Building, 3F
2-2-1 Hirakawa-cho
Chiyoda-ku, Tokyo 102
Japan

1 2 3 4 5 6 7 8 9 XXX 02 01 00 99 98 98 97 96

Library of Congress Cataloging-in-Publication Data

Parker, Richard, 1938-
Basic calculus with Maple / Richard Parker.
p. cm.
Includes index.
ISBN 0-8273-7408-9
1. Calculus--Computer-assisted instruction. 2. Maple (Computer file) I. Title.
QA303.5.C65P37 1997
515'.078'55369--dc21 97-4462
CIP

Prelude to Maple: Guide to Basic Commands

Objectives for This Chapter

This book is about using Maple to enhance your mathematical expertise. Each chapter shows you how to use a handful of Maple commands that bear upon the chapter topic. But before you start using Maple on these topics, you need to understand something about its use. That is the purpose of this introduction.

Maple, like mathematics, is a big undertaking. The best way to tackle such a large project is to start small and do a little more each day. That way, you move gradually ahead. You'll begin the process by getting an introduction to Maple here. It is definitely not meant to be a complete tutorial, it is only meant to get you started. The whole book is the tutorial! When you have completed it, you will be able to apply mathematics to a wide range of trigonometric topics and you will have the ability to verify the results of your calculations.

The specific objectives for this section are listed below.

1. Learn the purpose of a computer algebra system, such as Maple
2. Know how to issue a command to Maple
3. Know how to avoid some common pitfalls when issuing commands
4. Confirm your knowledge by taking a quiz

What Is Maple?

Maple is a kind of super-calculator called a *computer algebra system*. If you are using a computer with a graphical user interface, like the Macintosh®, Windows®, or an X-terminal, you will be using Maple's worksheet interface. The worksheet provides an integrated environment in which you can perform mathematical operations, use a simple word processor to explain your calculations, and include Maple plots, as well as allowing you to store formulas, equations, and other symbolic mathematical structures for use later on. Of course, you can do strictly numerical calculations as well. These may sometimes be a little more cumbersome than on your pocket calculator, but there are advantages. You can compute to many more decimal places than on your calculator, and you can save your results on disk or print them as part of a written report. This ability to store your results and use them again later on, perhaps with modifications, is a great advantage once you learn how it's done. Most important, there are techniques for verifying that your calculations are correct. This indispensable step is not nearly as easy on a calculator as it is with Maple.

Computer algebra systems are important tools for doing mathematics. They can help you use mathematics with confidence. The most important step in any problem is checking the answer. Does your answer really satisfy the problem you started out to solve? Using Maple, it is very easy to substitute the solution back into the original equations and see that everything checks, whereas doing this step by hand, even if you use a calculator, sometimes requires so many manipulations that the probability of a slipup becomes depressingly high. There is a temptation to say, "Oh, well, I must have gotten the right answer, so I'll go with it." However, later, after you leave school, there will be a lot more riding on the answer. You will want your calculations to be correct. That means you will want to use computational tools that permit you to record your work and allow you to verify your conclusions.

Mathematics has been under more or less continuous development for 2,500 years. There is an astonishing amount of accumulated knowledge, which the designers of Maple have tried to incorporate. Even so, you can use Maple effectively once you have learned only a handful of commands. This tutorial is meant to get you going. Further examples will be found throughout the book and you can use Maple's own help function to browse through the system's capabilities. Many people begin with the sections on plotting, especially the three-dimensional plot and animation packages. The best advice is—go ahead and enjoy!

Prelude: Before You Begin Maple

You are familiar with the pocket calculator, and you may be familiar with programming languages such as PASCAL, C, or C++. All these use approximations for real numbers. Typically, the accuracy of their numerical engines is about 10 or 12 significant figures. For instance, the difference

$$12345678901234567889 - 12345678901234567888$$

yields zero on my calculator, whereas the answer clearly is 1, as you can see by inspecting the two numbers. This occurs because all computations on these calculators are only approximate. Most calculators will refuse to accept 19-digit numbers, and those that do simply throw away anything after the 12th digit or so.

In contrast, Maple has been designed to treat numbers as *exact*. This will lead to confusion in first-time users, to whom it may seem that Maple is not working out an answer. As an example, take the mind-boggling easy problem of working out 1/3. If you type this into Maple as follows,

> 1/3;

$$\frac{1}{3}$$

you see that you get the same thing, just printed a little nicer as a fraction, the way you would see it printed in a textbook. If you think about this result for a little, you realize that one-third is the *exact* answer, while 0.33333333 is approximate, because in reality, in the decimal representation of one-third, the 3s go on forever.

Some Frequently Asked Questions

1. *Maple is great at algebra, but how do I work with decimal numbers?* One question that very quickly arises is, "How do you get the decimal equivalent to a number that Maple has specified exactly, such as 1/3, or sin(30)?" There is a way! In fact, there are many ways, and you will learn about them in the book. This means that you can use Maple as a calculator as well as doing symbolic manipulations with it. Don't be concerned that Maple tends to write results as fractions rather than decimals. It's easy to convert any of Maple's numbers to decimal form. You will be given many examples in this text.

2. *How do I issue commands to Maple?* Maple is an integrated package. You can use it as a simple word processor as well as a calculator. Therefore, the next question that needs answering is, "How do I tell Maple to compute something?" First, you ensure that you are in Maple input mode—that is, you see the Maple prompt (>) at the beginning of the line—then, you type in the expression you want to compute, and you remember to add a *semicolon* (;) at the end. The semicolon tells Maple to work on what you have typed in. If you don't put in the semicolon, Maple will do nothing! It will wait until you have typed in a semicolon before it attempts to work on what you have typed.

3. *What do I do when things seem to be going wrong?* The third thing to remember is this: if you have been progressing through the calculations yet Maple seems to be giving you the wrong answers or is doing nothing, the best thing to do is to start over. Quit Maple and restart it. Maple tries to remember everything you have done in order

to save time if you choose to repeat a calculation. But if you have typed something that has confused Maple, it will remember that too! This can cause all sorts of problems for new users. The best approach is to start over again rather than digging yourself into a deeper hole.

4. *I hate typing. Won't this slow me down?* The purpose of this book is to show you how to get reliable answers. An important part of the process is to keep typing to an absolute minimum. The book will show you how to use the copy and paste functions instead of retyping mathematical input. This eliminates the chance of error when you need to use the same expression again on a new input line.

5. The fifth and last general point concerns *naming objects in Maple*. Let's say you have typed in a quadratic equation. Later on, you want to use that equation again. Do you have to type it in again? Not if you gave it a name the first time. Once a complicated object has been assigned a name, all you have to do is use that name and Maple will replace the name with the object. Get in the habit of naming every expression you type into Maple. Here is an example of giving something a name. We will name the quadratic equation $6x^2 + 7x - 20 = 0$ by calling it Moe. Here's how:

> Moe := 6*x^2+7*x-20 = 0;

$$Moe := 6x^2 + 7x - 20 = 0$$

You name an object by using the *colon equal* assignment operator (:=). There must *not* be a space between the colon (:) and the equal sign (=). Think of this operator as one word. Furthermore, notice that because you have given a name to an equation, the equation contains an equal sign. The *colon equal* is an assignment operator, whereas the *equal sign* has its ordinary meaning in mathematics. The name *Moe* now stands for this equation. You can verify this by typing in this name on a Maple input line:

> Moe;

$$6x^2 + 7x - 20 = 0$$

Maple looks up the name *Moe* in its memory and sees that the name stands for an equation. It prints out the equation as the "value" of the name.

If you make a typing error and write *moe* instead of *Moe*, Maple responds as follows

> moe;

$$moe$$

because the name *moe* has not been defined. Be careful with all your typing, as Maple is

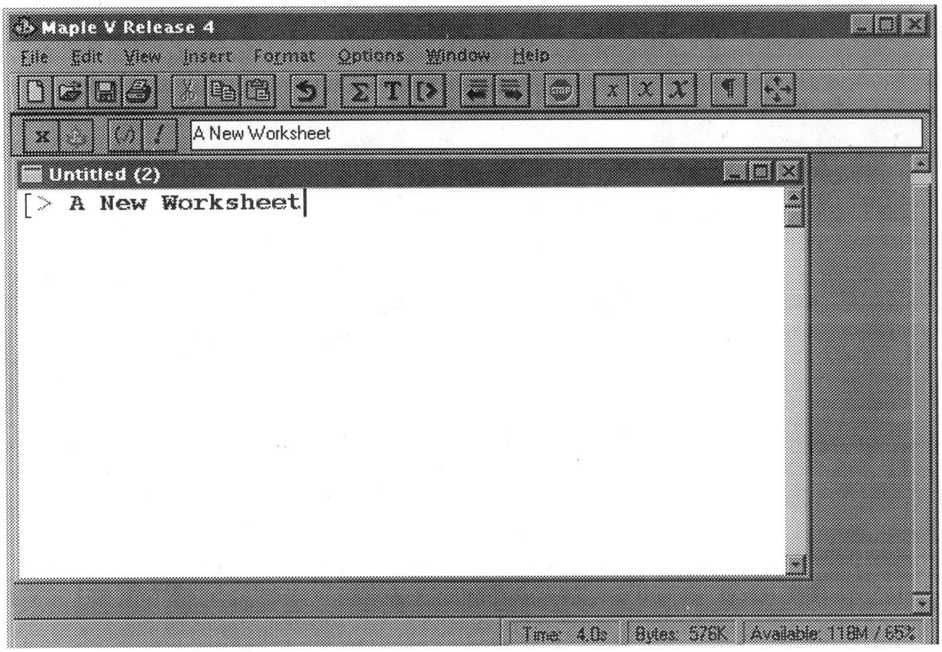

Figure A.1 The Worksheet Interface in Release 4

case sensitive. Clearly, naming objects makes it easy to refer to them again and again, so get in the habit of giving a name to any expression you might ever use again.[1]

The Workss eet

You'll probably start Maple by double clicking on the Maple icon. When you start Maple, the screen will look like Figure A.1 if you are using Release 4.

---

[1] You can reexecute a Maple command line by placing the cursor on a previously executed line and editing it. In particular, you can "wrap" another command "around" the command. Say your command was

>2*Pi*33.2;

You realize when you see the output of this command that you need to wrap an *evalf* around the command. Place the cursor before the *2* and type *evalf(*. Then move to the end of the command and close the brackets. You now have

>evalf(2*Pi*33.2);

Press Return, and the new command replaces the old. You can "build up" commands in this way, which allows you to check that intermediate results are what you want. (The *evalf* function is introduced in detail later in this Prelude.)

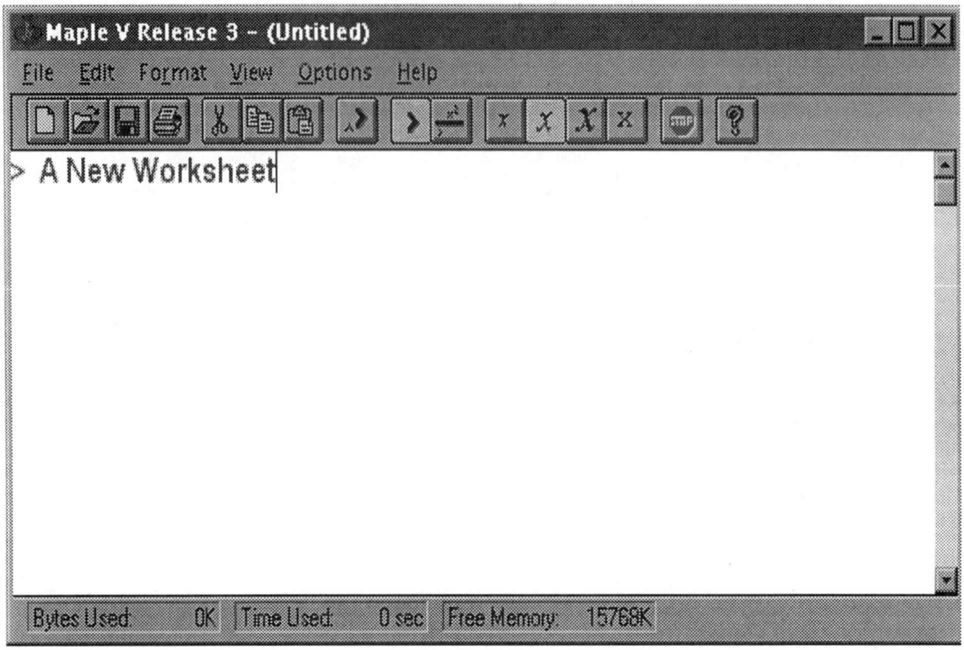

Figure A.2 The Worksheet Interface in Release 3

If you are using Release 3, the screen will look like Figure A.2.

We won't explain the function of all the buttons here. You'll become familiar with them as you work with Maple. If you have Release 4, there is one button that you should use from the beginning. It is the  button on the *context bar*. You have to type mathematical statements on a single line using the keyboard, but clicking on this button allows you to see what you have typed in true mathematical form. You can check your typing and make sure that you have input the correct expression by looking at the typeset result. Release 3 does not have this capability. If you are using Release 3, it is a good idea to type the expression you are going to work with on an input line (terminated by a semicolon and a return) *before* you apply any Maple operations to it. This technique permits you to see the typeset form of your input so you can check it for accuracy. Only when you are convinced that you are working with the right expression should you continue performing other operations on it. After all, if you begin wrong . . . !

Your First Maple Commands

Maple is a calculator, which you can use to do arithmetic. Type these expressions into Maple and observe the results.

Addition

> 3+4;

```
7
```

> 7+11;

```
18
```

> 12345678901234567890 +12345678901234567890;

```
24691357802469135780
```

Be careful when typing in this last command. Notice that the numbers are simply the numerals 1 to 10 typed in twice. In Maple, the result is computed to full accuracy. Pocket calculators don't give 20-digit accuracy.

Let's Practice. Use Maple to add the first 10 odd numbers. Here's the command

> 1+3+5+7+9+11+13+15+17+19; Answer:____________________

Now, find the sum of the first 11 odd numbers. *You don't need to retype!* Copy the previous command and paste it on a new input line. Place the insertion point between the *19* and the semicolon and type + *21*. By copying and pasting, you have used your previous work to eliminate almost all the typing for your new command.

> 1+3+5+7+9+11+13+15+17+19+21; Answer:____________________

Subtraction

> 1234567890123456789 - 1234567890123456788;

```
1
```

Here, the numbers differ by 1 in the units column; otherwise, the digits are the same. The result must be 1, since for any two numbers where one is one more than the other, $n + 1 - n = 1$.

Let's Practice. Subtract these numbers: 1 – (2 – 3) – (4 – 5) – (6 – 7). Check your result by mental arithmetic.

Answer: ____

Multiplication

> **3*4;**

12

> **25*34;**

850

> **1.2*7.4;**

8.88

This last example shows you that if you put a decimal point in the input numbers, Maple does give you a "real," or "floating," point (decimal) answer. More examples will follow under "Subtraction."

> **0.0000003 * 7;**

$.21\ 10^{-5}$

This is an example of how Maple writes an answer using powers of ten notation. Maple usually writes large or small decimal numbers as in this example, but it may use the float notation if the numbers get too small.

> **1234567890 * 1234567890;**

1524157875019052100

Let's Practice. Multiply the first six integers together. Remember to put in the multiplication sign (*).

Answer:____

Division

> **1/3;**

$$\frac{1}{3}$$

> **1/3.;**

.333333333

Do you see the difference in these commands? The first uses whole numbers, and the second uses a "real" number. There is a decimal point after the 3, so Maple interprets this number as 3.0, a decimal number. Once there is a decimal number in the expression, Maple switches to floating-point format rather than treating each number as exact. It gives its answer in decimal form, using ten digits as the default.

For example:

> **355/113; 355. / 113;**

$$\frac{355}{113}$$

.3141592920

The above fraction is an approximation to the number π, which is the ratio of the circumference of a circle to its diameter.

Let's Practice. Do this division problem in Maple: 1/2/3/4. If you are using Release 4, click on the *X* button at the far left of the context bar. Interpret what you see.

Answer: ____

Where are the implied brackets? Are they (a) ((1/2)/3)/4, or (b) 1/((2/3)/4)?

Answer: ____

Exponentiation

To raise one number to a power, use the "up arrow" (SHIFT+6) on the keyboard (or use **).

> **3^2;**

9

> **5^2.2;**

34.49324154

> **25^25;**

88817841970012523233890533447265625

These last two examples show that you can use a decimal number as the exponent, and that you can raise numbers to very large powers if you need to.

Let's Practice. Which is bigger, 22^{25} or 25^{24}? Answer: ____________________

Using *evalf* to Convert an Answer to Decimal Form

You know from some of the examples above that you can get Maple to work in floating point form by putting one decimal number in the expression. Just one decimal number will cause Maple to switch to decimal format. In that case, the answer will be approximate, but it will be in the familiar decimal notation. Another way to convert numbers is to use the *evalf* function. You will be using functional notation for many Maple operations.

The general format of a function call in Maple is shown by the *evalf* function. As an example, take the number π. Recall that this number is the ratio between the circumference and the diameter of a circle. If we input this number in Maple, we get:

> **Pi;**

$$\pi$$

Maple understands this input and writes the standard symbol for Pi. If we type:

> **pi;**

$$\pi$$

you might think that Maple treats this *pi* the same, but this is not true! It is important to note that Maple does make a distinction between upper-case letters and lower-case ones. This last result is "the Greek letter *pi*," not the ratio!

To find the decimal approximation for Pi, we use the *evalf* function:

> **evalf(Pi); evalf(pi);**

```
3.141592654
```

$$\pi$$

Pi has a value of approximately 3.14, whereas the Greek letter *pi* has no decimal value.

The Factorial

Maple knows how to find the factorial of a whole number:

> **3!;**

```
6
```

> **7!;**

```
5040
```

> **45!;**

```
119622220865480194561963161495657715064383733760000000000
```

The factorial function certainly grows quickly!

Integer Factorization: The *ifactor* Function

Maple contains many functions that help us work with whole numbers, among which is the *ifactor* function.

> ifactor(5670);

$$(2)\,(3)^4\,(5)\,(7)$$

It gives all the prime factors of a number.

Let's Practice. What is the decimal equivalent of Pi/3? Can you get the answer by typing:

>Pi/3.0?

Answer: ____

Remaindering

This is something that is not easy to do on a calculator. Say you want to find the remainder when 65,537 is divided by 13.

> irem(65537, 13);

4

The remainder is 4. One way to prove this is to take the quotient, using the function *iquo*:

> iquo(65537, 13);

5041

You can check the result by reversing the process:

> 5041*13 + 4;

65537

You can take the remainder of some very large numbers, a thing you cannot do easily using a calculator.

> irem(45!, 53);

32

Exercise: What is the remainder when you divide 45! by 11, by 17, by 44?

Maple's Help Facility. There are many other functions that you can use to investigate integers. This is a good place to introduce Maple's *Help* facility. The syntax is shown in Table A.1. Type a question mark and the name of the thing you want help for. You don't need to use a semicolon in this case. You will get a separate help window. If you don't get the help topic you want the first time, look at the "See Also" section of the Help screen. It may give you a good hint on where to look next. Try the command *?integer* to see the Help window for a list of integer functions (see Table A.1).

> ?integer

Table A.1

Maple Help Window for ?integer

Help For: Integers

Description:

An expression is of type integer if it is an (optionally signed) sequence of one or more digits of arbitrary length. The length limit of an integer is system dependent but generally much larger than users will encounter—typically greater than 500,000 decimal digits.

In addition to arithmetic operators, other basic functions of integers are

abs sign min max factorial

irem iquo modp mods mod

isqrt iroot isprime ifactor ifactors

igcd ilcm igcdex iratrecon rand

There are also many special functions for integers in the numtheory and combinat packages including the binomial coefficients, Fibonacci numbers, Stirling numbers, Jacobi symbol, Euler's totient function, etc.

See Also: type, numeric, constant, numtheory, combinat

Prime Numbers and Factoring

You may recognize some of these functions, while others may be a mystery to you. You can find out about each one by using the Help function again. One of the integer functions is *isprime*. If you ask for help on this topic, you will see that it gives a *true* or *false* answer, depending on whether the number you give it is prime (has no factors) or not:

> ?isprime

> isprime(65537);

```
true
```

The fact that 65,537 is prime is not too hard for you to figure out for yourself if you take a few minutes, but what about a really big number?

> isprime(65537^6 + 1);

```
false
```

How can you show that this number has factors? You would multiply it out, add 1, and then factor it:

> (65537^6 + 1);

```
7923541634588881603819457741 0
```

> ifactor(65537^6 + 1);

```
(2) (5) (18447869995091361793) (49477) (8681)
```

By looking at the expansion of the number, you see it is even. Therefore, it must have a factor of 2. Its digits end in 10, so it must also be divisible by 5. Don't make Maple a substitute for your own analysis of a problem. Use Maple to assist you in doing calculations, but first think about the problem carefully to see if you can grasp the answer directly, or at least make some significant simplifications.

Let's Practice. Which of these numbers is a prime? $2^2 + 1$, $2^3 + 1$, $2^4 + 1$, $2^5 + 1$, $2^6 + 1$, $2^7 + 1$, $2^8 + 1$

Answer: ____

Real Numbers

You have encountered the real number line before. Besides integers and fractions, the real number line contains the so-called irrational numbers like the square root of 2. In Maple, you can obtain a square root by using the *sqrt* function. Let's take the square root of 65,536.

> sqrt(65536);

```
256
```

The original number, 65,536, is a perfect square (the square of 256). What if we attempt to take the square root of 65,537?

> sqrt(65537);

$$\sqrt{65537}$$

Maple simply writes out the number again, using textbook mathematical notation. Note the difference between the *sqrt* function and the *isqrt* function:

> isqrt(65537);

```
256
```

The *isqrt* function gives you the whole number that squares nearest the given number. But what if you want the "answer" to the square root of 65,537? Remember what question you are asking! The square root of 65,537 is simply what Maple gave you: 65,537 under the square root sign. This is the completely precise answer. Anything else is an approximation!

Sometimes, however, it is the numerical approximation that you want, and in this case, remember that there are two ways to get it: either put in the number as a decimal number:

> sqrt(65537.);

```
256.0019531
```

or use the *evalf* function:

> evalf (sqrt(65537));

```
256.0019531
```

Notice that you can "nest" functions (put one function inside another) in Maple. The result of *sqrt* (65,537) has a value that is given to the *evalf* function. You always read such nested functions from the inside out. This is no different from other "functions of functions" that you have encountered earlier in mathematics, such as $f(g(x))$, which means that the function f is evaluated at the point $g(x)$.

Let's Practice. Find the decimal approximations to $\sqrt{3}$, $\sqrt{5}$, $\frac{1+\sqrt{5}}{2}$:

Answer:____

Assignments: Examples of Names in Maple

To use something like the square root of 65,537 repeatedly, you can give it a name. The name you give it will usually be something fairly short, but you must avoid using names that Maple itself uses. For instance, it is a bad idea to use "sqrt" as a name, because Maple uses it for the square root function. Instead use the letters *a, b, c*, and so forth. You can use capital letters as well, but don't use *D* or *I*. Maple uses these as names itself. You will see how to assign a name to an expression in the next section.

The Circumference of a Circle of Radius R. The formula for the circumference is $C = Pi * D$, where C stands for the circumference and D stands for the diameter. We could write, on a Maple input line,

> C = Pi *D;

$$C = \pi D$$

but we have broken the rule of not using capital D in an expression. Also, we have typed in an equation, *but we have not named anything*. To name $Pi * d$ (we are now using a legal variable for the diameter) as the circumference, we would type in:

> C := Pi * d;

$$C := \pi d$$

Notice the difference carefully. We have used the *colon-equals* sign to assign the value *Pi * d* to the name *C*. Read this last sentence repeatedly until the meaning becomes locked firmly in your mind. From now on, *C* is exactly the same as *Pi * d*. To see that this is the case, type *C* on a Maple input line:

> C;

$$\pi d$$

Maple responds by outputting *Pi * d*, because that is what *C* is "equal to." More accurately, *C* is a name for the expression *Pi * d*. Compare this with the equation for a straight line,

> y = m*x + b;

$$y = mx + b$$

> y;

$$y$$

Here, we have typed in an equation without naming it. In this case, *y* has not changed its value; it is still *y*, and is part of an *equation*, whereas *C* is a *name*. It has been *assigned* a value, *Pi * d*, by using the colon-equals assignment statement. Note as well that you can give an equation a name. Say we want to give the equation for a straight line a name:

> a := y = m*x + b ;

$$a := y = mx + b$$

We have used both the colon-equals and the equal signs in the same line. If we now ask Maple what *a* is, we get:

> a;

$$y = mx + b$$

The variable *a* is just the name for the equation, $y = m\,x + b$. To reuse the equation, we can either type out the whole equation or simply use the name.

The Substitute Command

If circle is defined as we have just done, how do we calculate a circumference given a particular diameter? We use the *subs* (substitute) command. In words, we substitute the given value of the diameter into the formula and evaluate. In Maple, we use the substitute command, *subs*.

Exercise: Calculate the circumference of a circle if the diameter is 6.

Solution:

```
> subs( d = 6. , C);
```

$$6.\pi$$

In this case, putting in a decimal value was not enough to get a numerical answer. Instead, we must use *evalf.*

```
> evalf ( subs( d = 6, C) );
```

$$18.84955592$$

The answer is 18.85 units (rounded to the nearest hundredth).

You will be tempted to think that the calculator method is a lot simpler for this problem, and of course you are right. But the method shown here is important in more complex situations, and you will be using it to solve many problems that are not possible to solve on the calculator.

Let's Practice. Given the polynomial $x^3 + 3.6111\,x^2 - 13.30893x + 1.153922$, find its value when $x = 0 - 10.0, -5.9, 0, +0.889, +2.2, 5, 10$. Define the expression, then use the *subs* command.

Answer:____

Algebra

Maple has an extensive set of commands for algebra. This book concentrates on results, not manipulations. One reason for this approach is that most people need to work within a given mathematical framework where the basic formulas are already given in a particular form. I believe that it is more important to be able to use a given expression (or equation) to accomplish goals, which are usually (a) computation, (b) solving equations, (c) plotting, or a combination of all three. Maple is an ideal assistant here. There are other software packages, like Theorist,® that give you control over each algebraic step in a transformation. Such packages are useful if you want to transform expressions from one form to an equivalent one, although Maple can be persuaded to do step-by-step calculations by experienced users.

The computation may be symbolic or numeric. You may need to find the solution to a quadratic equation, or you may want to picture the behavior of a function over a range of values. This means computing the value of a function at a point, or, if a set of points is required, a graph. You may want to get an overview of the function by plotting it and then computing the value of the function at a few points to high accuracy. A plot is the best way of appreciating the behavior of a function over a set of points.

The starting place for all these operations is the function (or expression). Pay extreme attention to brackets when you type in your formulas. Note the difference between what you want and what you get in the following examples.

Grouping—Brackets. In electronics, a quantity called the (magnitude of the) capacitive reactance is found using the formula $\frac{1}{2\pi fC}$. Enter this formula into Maple.

> **1/2*Pi*f*C;** *(Wrong!)*

$$\frac{1}{2}\pi fC$$

Paying proper attention to brackets, we type

> **1/(2*Pi*f*C);**

$$\frac{1}{2}\frac{1}{\pi fC}$$

The rule to remember is that the division operator (/) applies only to the next term and not to the rest of the multiplicative terms unless you enclose all of them in brackets.

Another need for brackets occurs when using fractional powers. For example, find the cube root of 27.

> **27^1/3;** *(Wrong!)*

$$9$$

This is an appealing answer, which might lure you into accepting it! However, what you really wanted is

> **27^(1/3);**

$$27^{1/3}$$

Now you see that you have typed the expression in properly, because Maple shows you how it interpreted your input. Unfortunately, it merely wrote it out again without simplifying it. The solution is to use the *simplify* command.

> **simplify(27^(1/3));**

$$3$$

Maple needs some encouragement in getting this result. The reason? There are three possible answers to a problem involving cube roots, so Maple preferred to keep the expression in its original form until it got some hint on how to proceed. The principal root was then found by using the *simplify* command.

Spaces—Where Not to Put Them. Give the name *eq1* to the equation $s = vt$. We will put a space between the colon and the equal sign. Look carefully!

> **s : = v*t;** (*Wrong!*)

```
Syntax error, `=` unexpected
```
[Release 4]

```
syntax error
> s : = v* t;
        ^
```
[Release 3]

Maple lets you know that a typing error has occurred. It even gives you some clue about where it has occurred. In Release 3, it puts an upward arrow (^) underneath the place on the input line where it detected the error. In Release 4, it puts the blinking cursor on the syntax error.

On the other hand, Maple does allow spaces as long as the meaning is preserved. You just can't put a space in the middle of a word (the colon-equal sign is a word).

> **sqrt (x+ y/ z);** (Spaces are allowed here because they are between the words.)

$$\sqrt{x + \frac{y}{z}}$$

There Is No Implied Multiplication in Maple. Many calculators allow you to omit the multiplication sign in certain cases, and it is usually omitted in math textbooks. This is a convention that works only if variable names are restricted to one letter. Maple allows variable names to be very long, so you must *always* put in the multiplication sign. If you had typed in "> s := vt;" for the previous formula, Maple would treat it as a legal expression.

> **s := vt;** (Forgot the multiplication sign!)

$$s := vt$$

Everything looks just fine, but you could be in for a nasty surprise. There is only one variable in the above formula and its name is *vt*. Rather than *v* multiplying *t*, there is the single variable whose name is *vt*. To ensure that you understand the difference, examine the two commands

> **s1 := vt; s2 := v*t;**

$$s1 := vt$$
$$s2 := v\ t$$

There is a slight difference in the output. In *s*2, there is a small space between the *v* and the *t*, indicating an implied multiplication sign. Let us try and substitute $v = 5$ and $t = 3$ in these expressions.

> **subs(v=5, t=3, s1); subs(v=5, t=3, s2);**

```
vt
15
```

Since there is no *v* (or *t*) in the first expression (s1), Maple leaves it unchanged. Maple does find a *v* and a *t* in expression s2; thus, the substitution works and Maple multiplies the two to get 15.

Watch the Caps-Lock Key. Maple is case sensitive, which means that you cannot type SQRT when you want to take the square root. Note the difference:

> **SQRT(16), sqrt(16);**

```
SQRT(16),4
```

The second command works, producing an answer of 4, but the first command is interpreted by Maple as an undefined function, SQRT. Since Maple does not know the definition of SQRT, it makes no attempt to evaluate it, but the command does not produce a syntax error. After all, you might define SQRT later on in the worksheet.

Terminate Commands with a Semicolon. A typical situation is represented here. Imagine that you typed a command and pressed Return without terminating the command with a semicolon. Release 4 will warn you to put in the semicolon. If you go back and put it in, everything will be fine.

In Release 3, nothing at all will happen. You look at the command and realize that you forgot to put the semicolon at the end. You go back to the command, type in the semicolon, and press Return. Maple indicates a syntax error. You examine the command and can't see anything wrong. The temptation is to make changes to the command, which makes things go from bad to worse.

> **3 + 4**

>

After getting this far, you realize that you forgot the semicolon. You move the cursor back, put in the semicolon, and press Return.

> **3+4;**

```
syntax error:
    3+4;
    ^
```

There is nothing wrong with the command or its syntax now! The error occurred because Maple remembered the first *3 + 4* that you typed in. When you went back and added the semicolon, the command became *3 + 4 3 + 4;* because Maple reads the line over again when you press Return. There is a syntax error in the command now, caused by the space between the *4* and the *3*. This is no longer a legal mathematical statement. To avoid the problem, when you notice that you have forgotten a semicolon, type it in on the next line (which will otherwise be blank) like this:

> **3 + 4**

>**;**

```
7
```

Trigonometry

The most important thing to remember when using Maple to do problems in trigonometry is that Maple uses radian measure, not degrees. Consequently, you must convert any angle into radians before using it in a Maple calculation. To convert from degrees to radians, multiply the angle in degrees by Pi and divide by 180.

Example: Convert 30 degrees to radians.

> **evalf(30*Pi/180);**

```
.5235987758
```

Thirty degrees is equivalent to 0.5236 radians.

You can also look up values for sin, cos, tan, and other trigonometric functions:

Example: Find the sine of 30 degrees.

> **sin(30);**

`sin(30)` (*Wrong!*)

You must convert to radians and use *evalf* if you want a decimal answer:

> **evalf(sin(30*Pi / 180));**

```
.5000000000
```

If you want the exact answer, simply type in the conversion factor between radians and degrees: Maple will reduce this to sin(Pi/6) and display the exact result.

> **sin(30*Pi / 180);**

$$\frac{1}{2}$$

Using exact numbers like Pi/6, Maple has given us the exact answer, 1/2, correct to "infinity" decimal places. When Maple gives an answer like .500000000, we cannot be sure whether the zeros go on forever or whether some other digits will show up later on in the expansion. However, when Maple gives the answer 1/2, we know it is exactly 1/2.

You will learn in this book that there are two units of measure for angles. In high school, you probably learned how to measure angles in *degrees*, so by now, measuring angles in degrees must be quite familiar to you. You may have encountered the other common unit, the *radian*. Radian measure is so important in advanced mathematics that most engineering texts freely mix radians and degrees, assuming that the reader will be able to determine from the context which unit is being used. However, some books exclusively use radian measure, and Maple does also. Since we will be using degrees frequently, we need to teach Maple to use these units as well, and here is a good place to start.

You'll see how to convert from radians to degrees and vice versa in Chapter 1. We will discuss here how to look up the basic trigonometric functions when an angle is given in degrees. If you want to use degrees as a unit of measure while you are using Maple, a simple method is to open a new Maple worksheet and issue these commands.

> **Sin := z -> sin(z*Pi/180);**

> **Cos := z -> cos(z*Pi/180);**

> **Tan := z -> tan(z*Pi/180);**

Save the worksheet and give it a name, such as *trig*. Whenever you want to do trigonometry with Maple, open this worksheet and execute the three commands. You may want to put in some more commands from Chapter 1 to help you convert from degrees to radians and back. An important distinction is in the name *Sin* (rather than *sin*). Case-sensitive Maple considers these two names to be totally different entities. It reserves the name *sin* (all lower-case letters) for the trigonometric function called the *sine*. To use the *sin* function, you must write the angle in radians. However, Maple does not define a *Sin* function, so you are free to define your own, as we have done in the previous commands.

To use either command, you write the statement just as you would in a mathematical essay. Here are a few examples. Take it as given that the sine of 30° is 1/2. This result follows from the definition of the sine function applied to an equilateral triangle, one of whose angles has been bisected to create the 30° angle. If you want to use Maple to look up the sine of 30°, you would issue the command:

> **Sin(30);**

$$\frac{1}{2}$$

This is the expected result. Note what happens if you use the lower-case *sin*:

> sin(30);

```
sin(30)
```

Mysteriously, Maple simple repeats the input line. You might decide that you should find the decimal approximation to *sin(30),* so you issue the command:

> evalf(sin(30));

```
-.9880316241
```

After you have made a slip like this a few times, you will realize that > *sin(30);* is the command for looking up the sine of 30 *radians* (approximately equal to 1,719 degrees), not 30 degrees. It's not surprising that the sine of 30 radians is a negative number. After all, 1,719 degrees is nearly 4¾ revolutions. The terminal side of this angle points nearly straight down, where the sine would be –1.

With these commands you have taught Maple how to work in degrees as well as radians. You could include a few more commands, such as:

> Csc := z -> csc(z*Pi/180); (The cofunction definitions)

> Sec := z -> sec(z*Pi/180);

> Cot := z -> cot(z*Pi/180);

> Arcsin := z -> arcsin(z)*180/Pi; (Definitions for the inverse trigonometric functions)

> Arccos := z -> arccos(z)*Pi/180;

> Arctan := z -> arctan(z)*Pi/180;

These definitions permit you to work with degrees in Maple as easily as with radians. Just make sure you use the correct function—the one that corresponds to the units in your problem!

Plotting

To plot a graph in Maple, you use the *plot* command. First you type in *plot (*. Then add the expression to be plotted (not the equation, just the right-hand side of the equation!), then a comma, then the range for the *x* axis. You finish off the command with a closing bracket and a semicolon. Here is an example:

To plot *sin(x)* for *x* ranging from –2*Pi to 2*Pi, type in the following (see also Figure A.3):

> plot (sin(x), x = -2*Pi .. 2*Pi);

Note especially how you specify the range for the *x* axis. You type *x* =, then give the left-hand (lower) starting place for the range, type in two dots (periods), and specify the right-hand (upper) boundary for the range.

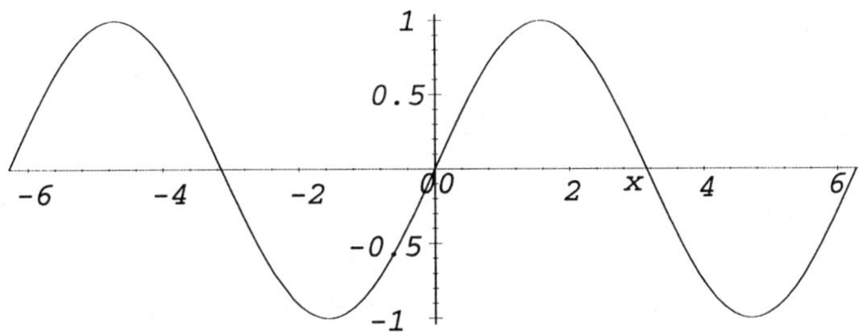

Figure A.3 Maple Result: The Graph of a Sine Wave

This brief introduction to Maple will enable you to get started. You will learn more commands as you work through the book.

Calculus

The two primary Maple commands for calculus are *diff* and *int*. *Diff* takes the derivative of a symbolic expression and expresses the answer in symbolic form. *Int* attempts to take the integral of a symbolic expression. Maple cannot find the integral for every combination of elementary functions, simply because a solution to this problem does not always exist. However, Maple is very good at finding most of the integrals you may happen to encounter in introductory calculus.

You will encounter other commands that are especially useful. Some, like the command called *middlesum*, are aids for learning calculus. Many of these commands are in a Maple package that has been called the *Student Package*, probably because it is primarily useful to students of calculus. The commands in this package can be loaded with the command

>with(student)

[D, Diff, Doubleint, Int, Limit, Lineint, Product, Sum, Tripleint, changevar, combine, completesquare, distance, equate, extrema, integrand, intercept, intparts, isolate, leftbox, leftsum, makeproc, maximize, middlebox, middlesum, midpoint, minimize, powsubs, rightbox, rightsum, showtangent, simpson, slope, trapezoid, value]

You see from the output of the command that a large number of new commands have been added to the ones in Maple's initial list. You will meet many of them as you work through this book.

Dos and Don'ts When Using Maple

Do end every Maple command with a semicolon.

Do start Maple over again if you get confusing results or no results at all. Maple tends to remember everything, even your mistakes!

Don't use double quotes to refer to the last result, assign names to Maple statements instead, or reuse the command by wrapping (nesting) commands around it.

Don't use:

> a := (1/x + 1/y); normal(");[2]

$$a := \frac{1}{x} + \frac{1}{y}$$

$$\frac{y + x}{xy}$$

Instead, you can use either one of the following two commands.

> normal(a), normal((1/x + 1/y));

$$\frac{y + x}{xy}, \frac{y + x}{xy}$$

The second of these commands is an illustration of "wrapping" a command around a previous command. On the other hand, you can't wrap a command around an assignment.

> normal(a := (1/x + 1/y));

```
syntax error:
normal(a := (1/x + 1/y));
          ^
```

Do use the *copy* and *paste* functions to edit Maple commands. Here's an example of simplifying an expression and evaluating a numerical result. The expression, $\frac{1-e^x}{1+e^{-x}} + \frac{1+e^x}{1-e^{-x}}$ is to be simplified and then evaluated at $x = 0.5$.

1. Type the expression into Maple. Use parentheses to maintain the grouping in the original expression:

---

[2] The *normal* command finds a common denominator and writes the expression in this form.

> (1 -exp(x)) / (1 + exp(-x)) + (1 + exp(x)) / (1 - exp(-x));

$$\frac{1 - e^x}{1 + e^{(-x)}} + \frac{1 + e^x}{1 - e^{(-x)}}$$

2. Check Maple's output and ensure that it is the same as the original problem. In Release 4, you can use the *X* button on the Context bar to see the typeset form of your input.

3. Now that you have ensured that the input is correct, highlight the Maple input line and copy it. Paste it on a new Maple input line.

> (1 - exp(x)) / (1 + exp(-x)) + (1 + exp(x)) / (1 - exp(-x));

4. Edit the line by wrapping the simplify command around the line you have copied. Use the *simplify* command.

> simplify((1 - exp(x)) / (1 + exp(-x)) + (1 + exp(x)) / (1- exp(-x)));

$$-4\frac{1}{(1 + e^{(-x)})\ (-1 + e^{(-x)})}$$

5. Copy this input line. You see from the output that Maple has not expanded the denominator. Maple will expand factors in the numerator, so you can force the expansion of the denominator by taking the reciprocal, expanding it, and taking the reciprocal again. Edit the line to make these changes and execute the command once more.

> 1/ (expand(1/ (simplify((1 - exp(x)) / (1 + exp(-x)) + (1+ exp(x)) / (1 - exp(-x))))));

$$\frac{1}{\frac{1}{4} - \frac{1}{4}\frac{1}{\left(e^x\right)^2}}$$

6. You want to evaluate the expression at a specific value of *x*. You can use the *subs* command on *any* of the previous lines since they are all equivalent. We will copy the first input line and substitute $x = 0.5$.

> subs(x = 0.5, (1 - exp(x)) / (1 + exp(-x)) + (1 + exp(x)) / (1 - exp(-x)));

$$\frac{1 - e^{.5}}{1 + e^{(-.5)}} + \frac{1 + e^{.5}}{1 - e^{(-.5)}}$$

7. One final copy and paste using the *evalf* command finishes the problem.

> evalf(subs(x = 0.5, (1 - exp(x)) / (1 + exp(-x)) + (1 +exp(x)) / (1 - exp(-x))));

6.327906828

8. We can check our work by evaluating the simplified expression at $x = 0.5$. Just wrap *evalf(subs(x* = 0.5 . . .) around the command in step 5. The result is the same as in step 7.

In this whole process, you have input the mathematical expression only once, and you have checked Maple's formatted output to make sure that it matches the problem. After seeing that the expression you are working with is correct, you "wrap" commands around it and observe the results in a step-by-step process.

Do use Maple's Help facility. Type *?*, followed by the name of the topic. For instance, to get help on the substitute command, type *?subs* after the Maple prompt.

Do be careful about the decimal point (the period). Maple uses the same character for the decimal point and the concatenation[3] operator. Thus, you must be very careful with the syntax, as shown in the following examples:

Range specifier: You want the range to be 0 to .3, so you type:
0...3

Maple interprets this as 0 to 3! Since there is no space between the double dots and the single dot, it assumes that you meant two dots and a 3.

The variable *x* has a value of 3 and you want to use 3.1, so you write
x.1,

thinking that Maple will substitute 3 for *x* and append the .1. Instead, Maple thinks you mean the variable *x1*.

Do be careful about the range specifier (two periods, one after the other). There must be no space between the periods in this case.

Don't write *sin(x)* as *sin x*, *log(x)* as *log x*, or *exp(x)* as *e^x*. (Even worse, don't write *sin*x* for *sin(x)*). Maple uses a consistent "functional" notation. A function depends on its variables. For instance, the *sin* function depends on the angle at which the sine is required. If the angle is theta, you write *sin(theta)* to evaluate the function. The expression *e^x* looks legal, and it is. The only drawback is that Maple treats *e* as the letter *e*, not as the base of the natural logarithms; thus, in Maple, *e* has no predefined value.

---

[3] See the help page. Type *?concatenation* or *?dot*.

Do practice wrapping functions around other functions. This is called "nesting." A common use of this technique is finding the decimal approximation for some quantity. Say you want to find the decimal value for the square root of 2. You could issue two Maple commands:

> A := sqrt(2); evalf(A);

or you could accomplish the same thing in one command by typing the command:

> evalf(sqrt(2));

An advantage of this technique is that you will acquire the habit of "thinking one step ahead."

Do execute all the Maple input lines when you open a Maple worksheet. When you use the File, Open command to use a previously stored worksheet, Maple does not recalculate the commands in the worksheet unless you tell it to. The file may contain Maple output, but that was the output that was stored with the file. It is not freshly computed. Thus, any names that appear on Maple input lines will not be known in the current Maple session. You must reexecute these commands to have these names defined. There are two ways of reexecuting Maple commands once you open a worksheet.

1. In Release 4, use the mouse and select Edit, Execute, Selection after highlighting the command you want executed. In Release 3, use the Format, Execute Worksheet command. This causes Maple to execute every command in the worksheet. This is the easiest way to make sure that every name used in the worksheet is defined. Also, the output lines will be recomputed and will replace whatever was in the original file. The drawback with this method is that it may take a long time to recompute every command if the worksheet is large or contains commands that take a lot of computing.

2. Place the cursor on the first Maple input line and press *Enter (Return)*. Maple will execute the command and place the cursor on the next Maple command line. There may be text or possibly plot outputs or other graphics between the two command lines. Maple will skip right over the text and graphics and place the cursor on the next command line. Don't be surprised when this happens. There may be more than one screen of text, so the first command and its output will not be visible. If you use the vertical scroll bar to go back to the first command, you will see it and its output. It may not look any different, but it has been recomputed. Now for another surprise: if you press *Return*, the screen will jump once more! Because you used the vertical scroll bar, the cursor remained where it was, on the second Maple command line. Pressing *Return* caused Maple to execute the second command—write the output of the command and go on—placing the cursor at the third command line in the file. A further complication is: Maple allows "program groups," so that what was described as a single command may actually be a group of commands, depending on how the

file was produced in the first place. The point to remember is that you can move through the worksheet, once you place the cursor on the first command, simply by pressing *Return* again and again.

3. You can be selective about which commands to run. Place the cursor on any Maple command line you wish to execute. Press *Return* and Maple will recompute the command line, or group of lines. It then places the cursor on the next command. You may execute that command if you wish, or you can move the cursor to some other command. This way, you can define only those names that you need for your current purpose. This last method may save you some time, but remember that you *must* execute the command lines if you want to use the names that have been mentioned in the worksheet. Just because you can see the names on the screen doesn't mean Maple knows what they are!

Using the Text Editor and Saving Your Worksheets

Maple has a built-in text editor (see Figure A.4) which you can use in describing your calculations and in the preparation of reports. In Release 4, you may select the text mode by clicking on the [T] button on the Tool bar (see Figure A.4). When you are finished typing text, you switch to a new input region by clicking on the [>] button.

In Release 3, you use the Format, Input Region command. You can also use a keyboard shortcut. Type F5 to switch between text input and command input.

To sum up, you can prepare a document containing text, Maple input lines, and Maple output and graphs, and you can save your worksheet on a disk. Saving to disk is accomplished either using the File, Save command, or by clicking the [save] button in either Release 4 or Release 3.

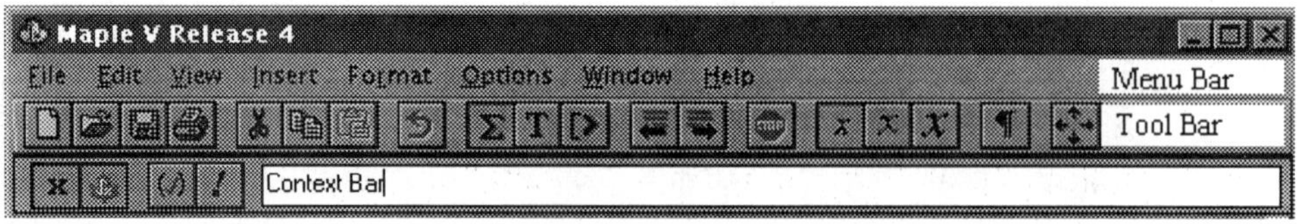

Figure A.4 The Menu Bar, Tool Bar, and Context Bar in Maple V, Release 4

Test Yourself

Answer these questions about Maple.

1. What are the syntax errors in these input lines? The parenthetical comment lines give what the Maple user intended.

(a) > eq1 = a*x^2 + b*x * c = y; Answer:________________

(b) > eq2 := a*x^2 + b*x +c := y; Answer:________________

(c) > eq3 := ax^2 +bx + c = y; (Quadratic formula) Answer:________________

(d) > eq4 := y = 1/1+x; (Reciprocal of 1 plus *x*) Answer:________________

(e) > eq5 := R*a, eq6 := P = V^2/R; Answer:________________

(f) > eq7 := SQRT(16); (Square root function) Answer:________________

(g) > eq8 := plot(x^2, x=0...5); (Plot x^2 from 0 to 0.5) Answer:________________

2. Answer these questions assuming that each group begins a new Maple session. In other words, assume that the *Restart* command has been given at the start of each section. Predict the results of each set of commands. Read the input line very carefully. These questions have been designed to illustrate subtle differences between pairs of similar-looking constructs in Maple.

(a) > a = 3; a + 3; Answer:________________

(b) y = m*x + b; subs(m = 3, b = –2, y); Answer:________________

(c) y := m*x + b; subs(m = 3, b = –2, y); Answer:________________

(d) A := Pi * r^2; subs(r = 2, A); Answer:________________

(e) A := pi * R^2; subs(R = 2, A); Answer:________________

(f) y := 1/3 + 1/4; z := 1./3 + 1/4; Answer:________________

3. State which commands are used to compute each of the following:

Example: Find all the factors of 1,155. Answer: ifactor (1,155)

(a) The exact square root of 65. Answer:________________

(b) The decimal approximation to the square root of 65. Answer:________________

(c) The hypotenuse, base = 7, height = 14. Answer:________________

(d) Solve the equation $3 + 4x = 5 + 6x$. Answer: ____________________

(e) Plot the line $y = 3x + 4$, for x from -2 to $+1$. Answer: ____________________

(f) All the factors of $6x^2 + 5x - 6$. Answer: ____________________

(g) Reduce $(x^2 - 1)/(x + 1)$ to simplest form. Answer: ____________________

(h) Plot the line segment from (2, 3) to (3, 5). Answer: ____________________

(i) Plot the curve $y = x^2$ for x from -2 to 2. Answer: ____________________

CHAPTER 1

Precalculus

Objectives for This Chapter

1. Review the order of operations and the basic laws of arithmetic
2. Understand how to apply the concept of a function in Maple using the arrow operator
3. Practice graphing functions in Maple
4. Review a summary of algebraic and trigonometric theorems for future reference
5. Manipulate radical, exponential, and logarithmic expressions in Maple

Maple Commands Used in This Chapter

binomial(*n, r*)	The binomial coefficient C_r^n.
evalf(Pi);	Convert values to their decimal approximations.
expand((x + y)^2;	Multiply expressions out in full.
factor((x^3 - 1));	Express a polynomial in its factored form.
if ... then ... fi;	Maple's *if* statement.
map(sqrt, *seq*);	Apply the operation to each member in a sequence.
normal(1/x + 1/y);	Put fractions over a common denominator.
plot(*expression, range*)	The basic form of the plot function.
radnormal	Simplify expressions with nested radicals, if possible.
rationalize(1/(1-sqrt(x)));	Express a radical expression in standard form.

readlib(radnormal)	*Radnormal* must be loaded from Maple's library before it can be used.
readlib(rationalize)	*Rationalize* must be loaded from Maple's library before it can be used.
simplify(cos(x)^2 + sin(x)^2);	Simplify an expression.
solve(*equation, variable*)	Find an exact solution to an equation.
sqrt	Maple's square root function.
Sum(1/x, x = 1 .. 100);	Write out the sum, but do not evaluate it.
sum(1/x, x = 1 .. 100);	Evaluate the sum.

Introduction

This review of algebra and trigonometry is provided for you to get some familiarity with Maple commands. The commands are applied to some basic formulas in this chapter. Read every command and study the output. If the output of the command is not shown, run Maple yourself and enter the command.[1] Predict what the result will be before hitting Return!

Order of Operations for the Basic Laws of Arithmetic

The rules of algebra are:

> a + b = b + a; a*b = b*a;	(commutative properties)
> a + (b + c) = (a + b) + c; a*(b*c) = (a*b)*c;	(associative properties)
> a*(b + c) = a*b + a*c;	(distributive property)
> a + 0 = a; 0 + a = a; a*1 = a; 1*a = a;	(identity properties)
> a + (-a) = 0; (-a) + a = 0;	(additive inverse)
> a*(1/a) = 1; (1/a)*a = 1;	(multiplicative inverse)

1. Note that in some cases, Maple applies the basic laws of algebra to reorganize your input. For instance, it always simplifies numerical sums and products, and it may change the ordering of terms in an expression. In the first command, it writes the output to > $a + b = b + a$ as $a + b = a + b$.

The rest of algebra, including the calculus, follows from these rules. The following facts can be deduced from them:[2]

> a * 0 = 0; 0 * a = 0;

> -(-a) = a;

> x := y; if a + x = a + y then x = y fi;

> (-a)*(-b) = a*b;

The last result is the famous "two negatives make a positive" when referring to multiplication of two negative numbers. You might want to type these lines into Maple and observe the results. The line containing the *if* statement may be confusing at this point, but you should be able to interpret Maple's output for all the others.

Maple handles rational numbers with ease. They can be added, subtracted, multiplied, divided, and raised to powers. Brackets are necessary in division and when raising a fraction to a power.

> 3/4 + 5/7, 3/4 - 5/7, (3/4) * (5/7), (3/4) / (5/7), (5/7)^3;

$$\frac{41}{28}, \frac{1}{28}, \frac{15}{28}, \frac{21}{20}, \frac{125}{343}$$

Decimals require a little more work. Realize that decimal numbers are often inexact. An approximation to the sequence of fractions above is produced by the *map* command.

> map(evalf, 3/4 + 5/7, 3/4 - 5/7, (3/4) * (5/7), (3/4) / (5/7), (5/7)^3);

```
1.464285714, .03571428571, .5357142857, 1.050000000,
                    .3644314869
```

The *evalf* command converts fractional or irrational numbers to their decimal approximation. (The *map* command is powerful because it can apply another command, like *eval*, to a whole set of values at once. We used it here to save space.) Here are some other useful forms of the evalf command:

> evalf(2/3), evalf(sqrt(2)), evalf(Pi), evalf(sqrt(2), 3);

```
.6666666667, 1.414213562, 3.141592654, 1.41
```

Note especially the last command in the sequence. The form *evalf(number, sig_figs)* allows you to specify how many digits to put in the answer. The results of this are sometimes surprising, since Maple sets the number of digits internally before doing the computation,

2. The third command is an example of Maple's programming language. You do not need to understand the syntax at this point. Just type the command on a Maple input line and observe the result. If you have some experience with programming and want a brief description of Maple's *if* statement, look up the Help page for *?if.*

unlike pocket calculators, which always use the same precision internally and perform the rounding only at the end.

Real numbers form a set. In Maple, a set is designated by enclosing values inside curly braces.

> {1, 2, 3, 4, 5, 1.1, -23, -2.3, -Pi, Pi/3, sqrt(2)};

$$\left\{1,\ 2,\ 3,\ 4,\ 5,\ -23,\ 1.1,\ -\pi,\ -2.3, \sqrt{2},\ \frac{1}{3}\pi\right\}$$

Notice that Maple does not order the set in the way in which it was typed in. A set is defined as a collection of objects, the order does not matter. We will be concerned primarily with the set of real numbers in this book, although complex numbers will be met with infrequently.

Polynomials

A polynomial is an expression consisting of a sum of terms. Each term is a product of a constant and one (or more) variables. The variables can appear as multiples of themselves, that is, they can be raised to nonnegative integral powers. Polynomials, like simple numbers, can be added, subtracted, multiplied, and raised to a power. Dividing one polynomial by another may result in a new entity, called a *rational fraction*. What follows are some operations with the polynomials $2x^2 - 7x - 13$ and $5x^2 + 9x - 12$.

Addition

> (2*x^2 - 7*x + 13) + (5*x^2 + 9*x - 12);

$$7x^2 + 2x + 1$$

Subtraction

> (2*x^2 - 7*x + 13) - (5*x^2 + 9*x - 12);

$$-3x^2 - 16x + 25$$

Multiplication

> (2*x^2 - 7*x + 13) * (5*x^2 + 9*x - 12);

$$(2y^2 - 7y + 13)\,(5y^2 + 9y - 12)$$

In the last result, Maple did not automatically multiply out the terms. The *expand* command must be used to perform the multiplications.

> expand((2*x^2 - 7*x + 13) * (5*x^2 + 9*x - 12));

$$10x^4 - 17x^3 - 22x^2 + 201x - 156$$

Special Products

Here are some products worth remembering:

> expand((a + b)^2); (sum of squares)

$$a^2 + 2ab + b^2$$

> expand((a - b)*(a + b)); (difference of squares)

$$a^2 - b^2$$

> factor(a^3 + b^3); (sum of cubes)

$$(a + b)(a^2 - ab + b^2)$$

> factor(a^3 - b^3); (difference of cubes)

$$(a - b)(a^2 + ab + b^2)$$

> expand((a + b)^3); (cube of a sum)

$$a^3 + 3a^2b + 3ab^2 + b^3$$

> expand((a - b)^3); (cube of a difference)

$$a^3 - 3a^2b + 3ab^2 - b^3$$

You can reverse the first three examples by factoring the Maple output to get the input statement. Likewise, you can expand the next two Maple outputs to get the input. The commands *factor* and *expand* are inverses of each other in this sense. In most cases the following is true:

> factor(expand (expression)) = expression;

For example,

> factor(expand(x - 1)^3) = (x-1)^3;

$$(x - 1)^3 = (x - 1)^3$$

The Binomial Theorem

The general formula for expanding $(a + b)^n$ is called the binomial theorem.

$$(a+b)^n=a^n+na^{n-1}b+\frac{n(n-1)}{2!}a^{n-2}b^2+\dots+b^n=\sum_{r=0}^{n}C_r^n\,a^{(n-r)}b^r$$

The coefficient C_r^n is called the *binomial coefficient.* If you need to review the binomial theorem, a discussion can be found in most algebra texts.[3]

Since n is indeterminate in this expression, Maple does nothing if you try > *expand((a + b)^n);*: it merely repeats the input. There is a formula for the expansion, although it may look intimidating if you are not used to summation notation.

> (a + b)^n = Sum(binomial(n, r)*a^(n-r)*b^r, r = 0 .. n);

$$(a+b)^n = \sum_{r=0}^{n} \mathrm{binomial}(n,\ r)\ a^{(n-r)}\ b^r$$

Maple does know the binomial coefficients, so you can evaluate any known coefficient: C_r^n =binomial(n,r). For instance,

> (a + b)^4 = sum(binomial(4, r)*a^(4-r)*b^r, r = 0 .. 4);

$$(a+b)^4 = a^4 + 4a^3b + 6a^2b^2 + 4ab^3 + b^4$$

which is the same as

> expand((a + b)^4);

$$a^4 + 4a^3b + 6a^2b^2 + 4ab^3 + b^4$$

Rational Fractions

The accepted way of adding two fractions is to find a common denominator. The Maple command for accomplishing this is *normal.* Maple does not automatically normalize rational fractions, although it does reduce the sum of two numerical fractions to a common denominator. An example should clarify the difference:

> p/q + r/q, 2/3 + 1/4;

$$\frac{p}{q} + \frac{r}{q},\ \frac{11}{12}$$

Whereas Maple left the first sum unchanged, it added the two numerical fractions by finding a common denominator. It did not reduce the fraction to a decimal, because that would not be an exact result.

If a common denominator of a fraction containing indeterminates is required, use the *normal* command:

3. See Parker, *Maple for Algebra* (Delmar Publishers) or Peterson, *Technical Mathematics with Calculus* (Delmar Publishers).

```
> normal(p/q + r/q );
```

$$\frac{p + r}{q}$$

The *normal* command works just as well if we start with two fractions whose denominators are different.

```
> normal( p/q + r/s );
```

$$\frac{ps + rq}{qs}$$

Thus, to reduce a fraction to its lowest terms, first express it as a single fraction by using *normal*. Then simplify it by using *simplify*. Finally, see if the result can be factored by using *factor*.

```
> 8*x/(x^2 - 9) + 4/(5*x - 15);
```

$$8\frac{x}{x^2 - 9} + 4\frac{1}{5x - 15}$$

```
> normal(8*x/(x^2 - 9) + 4/(5*x - 15));
```

$$\frac{4}{5}\frac{11x + 3}{(x - 3)\,(x + 3)}$$

The result is already simplified and factored, so no further work is required.

Radicals

Square roots, cube roots, and higher-order roots are written as $\sqrt{a}$, $\sqrt[3]{a}$, $\sqrt[4]{a}$, and so on. The way to express a root in Maple is to use a fractional exponent. For example, $\sqrt{16} = 16^{1/2}$, $\sqrt[3]{27} = 27^{1/3}$, $\sqrt[3]{4^2} = 4^{2/3}$, and so on. Since Maple treats the root as multivalued, it does not automatically simplify $\sqrt{64}$ to the principal square root of 64, which is 8.

```
> 64^(1/2);
```

$$\sqrt{64}$$

To reduce this to its principal root, use *simplify*.

```
> simplify(64^(1/2) );
```

$$8$$

The exception to this rule is the square root function, *sqrt*, which automatically simplifies perfect squares. For example,

> sqrt(64*a);

$$8\sqrt{a}$$

In calculus, it is sometimes necessary to rationalize the denominator containing a radical. If you need to remove all the radicals from the denominator of an expression like

> (2*sqrt(x^2 - 1) + sqrt(x^2 + 1)) / (3*sqrt(x^2 - 1) + sqrt(x^2 + 1));

$$\frac{2\sqrt{x^2-1}+\sqrt{x^2+1}}{3\sqrt{x^2-1}+\sqrt{x^2+1}}$$

you need more than *simplify*. The required procedure is not loaded in Maple's memory initially. You need to read this special procedure in, using the following commands

> readlib(rationalize);

```
proc(x:{ set,list,algebraic,relation} ) ... end
```

> rationalize((2*sqrt(x^2 - 1) + sqrt(x^2 + 1)) / (3*sqrt(x^2 - 1) + sqrt(x^2 + 1)));

$$\frac{(2\sqrt{x^2-1}+\sqrt{x^2+1})\,(\sqrt{x^2+1}+3\sqrt{x^2-1})}{8x^2-10}$$

If you wish, you could expand the numerator, but Maple would not preserve the common denominator. To expand the numerator *and* keep the common denominator, you would issue the following command:

> normal(expand(rationalize((2*sqrt(x^2 - 1) + sqrt(x^2 + 1)) / (3*sqrt(x^2 - 1) + sqrt(x^2 + 1)))));

$$\frac{1}{2}\,\frac{\sqrt{x^2-1}\sqrt{x^2+1}+5x^2-7}{4x^2-5}$$

You can see why naming expressions can be helpful in reducing the complexity in a nested command. In this way, many lines of paper and pencil work have been accomplished in a single line of Maple.

It is also sometimes necessary in calculus to rationalize the numerator of an expression. Let the expression be

> ex := (sqrt(x - 1 + h) - sqrt(x - 1)) / h;

$$ex := \frac{\sqrt{x-1+h}-\sqrt{x-1}}{h}$$

Here is the trick. Take the reciprocal of this expression and rationalize it.

> **ex1 := 1/ex;**

$$ex1 := \frac{h}{\sqrt{x-1+h} - \sqrt{x-1}}$$

> **ex2 := rationalize(ex1);**

$$ex2 := \sqrt{x-1} + \sqrt{x-1+h}$$

Now we take the reciprocal once more to get the required expression.

> **1/ ex2;**

$$\frac{1}{\sqrt{x-1} + \sqrt{x-1+h}}$$

Thus, using *rationalize,* we can rationalize either the numerator or the denominator of a given expression.

Solving Equations

The *solve* command is usually sufficient for most examples encountered in a first calculus course. This command comes in a couple of varieties, as shown below. The basic syntax of the command is *solve(equation, variable).*

> **solve(y = m*x +b, x)**; (solve a linear equation)

$$\frac{y-b}{m}$$

> **solve(34 - 2*x = 12*x + 6, {x})**; (solve, displaying result as an equation)

$$\{x = 2\}$$

Actually, the result is displayed as a set of equations, but since there is only one solution to this equation, that's all that is displayed. A quadratic equation has two solutions:

> **solve((x - 6)/(3*x + 4) - (2*x - 3)/(x + 2));**

$$0, \frac{-3}{5}$$

Note what we did *not* do in the previous command. We did not use an equation, and we did not tell Maple what variable to solve for. Maple automatically assumed that the expression should be set equal to zero, and it solved for the only indeterminate in the expression.

If Maple is given the general quadratic equation formula $ax^2 + bx + c = 0$,

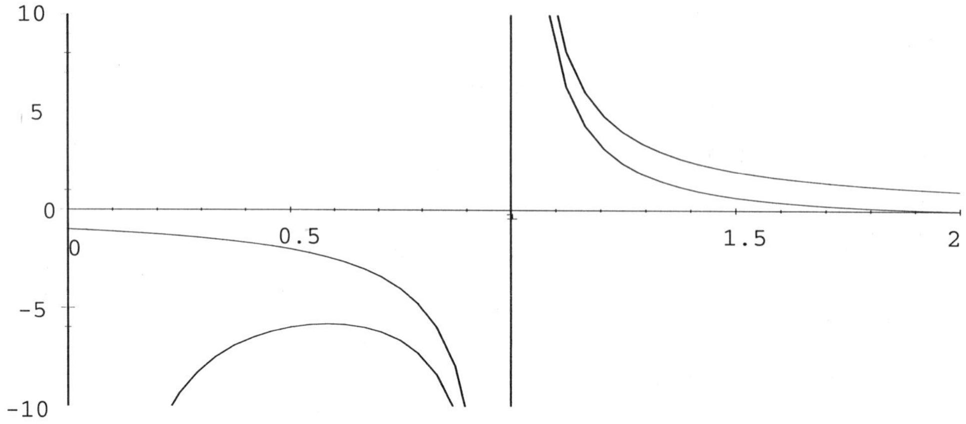

Figure 1.1 An Equation with No Solutions

> **solve(a*x^2 + b*x + c = 0, {x});**

$$\left\{x = \frac{1}{2}\,\frac{-b + \sqrt{b^2 - 4ac}}{a}\right\},\ \left\{x = \frac{1}{2}\,\frac{-b - \sqrt{b^2 - 4ac}}{a}\right\}$$

it solves it and gives back the general formula for the solution, including both roots.

What if there is no solution?

> **solve(1/(y*(y - 1) - 1/y = 1/(y - 1), {y});**

> *(nothing!)*

Maple returns no answer to this problem. It simply gives you a new prompt. A plot of both sides of the equation explains why. The two curves never cross, so there is no value of y common to the two expressions (see Figure 1.1).

Plotting

The Maple *plot* command will be of the utmost importance in the study of calculus. You need to visualize the functions you are working with, and it is really a simple procedure to plot them using Maple. You can also make approximate measurements on the plot and refine them if necessary. You can plot multiple expressions together.

The form of the *plot* command is *plot(expression, range)*. The thing to be plotted must be an *expression*, not an equation. If you want a graph of $y = 3x^2 - 2x - 2$, you would use only the right-hand side of the equation in the plot command. If you try to put in an equation, you'll get an error message. The *range* is the x axis region for the plot. You must decide what interval you want for your plot. You specify a range by typing $x = a \,..\, b$. The variable x must be the same as the variable in your plot expression, and the values a and b are the beginning and ending points for the x axis. Both a and b must be numerical values. Notice that a comma is used to separate the plot expression from its range. Maple will choose the scale for the y axis so that the whole graph is shown.[4] Here are some examples:

Plotting a Straight Line Function

Plot the straight line function $y = 3x - 2$ from $x = -2$ to $x = 3$ (See Figure 1.2).

> plot(3*x - 2, x = -2 .. 3);

Make special note of the way in which the plot command was formed. Ordinary brackets are used to enclose the plot expression and the range. A comma separates the expression from the range. Only the right-hand side of the equation is used. The lower limit of the range is separated from the upper limit by double dots (a pair of periods). There cannot be a space between the double dots. Be careful when you use decimal numbers! Make sure to type the lower limit, a space, the double dots, then another space, and then the upper limit.

Multiple Expressions on the Same Plot

Solve the system of equations, $y = 3x - 2$, $2x^2 - 1$, graphically for the positive values of x (see Figure 1.3).

> plot({ 3*x - 2, 2*x^2 -1 }, x = 0 .. 1.5);

When you want to plot a *set* of expressions (in other words, more than one expression), you enclose the expressions in curly braces and separate the expressions by commas. That's all there is to it. The resulting graph shows that the curves cross each other at $x = 0.5$ and $x = 1$. You can verify the solution by expanding the graph. Select a small range around $x = 0.5$ and have Maple redraw the graph. In this way, you will be able to measure the crossover point much more accurately.

> plot({ 3*x - 2, 2*x^2 -1 }, x = 0.49 .. 0.51);

You can find a graphical solution even when Maple cannot solve the system of equations algebraically. In this example, Maple can find the solution:

---

4. This causes problems if the graph goes off to infinity somewhere within the interval. In these cases, you will have to decide what interval you want for the vertical axis as well. The form of the plot command will be *plot(expression, x_range, y_range);*.

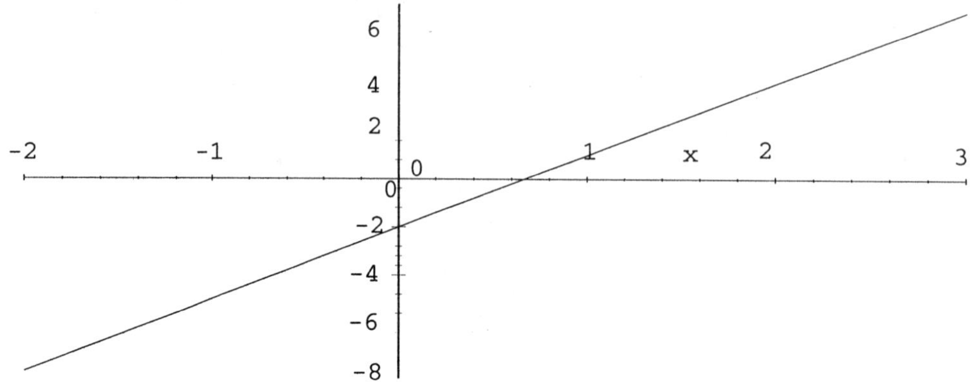

Figure 1.2 Plot of the Straight Line Function $y = 3x-2$

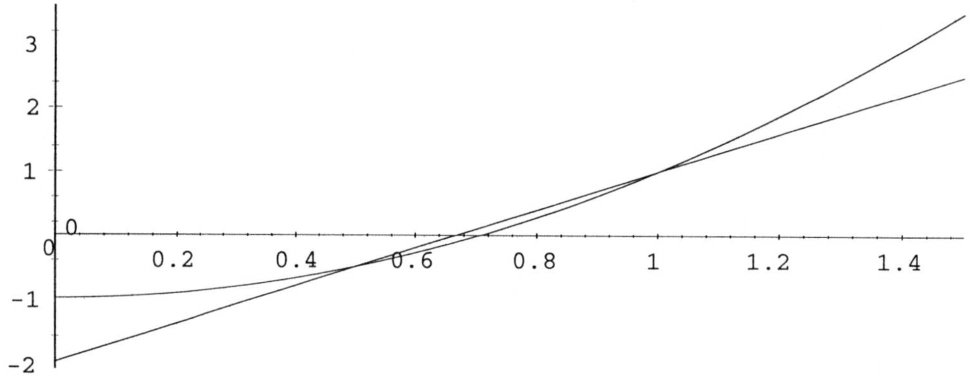

Figure 1.3 Solving Equations Graphically

> solve({ y = 3*x - 2, y = 2*x^2 -1 }, { x, y });

$$\left\{y = \frac{-1}{2}, x = \frac{1}{2}\right\}, \{y = 1, x = 1\}$$

Maple's exact solution confirms our earlier estimate.

There are many options to the *plot* command. You can look these up in Maple's Help menu. Select Help, *Keyword Search . . .*, and type *plot* in the *Search for string:* box. You will see many entries for *plot*, but you will need to scroll down the page a bit before they appear.

Summary of Algebraic Theorems

$$x^2 - y^2 = (x - y)(x + y) \quad \text{(difference of Squares)}$$

$$(x + y)^2 = x^2 + 2xy + y^2 \quad \text{(square of a Binomial [+])}$$

$$(x - y)^2 = x^2 - 2xy + y^2 \quad \text{(square of a Binomial [–])}$$

Manipulation of Radical, Exponential, and Logarithmic Expressions

Fundamental Exponential Identities

$$a^m \cdot a^n = a^{(m+n)}$$

$$a^0 = 1,\ a^1 = a$$

$$\frac{a^n}{a^m} = a^{(n-m)}$$

$$a^{-n} = \frac{1}{a^n}$$

Fundamental Logarithmic Identities

$$\text{Let } x = b^u$$

$$\text{Then } \log_a(b^u) = \log_a(x)$$

$$\text{and } u\log_a(b) = \log_a(x)$$

$$\text{Therefore } u = \frac{\log_a(x)}{\log_a(b)}$$

$$\text{change of base formula } \log_b x = \frac{\ln x}{\ln b}$$

Solved Problems

Example 1-1: Calculation

Evaluate 12 factorial (12!) using Stirling's formula for approximating $n!$. Compare the result of using Stirling's formula to the exact result. What is the percentage error? The formula is

$$n! \approx \sqrt{2\pi n}\, n^n e^{-n}$$

and the approximation gets better as n gets larger.

Solution:

```
> f1 := sqrt(2*Pi*n)*n^n*exp(-n);
```

$$f1 := \sqrt{2}\sqrt{\pi n}\, n^n e^{(-n)}$$

```
> f1a := subs(n = 12, f1);
```

$$f1a := 8916100448256\sqrt{2}\sqrt{12}\sqrt{\pi}\, e^{(-12)}$$

Maple has evaluated some of the numbers fully and has left others alone. The ones it leaves alone cannot be simplified without losing some accuracy. The approximate result in decimal form is

```
> evalf( f1a );
```

$$.4756874863\ 10^9$$

Compare this to 12 factorial:

```
> 12!;
```

$$479001600$$

Examine the last two results and mentally compute the approximate difference. The error is seen to be about 3 parts in 500. More accurately, the percentage error is

```
> evalf( 100*(12! - f1a) / 12! );
```

$$.69187948$$

Our mental estimate was 0.6%, and the calculation using Maple is 0.69%. Always try and make mental estimates as you go along. It will keep you on-track while solving the problem.

Example 1-2

Simplify the rational fraction $\frac{2x}{x+3} - \frac{x-4}{x^2-9}$.

Solution: Type the expression in Maple and check that the formatted output matches the original problem statement:

```
> 2*x / (x + 3) - (x - 4) / (x^2 - 9);
```

$$2\frac{x}{x+3} - \frac{x-4}{x^2-9}$$

Simplify the expression using either *normal*, which brings both fractions under a common denominator, or the *simplify* command, which applies a number of simplification rules. Don't retype the line. Highlight it, copy it, and paste it on a new Maple input line. Wrap the *simplify* command around the expression by editing it. Make sure to put in the closing bracket of the simplify command just before the semicolon.

> simplify(2*x / (x + 3) - (x - 4) / (x^2 - 9));

$$\frac{2x^2 - 7x + 4}{x^2 - 9}$$

Example 1-3

Solve the following system of equations.

$$\frac{3}{x}+\frac{2}{y}-\frac{1}{z}=\frac{11}{6}$$

$$\frac{1}{x}-\frac{1}{y}+\frac{3}{z}=-\frac{11}{12}$$

$$\frac{2}{x}+\frac{1}{y}+\frac{1}{z}=\frac{7}{12}$$

Solution: Write the three equations, giving them the names *f3a*, *f3b*, and *f3c*. Maple has a powerful equation solver. You do not have to isolate the variables all on one side or perform other simplifications. Just type the equations in as they appear.

> f3a := 3/x + 2/y - 1/z = 11/6; f3b := 1/x - 1/y + 3/z = -11/12; f3c := 2/x +1/y + 1/z = 7/12;

$$f3a := 3\frac{1}{x} + 2\frac{1}{y} - \frac{1}{z} = \frac{11}{6}$$

$$f3b := \frac{1}{x} - \frac{1}{y} + 3\frac{1}{z} = -\frac{11}{12}$$

$$f3c := 2\frac{1}{x} + \frac{1}{y} + \frac{1}{z} = \frac{7}{12}$$

Check the typing for accuracy. Use Maple's *solve* command.

> a3 := solve({ f3a, f3b, f3c }, { x, y, z });

$$a3 := \{x = 3,\ y = 4,\ z = -3\}$$

Check the solution.

> assign(a3); f3a, f3b, f3c;

$$\frac{11}{6} = \frac{11}{6},\ \frac{-11}{12} = \frac{-11}{12},\ \frac{7}{12} = \frac{7}{12}$$

The *assign* command causes Maple to remember the solution. More accurately, the variables *x*, *y*, and *z* now have the values 3, 4, and -3 assigned to them. All you have to do to recall them after that is to mention the names for the three equations. We have compressed all this work on one line to save space. Maple substitutes the values of x, y, and z in equations *f3a*, *f3b*, and *f3c*. Since the equations now contain only numbers, Maple does the arithmetic and reports the results. All you have to do is to check that the number on the right-hand side equals the number on the left-hand side.

If you want to use x, y, and z again as variables, you must *unassign* them. Once the values have been assigned, Maple remembers that x, y, and z have specific values until it is told to forget them. You can reassign the value of a variable. The command > *x := 10;* will cause x to acquire the value of 10. One way of getting Maple to unassign these variables is to reassign them to themselves! This is done by the following command. The variable on the right-hand side of the assignment is enclosed in forward single quotes. (The *single quote* is usually found on the same key as the double-quote character.)

```
> x := 'x'; y := 'y'; z := 'z';
```

Example 1-4

The *differential quotient* is defined as $\frac{f(x+\Delta x)-f(x)}{\Delta x}$. Evaluate the difference quotient for the function $f(x) = 3x^2 + 4$.

Solution: The difference quotient contains a symbol that Maple does not have. Rename Δx for Maple. Call it *dx*.

```
> DQ := ( f(x + dx) - f(x) ) / dx;
```

$$DQ := \frac{f(x + dx) - f(x)}{dx}$$

```
> f := x -> 3*x^2 + 4;
```

$$f := x \to 3x^2 + 4$$

```
> DQ;
```

$$\frac{3(x + dx)^2 - 3x^2}{dx}$$

```
> expand( DQ );
```

$$6x + 3dx$$

Here, we have defined the difference quotient as *DQ*. *DQ* has been defined in terms of an *unknown* function, *f*. Then the function *f* is defined. If you name *DQ* again and ask Maple to evaluate it, Maple will substitute the specific value of the function *f* in *DQ* and evaluate it at $x + dx$ and at x. We *expand* *DQ* to simplify the result.

Maple Lab

ML1-1

Find the sum of the first 100 terms of the series

$$\frac{1}{2\cdot 3}+\frac{1}{3\cdot 4}+\frac{1}{4\cdot 5}+\ldots+\frac{1}{(n+1)(n+2)}+\ldots+\frac{1}{102\cdot 103}$$

Hint: Maple's *sum* command can be used. Express the sum as

> sum(1 / ((n+1)*(n+2)), n = 1 .. 100);

Answer: ______________________________

ML1-2

Simplify the radical expression

$$\sqrt{2(3-\sqrt{2}-\sqrt{3}+\sqrt{6})}$$

Type in the expression:

> m2 := sqrt(2* (3 - sqrt(2) - sqrt(3) + sqrt(6)));

$$m2 := \sqrt{6 - 2\sqrt{2} - 2\sqrt{3} + 2\sqrt{6}}$$

Use the Maple library function *radnormal* to remove one layer of radical signs:

> readlib(radnormal); radnormal(m2);

Answer: ______________________________

ML1-3

Solve the equation $i = 150\cos(377\pi t + \frac{\pi}{3})$ for the smallest *positive* time for which $i = 100A$.

Answer: ______________________________

ML1-4

Solve the radical equation $\sqrt{2r+5} = \sqrt{r} + \sqrt{5}$.

Answer: ______________________________

ML1-5

Use Maple to determine if the trigonometric equation is an identity

$$\frac{\cos(\theta)}{1-\sin(\theta)}+\frac{\sin(\theta)-1}{\cos(\theta)}=2\tan(\theta)$$

Hint: Use Maple to subtract the right-hand side from the left-hand side. Use the *simplify* command. If the result is zero, the equation is an identity. Is the equation an identity?

Answer: ______________________________

ML1-6

Find A in the formula, given $W_o = 50.6$, $W = 30$, $K = 0.925$, $A_o = 240$.

$$W = W_o\sqrt{K^2+\log_{10}\left(\frac{A}{A_o}-1\right)}$$

Answer: A = ______________________________

ML1-7

Solve for x:

$$8.03^{2x} = -.731^{1-3x}$$

Answer: x = ______________________________

ML1-8

Use Maple to investigate the deflection of a steel beam 24.5 feet long. The beam is supported at both ends. The beam is holding up a part of a building, which puts a uniform load on all parts of the beam. You are given the formula for the deflection of the beam (d is in inches, x is in feet):

$$d = 3.23\times 10^{-6}x^2(245-x)^2$$

(a) Type the equation into Maple and give it a name, say *m1*.

Answer: The Maple equation is ______________________________

(b) Where is the maximum deflection of the beam? To find out, plot the right-hand side of d.

Answer: The maximum deflection is at ______________________________

(c) It is later found that the load is not quite uniform. Careful measurements show that the deflection equation is given by

$$d = 3.23 \times 10^{-6}x^2(245 - x)^2 + 0.256 \times 10^{-6}x^3 - 0.974 \times 10^{-5}x^2 + 0.795 \times 10^{-4}x$$

Type this equation into Maple and give it a name. *Hint:* Just type the additional term and add it to the right-hand side of *m1*. Call this new equation *m2*.

Answer: Equation *m2* is ______________________________

(d) The point of maximum deflection has shifted. Find the new point of maximum deflection by plotting the right-hand side of *m2*.

Answer: The point of maximum deflection is at $x =$ ____________________

(e) The maximum permissible deflection for a beam of this material is 0.1 inches. Is this beam within its tolerance value?

Answer: The maximum deflection is __________, so the beam is (within/not within) tolerance: __________. (Support your answers by sketching the graphs or, if you have access to a printer, attach them to your lab report.)

ML1-9

You can get Maple to switch to decimal mode by placing a decimal point in one of the numbers in the input. Does Maple reduce this command to a decimal number?

```
> 17.3*sin(2*Pi*377.1 + 1.04);
```

Answer: ____

ML1-10

What's wrong with this command?

```
> 3x^2 + 2x + 7;
```

Answer: ____

ML1-11

Find the typing error.

```
> subs( x = 5, (3*x^2 + 1);
```

Answer: ____

ML1-12

The expression $\frac{1}{1-x}$ is to be typed into Maple. Find the typing error.

> **1/1-x;**

Answer: ____________________

ML1-13

The expression e^{x} is to be typed into Maple. Find both mistakes.

> **e^x; exp*(-x);**

Answer: ____________________

ML1-14

(a) Try and add the reciprocals of the first hundred numbers. You would probably use the command

> **sum(1/x, x = 1 .. 100);**

(b) What is Maple's answer to this problem?

Answer: ____________________

(c) How do you express the answer in fractional form?

Answer: ____________________

Hint: You should suspect that Maple is applying some special rule for the sum of reciprocals. It probably recognizes a sum of reciprocals that starts at one.

CHAPTER 2

The Difference Quotient, Limits, and Continuity

Objectives for This Chapter

1. Define the difference quotient
2. Define a limit and apply Maple's *limit* command
3. Know the properties of limits
4. Use Maple's calculating power to explore the numerical concept of a limit
5. State the squeeze theorem
6. State the intermediate value theorem
7. Know the definition of continuity and apply Maple's *iscont* command

Maple Commands Used in This Chapter

->	Functional notation. *Example: sq := x -> x^2* is the squaring function.
(f@g)(x)	Functional composition: $(f@g)(x) = f(g(x))$.
@@	The repeated composition operator: $(f@@3)(x) = f(f(f(x)))$.
[[x1, y1], [x2, y2]]	A list of *x*, *y* coordinate points. Used in the *plot* command.
abs(x)	The absolute value function in Maple: $abs(x) = \|x\|$.
color = black	A *plot* option. It ensures that the color of the curve is black.

Digits	The command *Digits := 20* tells Maple to work to 20-digit accuracy in numerical calculations.
evalf	Evaluates expressions, reducing them to decimal form.
Labels = [` x-label ` , ` y-label `]	A *plot* option. The axes are labeled with the names listed.
Limit(f(x), x = a);	Inert form of the limit command.
Limit(f(x), x = a, right);	Inert form of limit as the *x* value approaches *a* from the right.
Limit(f(x), x = a, left);	Inert form of limit as the *x* value approaches *a* from the left.
assume(x < 1);	Place restrictions on the variable x.
is(x > 3);	Test the inequality; the command returns *true*, *false*, or *FAIL*.
iscont(1/x, x = -1 .. 1);	Tests the continuity of the function 1/*x* over the interval -1 .. 1.
expand(f(x));	Expand the expression *f*(*x*), which is usually a polynomial or a rational fraction.
f1 := (x) -> if is(x>3) then sqrt(x) else 0 fi;	Discontinuous function definition.
fsolve (f(x) = 0, x, 3/4 .. 7/8);	Numerical solver, giving interval in which to look for solution.
limit(f(x), x = a);	Compute the limiting value of the function as *x* approaches *a*.
limit(f(x), x = a, right);	Compute the limiting value as *x* approaches *a* from the right.
limit(f(x), x = a, left);	Compute the limiting value as *x* approaches *a* from the left.
simplify(expression)	Apply simplification rules to an expression.
sin(x)	The trigonometric sine function.
subs(x = 3, x + 1)	The substitute command. Here, the variable *x* is given the value 3.
value(expression)	Evaluate inert functions, like *limit*.

The Origins of the Difference Quotient

The calculus you study in an introductory course is concerned mostly with *smooth* functions. You can picture a smooth function as one that can be drawn without taking your pen off the paper. There are many examples of smooth functions in science and engineering, such as the following two. The temperature in a museum is usually monitored by a type of thermometer that traces the weekly variation on a chart. The chart shows that the temperature goes up and down smoothly. Similarly, even the sudden stops of colliding automobiles (at a test site, we hope) can be measured by instrumentation or by high-speed photography. These show that the cars do not stop suddenly. Instead, the speed goes smoothly to zero from whatever it was before the crash, but over a very short period of time. At no time does the speed suddenly jump from one value to another without going through all the intermediate values.

Let us pursue the idea of the speeding car a little further. Instead of examining the drama of a car crash, we will investigate a familiar daily occurrence, that of getting in a car and driving off. As usual in math courses, we make some simplifications. The car is facing directly in the direction you want to go. You have a clear path ahead. You start the car, press on the accelerator, and the car begins to move. You note that the speedometer changes smoothly from 0 to 30 miles per hour (mi/hr) in 15 seconds.

This seemingly simple situation contains some rather abstract ideas. What is meant by "the car begins to move"? How do you measure moving objects? Your probable response is: "Well, the car was stopped. When I started the engine and pressed on the accelerator, the car sped up." Aha! We have a new concept, that of *speed.* Obviously, speed is related to motion. You would say that speed is the measure of motion. The higher the speed, the more motion you have. Unfortunately, this seems like a circular argument. Speed and motion seem to express the same idea. What's the next step?

Science requires measurement. To refine the notion of speed, we must figure out a way of calculating a numerical value. After all, the auto manufacturers have done that, and called it a speedometer. What definition of speed did they use? Speed is defined as the ratio of the distance traveled to the time taken. If you are driving along at 60 miles per hour (mi/hr) for 1 minute, you must have covered 1 mile $\left(60\frac{mi}{hr} \times \frac{1}{60}\, hr = 1\ mi\right)$. You may raise an objection, saying that you need to wait one minute before you can measure the speed. Moreover, the speed must be constant during the whole minute.

Now we get to the reason for the appearance of this example in a chapter on limits. If we try and refine the definition of speed to meet these objections, we will touch on all the major ideas of the calculus. In fact, the calculus was invented by Sir Isaac Newton and Gottfried Wilhelm Leibniz in the mid seventeenth century as a way of dealing with the concepts of motion. Let us go back to the example of the car. We will devise an experiment that allows us to make precise measurements of its motion. We put a large tape measure along the track so that we will be able to measure the distance traveled at any time. We will use a video camera that puts an accurate time stamp on every frame. The video camera records 15 frames per second, thus, when we replay the motion, we will be able to measure the position of the car to the nearest 1/15th of a second. Now we can invent a new definition for the speed of the car: it is the distance traveled divided within the time interval, which works out in this case to the distance traveled in 1/15th of a second multiplied by 15. If we want to know the speed of the car at any time, say at the end of 10 seconds (s), we play the tape up to the 10-second mark and measure the position of the car. We advance one frame and measure the position of the car again. Here is our formula:

Speed of the car at 10 s = ((Position of car at 10 s + one frame)
– (Position of car at 10 s)) x 15

Now we make up a mathematical formula for the speed. Let the position of the car at any time be given by the function $s(t)$. We will call the speed v. We can plot the function at intervals of 1/15th of a second using our videotape. The speed at 10 seconds is

$$v_{t=10sec} = \frac{s\left(10 + \frac{1}{15}\right) - s(10)}{\frac{1}{15}} \qquad (2\text{-}1)$$

Suppose the tape measure's markings are in feet and each foot is divided into hundredths. You measure the distance traveled by checking the position of the car on the videotape frames at 10 s and at 10 + 1/15 s. Suppose the readings are 146.67 ft and 148.63 ft, respectively. Putting these values into the formula, we have for the speed at 10 seconds

$$v_{t=10\ sec} = \frac{148.63\,ft - 146.67\,ft}{\frac{1}{15}} = 1.96 \times 15 = 29.4\,ft/s \qquad (2\text{-}2)$$

Impressed with your result, you explain it to a friend, who raises a subtle point you had not considered. The car's speed is continually changing. How can you say that it was going 29.4 ft/s during the 1/15th-second interval? To drive (no pun intended) her point home, your friend measures the position of the car at 10 + 2/15 second and finds it to be 150.60 ft. She calculates the speed using the formula, getting

$$v_{t=10+\frac{1}{15}sec} = \frac{150.60\,ft - 148.63\,ft}{\frac{1}{15}} = 1.97 \times 15 = 29.55\,ft/s \qquad (2\text{-}3)$$

Now she poses her trick question. At 10 seconds, the car was going 29.4 ft/s. At 1/15th second later, the car was going at 29.55 ft/s. Does the speed jump from one reading to another? It would make for a bumpy ride! Worse, the jumps in speed seem to be dependent on the time interval used in the videotape frames. What if the camcorder had a feature that allowed you to record in slow-motion? If each frame of the slow-motion recording had a time stamp 1/100th of a second apart, the distance readings at 10 s would be the same, namely, 146.67 ft. But the next frame, a hundredth of a second later, would show the car at 146.96 ft. The same calculation for these numbers gives a speed of 29 ft/s.

Admittedly, the calculations give similar results. You can state with confidence that the speed of the car when the clock reads 10 seconds is *around* 29 ft/s. But is this good enough? Science and engineering require precise concepts as well as precise measurements. We will leave the question of a precise definition of speed for a moment to look at its relation to the distance traveled.

We could work out the speed of the car between any two successive frames on the videotape. Let us think of doing it for *all* frames by pairing them. We could make up a table of speeds for each 1/15th-second time interval, from $t = 0$ to $t = 15$ seconds. Now we risk a circular argument once again by using these speeds to calculate the distance traveled by the car in each time interval. We start with the two frames, $t = 0$ and $t = 1/15$ s. The distance traveled at $t = 0$ is $s_o = 0$ because the car hasn't moved yet. The distance measurement at $t = 1/15$ s is 0.0065 ft, too small to be measured on our tape, but let's accept the reading nonetheless. Using these values, we calculate the distance traveled by substituting them in the formula

$$s = v \times \frac{1}{15} + s_o \tag{2-4}$$

When we calculate the distance traveled using the speed formula, it gives us the *distance traveled during a specific 1/15th of a second,* not the *total* distance traveled. To get the total distance from the beginning, we must add on the distance traveled up to the time interval in question. That is the reason for the s_o term. It is the accumulated distance. We add all the incremental distances together to get the total. Does this total, which was obtained by adding all incremental distances, agree with the last frame on the videotape? The last frame shows the car at the final distance. There is no arguing with the position of the car as shown by that frame, which *must* show the total distance traveled. Our logic tells us that if we measure the distance traveled during each pair of frames and add them all up, we must also arrive at the same total distance. (Note that even if the distance measurements were made to the nearest millionth of an inch, the incremental distances still wouldn't add up to the total distance. The reason has to do with the concepts of "jumps" mentioned earlier. The speed is changing smoothly as the car accelerates, but the formula calculates a new speed at intervals of 1/15th s. What about the times in between? What was the speed during these in-between times?)

In the first time interval, the distance calculation gives us $s_o = 0.1$ ft. This is believable, but a small problem remains. The speed calculation gives us a nonzero speed at $t = 0$, yet we know that the car was stopped at t = 0. It started to move *right after* time t = 0. We can remove this difficulty by observing that the formula we developed is really an *average* speed. At the beginning of the time interval, the car was going slower than it was at the end of the time interval. At some time in between, the average speed would occur. Now we seem to be sliding away from our objective! We encountered conceptual difficulties when we tried to define a speed that was changing. We retreated to a concept of *average speed* rather than achieving our goal, which was to define *speed*, not average speed. Is such a definition possible?

The definition is given later on in this chapter, but let us review what we have observed. Every major concept (and every major problem) in calculus has been exposed by this seemingly everyday example. The questions that need to be answered are:

1. How do you make exact something that is continuously changing?
2. How do you define a changing quantity, like speed, so that it does not depend on the interval?
3. How do you add up all the pieces (like incremental distances) and get the correct total?

Even though scientists like Newton and Leibniz discovered the calculus, it took mathematicians many years to solve the logical pitfalls they encountered as they worked through problems similar to our example. They succeeded, and so can you! The first task is to explore the notion of smoothness. This can be done by a process of subdivision. Mathematicians have discovered that there are two related concepts here. They are called *limits* and *continuity*. You'll begin with these studies in this chapter. Armed with these concepts, you can begin the study of point 2, which is the study of differential calculus. You will find that it can be pictured as a tangent problem. The calculation of the speed of a car can be related to the problem of finding the tangent to a curve. In this case, the curve is a distance-time graph. These concepts are presented in detail in Chapters 3 and 4. Point 3 leads to the integral calculus, which can be pictured as the area under a curve. The method of calculating the total distance traveled by adding all the incremental distances can be related to finding the area under a speed-time graph. The ideas in point 3 will be investigated in Chapters 5 and 6.

Fractional Change and Percentage Change

Don't be intimidated by the calculus. Its basic ideas are already familiar to you. The differential calculus is related very closely to such topics as average speed, rate of change, fractional change, and percent change. You've seen the definition of average speed in the last section, now look at the concepts of fractional change and percentage change. The reason for calculating percentage change, or fractional change for that matter, is to get a measure of the relative increase, or decrease, in some quantity. If you work for a company and you need to ship an extra 400 pounds of product, will you encounter a problem with your shipping contractor? You will probably not if the overall shipment is 400,000 pounds and it is going on cargo ship. It probably will be a problem, however, if the original shipment was itself 400 pounds and it was to go in a van. In the first case, the additional 400 pounds is merely a drop in the bucket. In the second case, the extra 400 pounds doubles the weight of the shipment. The percentage change caused by the additional 400 pounds, which will be very different in the two cases, provides a good measure of the effect of adding the extra 400 pounds to the original amount.

The definition of fractional change is fairly obvious. Consider some quantity whose value changes. The quantity could be the cost of a foreign currency on any given day, the cost of gas at the service station pump, or the change in the number in the number of atoms in your body caused by eating a meal. Call the base value of some quantity b and the new value of the quantity n. The fractional change is defined as

$$\text{Fraction change} = \frac{n-b}{b} \tag{2-5}$$

The percentage change is just the fractional change multiplied by 100.

$$\text{Percentage change} = \frac{n-b}{b} \times 100 \tag{2-6}$$

We'll begin with an example of percentage change.

Example 2-1: Where Are the Sales?

Maeve Mobeley wanted to purchase a new pair of running shoes, but she was waiting for her favorite store to put them on sale. Unfortunately for her, inflation caused the price of the shoes, which previously sold for \$113.95, to go up to \$119.95. What was the percentage increase in the cost of the running shoes?

Solution: Use Equation 2-6 with $b = 113.95$ and $n = 119.95$.

$$\text{Percentage change} = \frac{n-b}{b} \times 100 = \frac{119.95 - 113.95}{113.95} \times 100 = 5.27\% \tag{2-7}$$

Naturally, the calculation can be done using Maple:

```
> (119.95-113.95)/113.95*100;
                        5.265467310
```

Since the fractional change in the price of the shoes is 0.0527, the percentage change is 5.27%. Maeve could say, "Oh, the price of the shoes that I wanted to buy went up by about 5%." Can this value of the percentage increase permit you to calculate the new selling price based on the original price? You would take the original price, \$113.95, find 5.26546731% of this figure, and add it to \$113.95.

```
> 113.95*0.05265467310 + 113.95;
```

```
119.9500000
```

Applying the percentage increase to the base price does indeed produce the new selling price.

Your Turn

(a) In economics, the basic idea of inflation is related to the increase in cost of a "shopping basket" of consumer items. In 1995, the cost of these items was \$456.78. If the same items cost \$473.12 in 1996, what is the rate of inflation for 1995–1996? Express your answer in percent.

Answer: ______________________________

(b) If this rate of inflation stays the same in 1997, how much will these consumer items cost then?

Answer: ______________________________

In the study of calculus, it's more common to use fractional changes rather than converting to percentage change. Most people use percentages in normal conversation, in news reports, and in many other types of reports. You should be comfortable switching between the two ways of saying the same thing. If you take the time to thoroughly understand the concept of fractional change, you will have no trouble with the differential calculus. The *derivative*, which you will study shortly (in Chapter 3), is closely related to the notion of fractional change, with the concept of a *limit* thrown in.

Rate of Change

Whenever something changes, it is natural to ask the question, "How fast is this quantity changing?" The thing being measured might be height of a child in his or her growing years, the number of automobiles sold in the United States as time goes on, or the distance apart of two continents as the tectonic plates ever so slowly drift under the influence of currents in the earth's molten core. The numbers being measured may be big or small, so it is useful to use the formula for fractional change, Equation 2-5, in most cases. By forming a ratio, the result can usually be made more meaningful than simply stating the change in the variable. For example, if you were told that the population of a certain city would increase by 532 because a plant was relocating its workers, would you be able to conclude that there would likely be a housing shortage and prices would go up? You could not do so until you knew the original population of the city. If the city was New York, where there are thousands of houses on the market at any given time, 532 new house buyers might not be noticed. But if the city was Smallville, population 5,367, the effect might be a sellers' paradise. Actual numbers are usually less meaningful than relative numbers, and the formula for fractional change provides the needed comparison.

Consider the problem of measuring the number of bacteria in a controlled laboratory experiment. The number of bacteria in a culture container is measured every hour. The tank can hold a maximum of N bacteria. Let the population at any time, t, be $P(t)$. Then the *normalized population* can be defined as $p(t) = P(t)/N$. Thus, $p(t)$ has domain $0 \le p(t) \le 1$. It is much easier to deal with $p(t)$, which represents the fraction of the total possible population, rather than talking about $P(t)$, a number that may be in the hundreds of billions, since it is bacteria we are talking about and they can be quite numerous. The population growth rate can be defined as $p_{t+1} - p_t$, and the fractional growth rate as

$$\frac{p^{t+1} - p_t}{p_t}$$

The Difference Quotient

The concept of rate of change and its relation to fractional change can be studied by examining some models of population growth. The *logistic equation* is $y = x + rx(1 - x)$. We will derive this equation shortly.

Biologists frequently measure populations at definite time periods. For instance, animal populations may be measured every year, after the young are born. Let us say that the population in the bacterial culture was measured every hour. In that case, the population growth from one hour to the next is given by $p_{t+1} - p_t$, and the fractional growth rate is

$$\textit{Fractional Rate of Change in One Time Period} = (p_{t+1} - p_t)/p_t. \qquad (2\text{-}8)$$

Example 2-2. Calculation of a Specific Growth Rate

At the beginning of an experiment, the population of *E. coli* bacteria was 17% of the maximum number the tank can hold. A bacterial count is made every hour after that. The next count, made 1 hour after, showed the population to be $p_{t+1} = 17.67\%$. What was the fractional growth rate per hour?

Solution: Apply the formula, $(.1767 - .17)/.17 = 0.0394$ per hour. Thus, the bacterial culture in the tank was growing at about 4% per hour.

Your Turn: At a certain time, the number of minnows in a tank was 500,000. A day later, the population was measured to be 450,000. What is the fractional growth rate per day? Note that in this case, the fractional growth rate is negative.

Answer: ______________________________

A Population Model

A simple model of population dynamics can be derived from the assumption that the population growth rate is proportional to "growing room," that is, the difference between the population size as it is currently and the maximum possible population. This quantity is called the *population deficit.*

From Equation 2-8, the population growth rate is $(p_{t+1} - p_t)/p_t$. The population deficit is the difference between the population at any given time and the maximum population. Since this is a normalized population, the maximum population is 1 and the population deficit is $1 - p_t$. The equation for this population model is $(p_{t+1} - p_t)/p_t = r(1 - p_t)$. In this model, the growth rate is dependent on two things, growing room and the number r, which represents a measure of the reproduction rate of the species.

In Example 2-2, we found that the growth rate for the situation was 0.0394 per hour. Does that mean that $r = 0.0394$? Think carefully before answering! Putting the given numbers into the population model yields $(0.1767 - 0.17)/0.17 = r(1 - 0.17)$. Solving for r yields the equations

$$0.0394 = 0.83r$$

$$r = 0.0394/0.83 = 0.475$$

The growth rate is indeed 0.0394, but the population deficit must be taken into account to calculate r.

We could make up a table of values to tabulate the population each hour and use the values to plot variation in population over time. We will define a Maple function to help us do the job. Solve Equation 2-8 for p_{t+1}. Accordingly, $p_{t+1} = p_t + p_t r(1 - p_t)$. Let $x = p_t$ and call the function that calculates p_{t+1} simply p. The Maple function is

```
> p := x -> x + r*x*(1-x);
```

$$p := x \rightarrow x + rx(1 - x)$$

We have a value for r in this example, so we can define a new function *p1* with this value of r:

```
> p1 := x -> x +0.0475*x*(1-x);
```

$$p1 := x \rightarrow x + .0475x(1 - x)$$

Let's test the function by supplying it with the initial population value of 0.17 and seeing what it predicts for the population at the next time period:

> p1(0.17);

```
.17670225
```

The model predicts the same population at the end of an hour as was observed.

There are a number of different choices for creating a table. You could proceed step by step. Calculate $p1(0.17)$, copy down the result, and use $p1$ again by using the result, which is 0.1767. You could continue in this fashion until you have enough points to plot a graph. Maple permits you to "automate" the steps once you have decided what to do. The command for producing a sequence of calculations is simply *seq*. For example, the command to produce a sequence of numbers from 2 to 13 is

> seq(ii, ii=2..13);

```
2, 3, 4, 5, 6, 7, 8, 9, 10, 11, 12, 13
```

Notice that *ii* was used for the variable. It is better to use double letters as the variable in the sequence command because you might have used the single variable, *i*, elsewhere in the same worksheet, while it is unlikely that you would have used *ii*. Now construct the command to produce a sequence of population figures. The function $p1$ calculates the population at the next time interval given the current population. You can see that you need to apply the function to itself in a repetitive fashion. Maple also has a command to do this! To find the population at the third time interval, issue the command

> (p1@@3)(0.17);

```
.1907326814
```

Test this command yourself. Change the 3 to a 1 and calculate the population at the first time interval. You will find that it is 0.1767. You can even check it at $t = 0$ by issuing the command *> (p1@@0)(0.17);* and you will find that it is 0.17, as it should be. Thus you could edit the command, changing the 3 to all the numbers in the sequence 1, 2, 3, 4, 5 ... and recording the results until you have enough points for your graph.

Maple allows you to automate the process even further. A point is described to Maple by enclosing the x and y coordinates in square brackets. Thus, our first two points in the sequence are [0, 0.17] and $[1, p1(0.17)]$. The next point would be $[2, p1(p1(0.17))]$. You have seen that it can be written as [2, ($p1$@@2)(0.17)]. Note carefully the brackets around $p1$@@2 and around 0.17. It is essential that the brackets appear in this way. Now form the sequence command.

> P := seq([n,(p1@@n)(.17)], n=0..100):

The command has calculated the population up to the 100th time period, and it has recorded all the data points in the variable P. All you have to do is plot P to see the population growth.

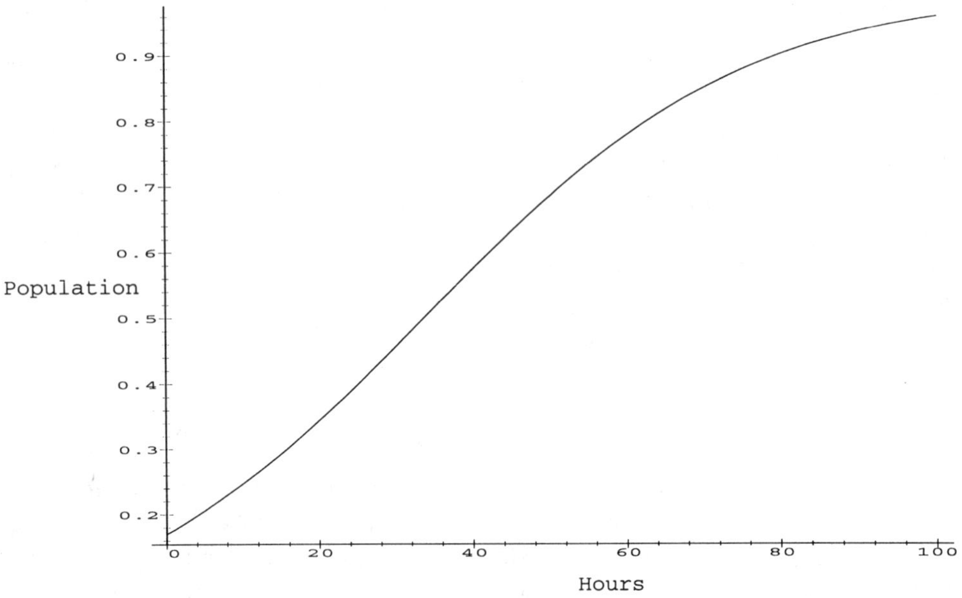

Figure 2.1 Bacterial Population as a Function of Time

> **plot([P]);**

Note that P is enclosed in square brackets, too. This is essential for the plot command to work. Look at the shape of the graph (Figure 2.1). At first, the graph curves upwards. This is the typical shape for a population expanding into a new territory with abundant food sources. As the population grows, it makes more and more demands on the food source. This model shows that a balance is achieved between the tendency for populations to increase and the limit imposed by the food source. The model can be used to predict when the population will reach certain levels. For instance, you might be interested in how long it will take for the population to reach 50% of its maximum value. The graph shows that it takes about 34 time periods. You can check by issuing the command that directly calculates the population at time period number 34.

> **(p1@@34)(0.17);**

```
.5039286875
```

Direct calculation verifies that the population reaches the 50% level after 34 time periods.

The Difference Quotient Revisited

You have seen how to calculate fractional change and percent change and how to represent quantities as a fraction of their maximum value. A natural question to ask about the previous example is, "How fast is the population growing at any given time?"

Let us attempt an answer for the initial period. We will use this example to construct general formulas that will form the foundation of the differential calculus. Our first definition of the *growth rate* is the change in population per unit time.

Definition 1: Growth rate = $(p_{t+1} - p_t)/1$

This formula works well for any situation whose natural time unit is 1. However, it is better to refine the definition of the growth rate by allowing the time interval to be adjustable. The definition remains the same, but the formula is different:

Definition 2: Growth rate = $(p_{t+\mathrm{T}} - p_t)/T$

In our current example, T would naturally be 1, but we could choose T to be any other number. If we choose to measure the population at 2-hour intervals, T would be 2. If we evaluate the growth rate with $T = 2$, we have

Growth rate over a period of 2 hours = (0.1836 – 0.17)/2 = 0.0068

whereas the growth rate based on a 1-hour interval is

Growth rate for the first hour = (0.1767 – 0.17)/1 = 0.0067

You see that the growth rates are very nearly the same, but not identical. This is a bit unsatisfactory, since mathematicians like to give a definite rule that gives definite results. Note the distinction between the definition of *growth rate* and the definition of *fractional change*. The numerators of the fractions are the same in the two definitions, but the denominators differ. In the case of fractional change, the denominator is the same as one of the terms in the numerator. In the case of rate of change, the denominator is *time*. When we calculated the fractional change, the time interval was stated as an ancillary part of the description. The time variable is explicitly mentioned in the growth rate formula. Fractional change formulas are most useful when the time for the change is obvious to anyone. Growth rate formulas are useful when the time over which the change occurs is more flexible. Calculus deals with the latter.

Let us look at the graph for the two times mentioned in definition 2 (Figure 2.2). If we plot the points for hours 0, 1, and 2, the Maple command is:

```
> plot([seq([n,(p1@@n)(.17)], n=0..2)]);
```

Let us draw in the lines to form the triangles that show the difference quotient graphically.

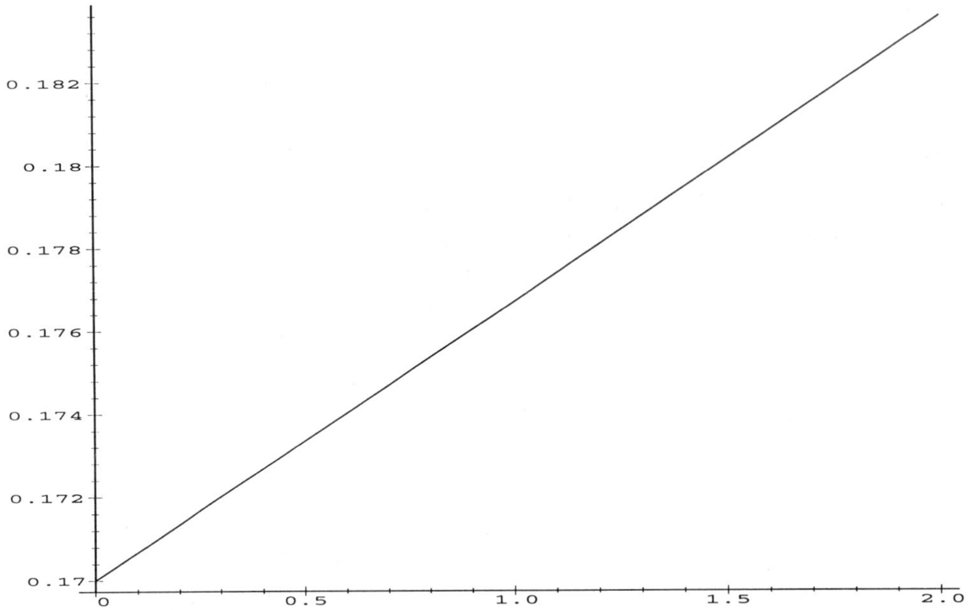

Figure 2.2 Population Line Joining the First Measurement to the Last

```
> plot({[[1,0.17],[1,p1(.17)]], [[2,0.17],[2,(p1@@2)(.17)]],[seq([n,(p1@@n)(.17)],
n=0..2)]}, color = black);
```

The two triangles look similar, which indicates that the ratios of their sides are the same (Figure 2.3). If you sight along the population line, you will see a very slight break at $t = 1$. We have seen that the ratios of the height to the base in the two triangles are not quite identical, since the ratio of the sides for the larger triangle is 0.68, while it is 0.67 for the smaller triangle.

Let us go over much of the same ground once more, this time striving for a more general formula. We will use the usual definition for a function, $y = f(x)$.

We want to find a way of answering the question, "How does y change when x gets slightly larger, relative to the increase in x?" This is another way of asking for the growth rate. We call the formula for the rate of change the difference quotient.

Definition: The difference quotient is

$$DQ = \frac{f(x + dx) - f(x)}{dx} \tag{2-9}$$

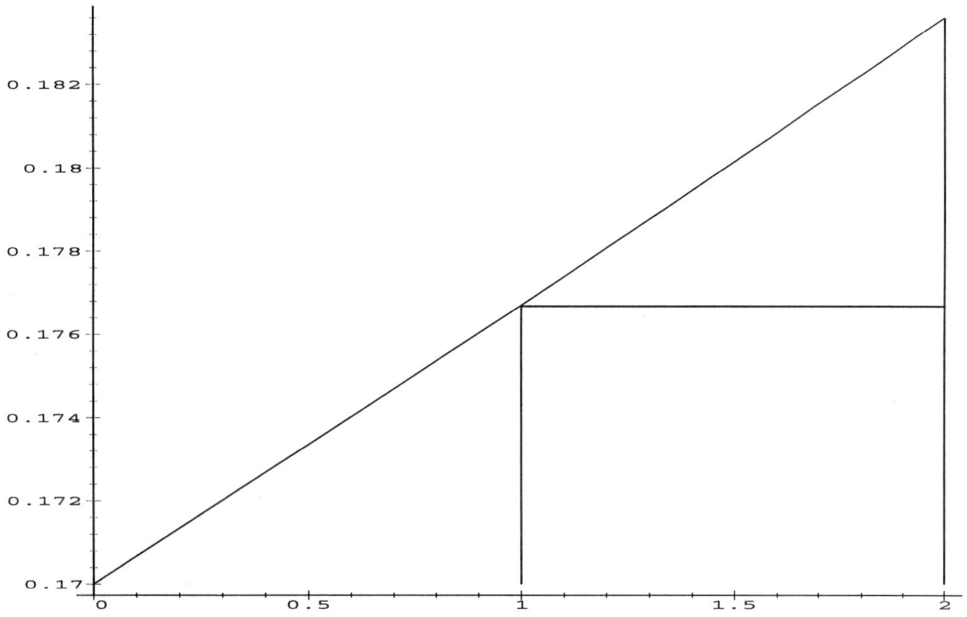

Figure 2.3 Population from 0 to 1 Hours, Then from 1 to 2 Hours

From the definition, observe that: (1) This is exactly the same formula as the previous growth rate formula, with $T = dx, f = p$, and $x = t$. The variables have different names. Apart from that, the formulas are identical. (2) The formula can be written in symmetric form

$$DQ = \frac{f(x_2) - f(x_1)}{x_2 - x_1} \tag{2-10}$$

by substituting $x = x1$, $dx = x2 - x1$, and $x + dx = x2$ in Equation 2-9. It is useful to display the difference quotient graphically. The specific example will be the growth of the function $y = x^2$ from $x = 1.5$ to $x = 2.5$.

> plot({x^2, [[1.5, 1.5^2],[2.5, 1.5^2],[2.5, 2.5^2],[1.5, 1.5^2]]}, x = 1..3, color=black);

If the function is fairly smooth, like this one, the hypotenuse of the triangle will lie close to the curve $y = f(x)$, but will be clearly different from the curve. We have taken the two points fairly far apart, at $x = 1.5$ and $x = 2.5$ (Figure 2.4). If the points are taken closer together, the hypotenuse will be almost indistinguishable from the curve. For example, choose points $x = 1.95$ and $x = 2.05$. The plot command for these values is

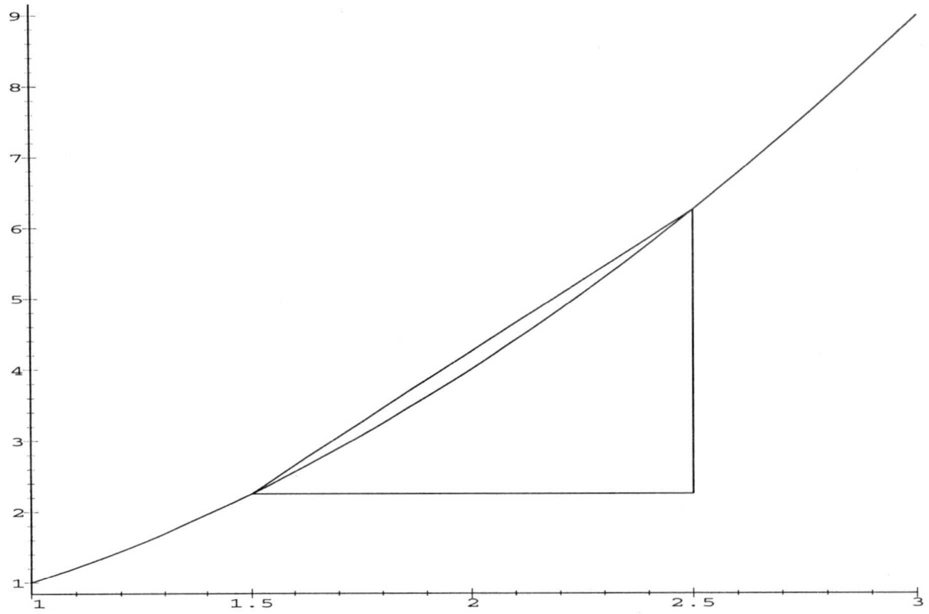

Figure 2.4 The Symmetric Difference Quotient for the Function x^2 at $x = 2$

```
> plot({x^2, [[1.95, 1.95^2],[2.05, 1.95^2],[2.05, 2.05^2],[1.95, 1.95^2]]}, x = 1.9..2.1,
color=black);
```

and there is only a hint of a difference between the hypotenuse and the curve $y = x^2$ (Figure 2.5).

These investigations lead to the conclusion that a small triangle can be constructed anywhere on a smooth curve and that the hypotenuse will be almost indistinguishable from the curve. Remember: the triangle is a result of the formula for the difference quotient. The base of the triangle is dx, and the altitude is $f(x + dx) - f(x)$. The differential calculus begins by calculating formulas for the difference quotient for various functions. We will look at the constant function first.

The Difference Quotient for Constant, Linear, and Quadratic Variation

The constant function is really just a number. In Example 2-3, the value of the function f is 5, no matter what the value of x. Constant functions are quite common in real-life situations. For example, consider the height of an adult human. Let the height function be $h(t)$. As a

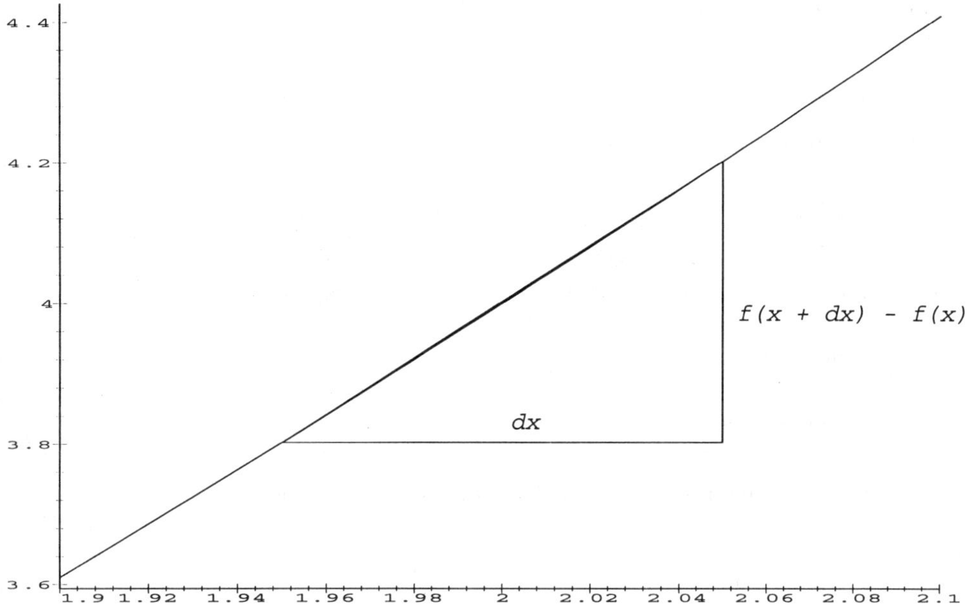

Figure 2.5 The Symmetric Difference Quotient for $x = 2$ with a Smaller Interval Δx.

child, the person's height increased from year to year. Upon reaching adulthood, the person's height remains the same for many years. During that time, the person's height can be measured every year. The fact that it remains the same is stated in functional form as $h(t) =$ constant.

Example 2.3: The Difference Quotient of a Constant Function

Evaluate the difference quotient for $f(x) = 5$.

Solution: The graph of this function is a horizontal straight line at $y = 5$. At any given value of x, the function evaluates to 5. Thus, the difference quotient is

$$(f(x + dx) - f(x)) / dx = (5 - 5)/dx = 0 \qquad (2\text{-}11)$$

If you try to form a triangle on the function $f(x) = 5$, no matter what value for dx you choose, the y side will be 0. Clearly, the fact that the constant in this example was 5 makes no difference. The result would be the same for any given value of the constant.

General Rule 1: The difference quotient of the constant function is zero.
If $f(x) =$ constant, $DQ = 0$.

A constant function graphs as a horizontal straight line. However, what if the graph is a straight line but makes some angle with the x axis? This is called the linear function.

Example 2-4: Linear Variation

A car is traveling along a straight highway at 117 kilometers per hour (32.5 meters per second). Its distance from a certain marker post as a function of time is given by $s = 32.5t$, where s is the distance in meters and t is the time in seconds.

(a) Graph the situation.

(b) Find the difference quotient DQ.

(c) Illustrate the difference quotient on the graph for $t = 2$ seconds.

Solution:

(a) The graph is a straight line, as shown by the *plot* command (see Figure 2.6).

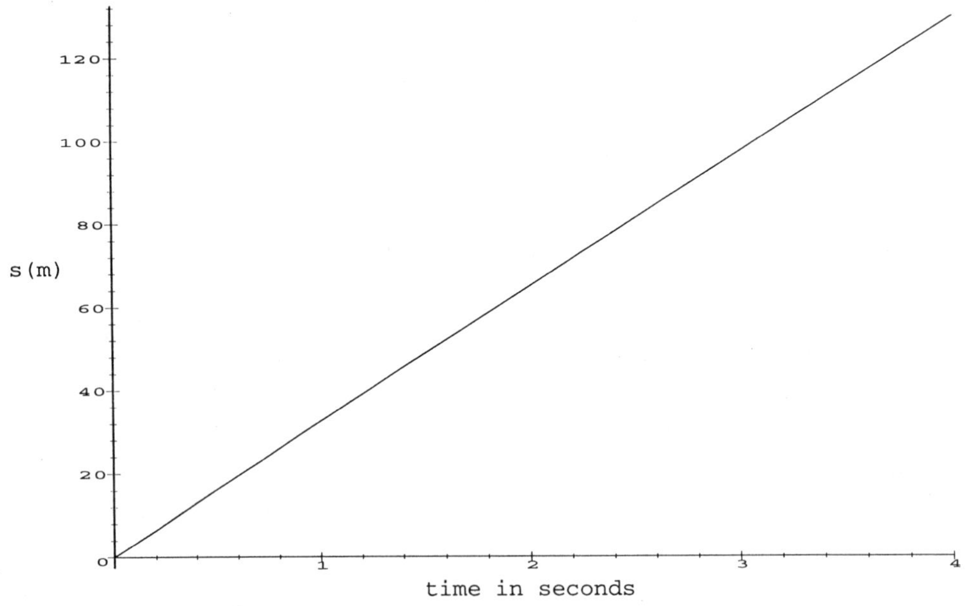

Figure 2.6 Distance-Time Graph for Example 2-4

```
> plot(32.5*t, t = 0 .. 4, labels = [`time in seconds`, `s (m)`]);
```

(b) To apply the formula for the difference quotient, a function is needed. In this case, you can define the distance function as

```
> s := t -> 32.5*t;
```

$$s := t \to 32.5t$$

The distance at any time t is given by $s(t)$. For instance, the distance at $t = 10$ seconds is

```
> s(10);
```

$$325.0$$

Since we are not given any specific time at which to calculate the difference quotient, we use the general form of the formula

```
> DQ = (s(t +dt) - s(t))/dt;
```

$$DQ = \frac{s(t + dt) - s(t)}{dt} \tag{2-12}$$

You may be surprised at the result. The difference quotient is a constant, equal to 32.5. If we perform the calculation longhand, we have

$$DQ = (32.5(t + dt) - 32.5t)/dt = (32.5t + 32.5dt - 32.5t)/dt = 32.5dt/dt = 32.5$$

The terms in t and dt drop out of the calculation, leaving a constant result. Once again, the result is independent of the size of dt.

Your turn: Calculate the difference quotient using Maple.

Answer: The appropriate Maple command is ________________________________

(c) To plot the graph, use the command

```
> plot({s, [[1.5, s(1.5)],[2.5, s(1.5)],[2.5, s(2.5)]]}, 0 .. 4, color = black, labels = [`t (s)`, `s (m)`]);
```

Since the graph is a straight line, all triangles built on this line will be similar (see Figure 2.7). The ratio of their sides are the same. In analytic geometry you learned that this ratio is called the *slope* of the line. In symbols, $dy/dx = m$.

General Rule 2: The difference quotient of a linear function is a constant.
If $f(x) = cx$, then $DQ = c$.

Before investigating the difference quotient for other types of functions, let us ask about the difference quotient of combinations of functions. In general, what is the difference quotient

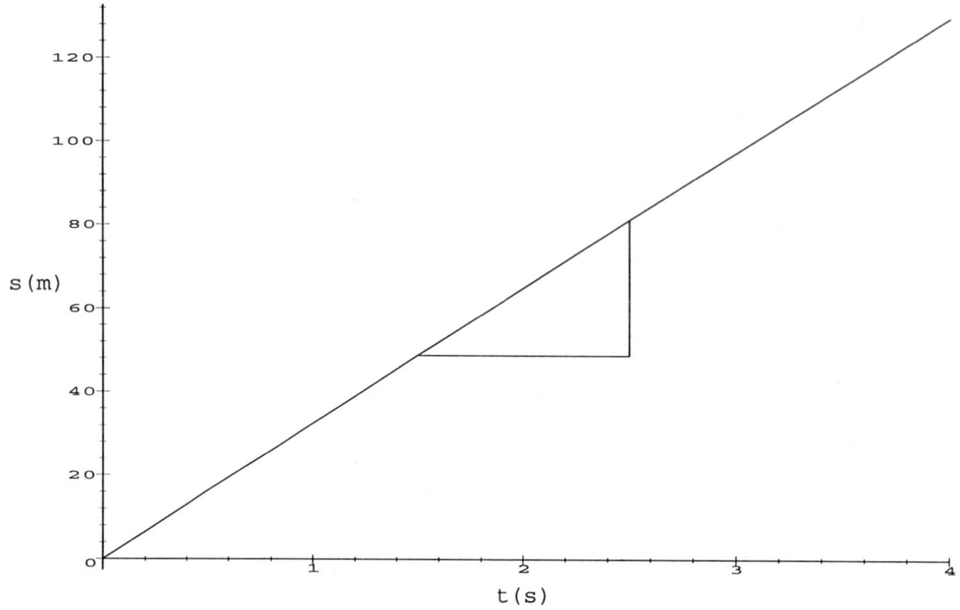

Figure 2.7 Calculating the Difference Quotient from the Graph

of sums and differences of functions, and what is the difference quotient of a function multiplied by a constant?

> General Rule 3: The differential quotient of a sum of functions is the sum of the difference quotients: $DQ(f+g) = DQ(f) + DQ(g)$.

We want to show that $DQ(f+g) = Dq(f) + DQ(g)$. In other words, you can add the functions first and then take the difference quotient or you can take the difference quotients of both functions and then add the results. Either way, the answer is the same.

Proof: This is a good place to illustrate how to tell Maple about sums of functions. If f and g are two functions, their sum is simply $f+g$. If you want to evaluate this combined function at x, the syntax of the command is

```
> (f + g)(x);
```

$$f(x) + g(x) \qquad \text{Form 1}$$

Note the brackets around $(f+g)$. These tell Maple that both f and g are to be evaluated at x. Now we form the difference quotient for $f+g$.

> ((f+g)(x+dx)-(f+g)(x))/dx;

$$\frac{f(x+dx)+g(x+dx)-f(x)-g(x)}{dx} \qquad \text{Form 2}$$

This is easily seen to be the same as the difference quotient of f added to the difference quotient of g.

> (f(x + dx) - f(x))/dx + (g(x + dx) - g(x))/dx;

$$\frac{f(x+dx)-f(x)}{dx}+\frac{g(x+dx)-g(x)}{dx} \qquad \text{Form 3}$$

This last result can be brought to a common denominator by using the *normal* command.

> normal((f(x + dx) - f(x))/dx + (g(x + dx) - g(x))/dx);

$$\frac{f(x+dx)+g(x+dx)-f(x)-g(x)}{dx} \qquad \text{Form 4}$$

Since Form 2 is identical to Form 4, General Rule 3 is established.

Your Turn: Show that the difference quotient of $f-g$ is the same as the difference quotient of f minus the difference quotient of g.

Answer: ____

General Rule 4: The difference quotient of a function multiplied by a constant is the same as the constant multiplied by the difference quotient of a function: $DQ(cf(x)) = cDQ(f(x))$.

Proof: Use Maple to form the difference quotients.

> (c*f(x + dx) - c*f(x))/dx = c*(f(x + dx) - f(x))/dx;

$$\frac{cf(x+dx)-cf(x)}{dx}=\frac{c(f(x+dx)-f(x))}{dx}$$

You may be able to see that the two sides of the equation are the same. If you need further proof, expand both sides.

> expand((c*f(x + dx) - c*f(x))/dx) = expand(c*(f(x + dx) - f(x))/dx);

$$\frac{cf(x+dx)}{dx}-\frac{cf(x)}{dx}=\frac{cf(x+dx)}{dx}-\frac{cf(x)}{dx}$$

The two sides are identical and the proof is complete. Now we will continue to investigate the difference quotient of the quadratic function. We will show two examples.

Example 2-5: Basic Quadratic Variation

Find the difference quotient of the squaring function.

Solution: The squaring function takes a number and multiplies it by itself. In equation form, it is $y = x^2$. In functional notation it is

```
> sq := x -> x^2;
```

$$sq := x \rightarrow x^2$$

The differential quotient for the squaring function is

```
> (sq(x + dx) - sq(x))/dx;
```

$$\frac{(x + dx)^2 - x^2}{dx}$$

Make sure you understand what was done in this command. We have used *sq* instead of *f* because *sq* is the name of this particular function. Then we evaluated the function *sq* at two points, $x + dx$, and x, just as we did in Equation 2-9. This command is *exactly* the same as the formula for the difference quotient, using *sq* as the specific name for the function *f* in Equation 2-8.

This result can be simplified by using *expand*.

```
> expand((sq(x + dx) - sq(x))/dx);
```

$$2x + dx$$

The term *dx* dropped out of the difference quotient for constant and linear variation. In the case of quadratic variation, the expression for the difference quotient contains *dx*. This means, of course, that the result depends on the size of *dx*. Review Figures 2.3 and 2.4. There you saw that the size of *dx* affects how well the hypotenuse matches the curve that represents the function. The smaller *dx*, the better the fit.

Example 2-6: Linear and Quadratic Variation Combined— The Logistic Equation

Find the difference quotient for the logistic equation $y = f(x) = kx(1 - x)$. How fast is y increasing when $x = 1/3$ and $k = 3$?

Solution:

```
> f := x -> k*x*(1 - x);
```

$$f := x \to kx(1-x)$$

```
> DQ = (f(x + dx) - f(x))/dx;
```

$$DQ = \frac{k(x+dx)(1-x-dx) - kx(1-x)}{dx}$$

```
> simplify(DQ = (f(x + dx) - f(x))/dx );
```

$$DQ = -k(2x - 1 + dx)$$

```
> expand(simplify(DQ = (f(x + dx) - f(x))/dx ));
```

$$DQ = -2kx + k - kdx$$

Remember that you do not have to retype all the commands. You do have to type the definition of the logistic function and the formula for the difference quotient. After that, type the simplify command as *simplify();* and then copy the formula for the difference quotient from the Maple input line you already typed and paste it between the brackets in the *simplify* command. Similarly, you copy the entire *simplify* command (without the final semicolon) and paste it between the brackets of *expand();*. In this way, you can see all of the intermediate results with just a little extra typing.

Since the logistic formula contains a quadratic term, there is a *dx* term in the result.

(b) To evaluate the difference quotient for given values of *k* and *x*, simply substitute the values in the result.

```
> subs( x = 1/3, k = 3, expand(simplify(DQ = (f(x + dx) - f(x))/dx )));
```

$$DQ = 1 - 3dx$$

The value of the difference quotient for the given values of *k* and *x* is $1 - 3dx$.

When the difference quotient contains *dx*, you would naturally ask, "How big is *dx*?" This question gets at the heart of the differential calculus. To help you decide on the size of *dx*, let's graph the logistic equation $y = f(x) = 3x(1 - x)$ (Figure 2.8).

```
> plot( { 3*x*(1 - x), [[1/3, 0],[1/3, subs(x=1/3, 3*x*(1 - x))]] }, x = 0 .. 1, color=black);
```

We wish to calculate the increase in *y* when we let *x* increase from 1/3 to "a little bit" greater than 1/3. We will call this little bit *dx*. We will plot the graph for a small interval around $x = 1/3$, and we will construct the triangle whose sides are *dx* and $y(x + dx) - y(x)$.

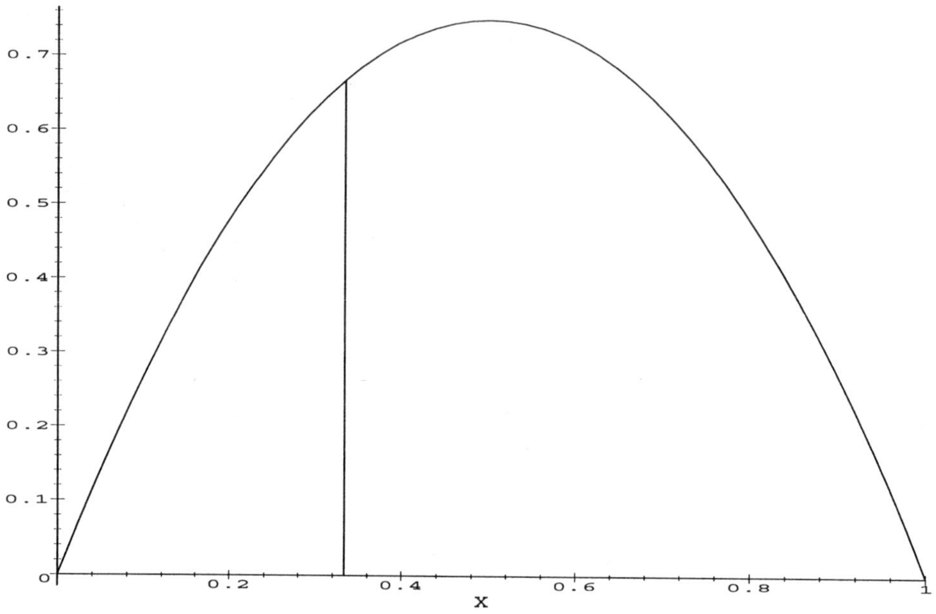

Figure 2.8 The Value of the Logistic Equation $y = 3x(1 - x)$ at $x = 1/3$

> y := x -> 3*x*(1 - x); dx := 0.01;

$$y := x \to 3x(1 - x)$$

$$dx := .01$$

> plot({ y, [[1/3, y(1/3)],[1/3+dx, y(1/3)], [1/3+dx, y(1/3+dx)]] }, 1/3 - dx .. 1/3 + 2*dx, color=black);

The base of the triangle in Figure 2.9 is $dx = 0.01$. The altitude is

> y(1/3 + dx) - y(1/3);

$$.0096999998$$

Therefore the difference quotient is (approximately)

> .0097/.01;

$$.9700000000$$

The differential quotient is a number just a little less than 1.

Note that the hypotenuse of the triangle seems to lie right over the curve y. We have chosen a value of dx small enough that the graph of $y = 3x(1 - x)$ looks like a straight line over the interval $x = 0.33 .. 0.35$. This is an important observation. For one thing, it means that we could replace the portion of the curve between $x = 1/3$ and $x = 1/3 + 0.01$ with the hypotenuse of the triangle formed from the difference quotient without making much of an error. This notion will be expanded in the next section, where we begin the discussion of limits.

The Idea of a Limit

You can see by looking at Figure 2.9 that, as dx is made smaller and smaller, the hypotenuse will almost coincide with the function. The difference quotient approaches $3 - 6x$, or in the case of the general logistic equation, the difference quotient approaches $k - 2kx$ (see Example 2-6, fourth Maple output line).

You could arrive at this result for the difference quotient simply by setting dx equal to zero, but then you would be dividing by zero in the formula for the difference quotient. How small can dx get? Let us try various values for dx and see what the difference quotient evaluates to. We start with the definition of the difference quotient, expressed in functional notation:

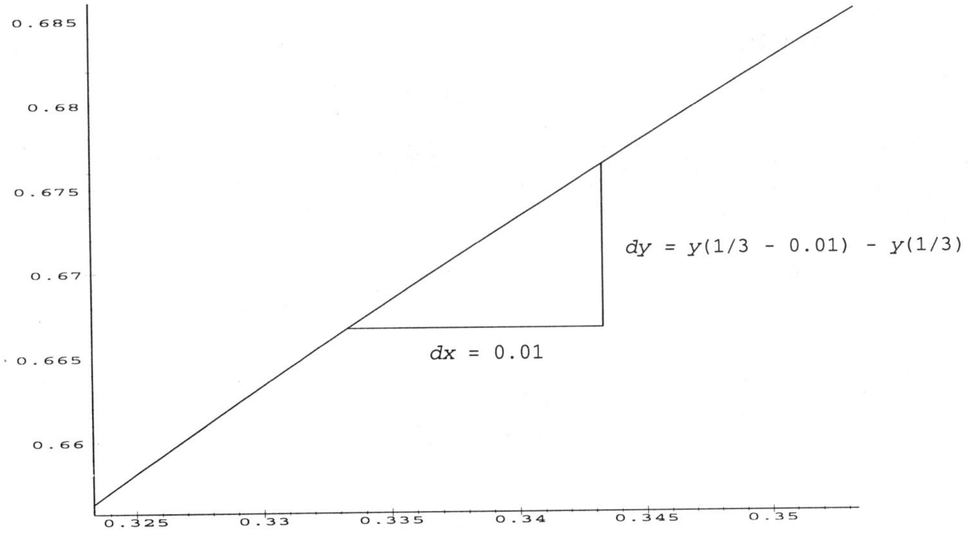

Figure 2.9 The Differential Quotient for the Logistic Equation at $x = 1/3$ and $dx = 0.01$

```
> dq1 :=f -> (f(x + dx) - f(x))/dx;
```

$$dq1 := f \to \frac{f(x + dx) - f(x)}{dx}$$

Make sure that *dx* does not have a numerical value by unassigning it.

```
> dx := 'dx':
```

Define the particular function we want to investigate: the logistic function.

```
> f := x -> 3*x*(1 - x);
```

$$f := x \to 3x(1 - x)$$

Find its differential quotient for the case $x = 1/3$ and give the result a name. Call it *R*. Note the use of the *subs* command to give *x* the value 3.

```
> R := subs(x = 1/3, dq1(f));
```

$$R := \frac{3\left(\frac{1}{3} + dx\right)\left(\frac{2}{3} - dx\right) - \frac{2}{3}}{dx}$$

We'll use the *subs* command again to calculate the difference quotient for various values of *dx*.

```
> subs(dx = 0.01, R);
```

$$.969999800$$

This result has already been calculated. We have included it here as a check on our work. Now, try smaller values for *dx*. Copy and paste the equation, and then edit the number 0.01 by putting in an extra zero.

```
> subs(dx = 0.001, R);
```

$$.9969998000$$

The result is much closer to 1. We'll try making *dx* smaller yet to see if we get even closer to 1.

```
> subs(dx = 0.0001, R);
```

$$.9996980000$$

The technique seems to be working. The smaller the value of *dx*, the closer the difference quotient gets to 1. Let's try a much smaller value for *dx*.

> subs(dx = 0.0000001, R);

```
.9980000000
```

This number is farther away from 1 than the previous result! Moreover, if we add even more zeros, making *dx* very small indeed, the result is completely wrong.

> subs(dx = 0.0000000001, R);

```
-1.000000000
```

There is no way the result can be negative as both *dy* and *dx* are positive numbers. Thus, the ratio *dy*/*dx* must be a positive number. What has Maple done to give us such an impossible answer? To find out, count the number of zeros after the decimal point in *dx*. There are 9 zeros, and then the 1. By default, Maple works to 10-digit accuracy:

> Digits;

```
10
```

Thus, our number *dx* is too small for the accuracy to which Maple is working. Once this happens, you cannot rely on the computation. The solution is to increase the accuracy to which Maple works when it does (approximate) numerical calculations and then recalculate the difference quotient.

> Digits := 20; subs(dx = 0.0000000001, R);

```
Digits := 20
.99999999950000000000
```

We have regained the lost accuracy, and the result is a number very close to 1.

You can conclude that the formula for the difference quotient can be applied to very small values of *dx*, as long as you are careful with the numerical calculations. If you are working to 10-digit accuracy, you can't set *dx* smaller than about 0.000001 if you want a reliable result. If you do increase Maple's accuracy, you can work to just about any reasonable number of digits. Since Maple is doing the calculations, set *Digits* to 100 and calculate the difference quotient.

> Digits := 100; subs(dx = 0.00000000000000000001, R);

```
Digits := 100

9999999999999999999699999999999999999999999999999999999999999
999999999999999999800000000000000000000
```

This number differs from 1 somewhere after the 50th digit. Thus, the conclusion that the difference quotient approaches 1 as *dx* gets smaller and smaller is inescapable.

We don't want to work to 100 digits any longer, so we set the number of digits back to 10 with the command:

> **Digits := 10;**

```
Digits := 10
```

(Instead, we could have simplified the difference quotient before substituting values for *dx*.)

> **simplify(dq1(f));**

$$-6x + 3 - 3dx$$

Now the setting for Maple's *Digits* variable is no longer important. In fact, if we make *dx* very small and substitute it in the expression above, the result will be 1.

> **subs(dx = 0.00000000001, subs(x = 1/3, -6*x +3 - 3*dx));**

```
1.000000000
```

Why did we get the correct result this time? Notice that we are no longer dividing by *dx*. If the value of *dx* in the above command is zero, the term containing *dx* simply "drops out" of the expression entirely. In other words, setting *dx* to be zero gives the correct answer. In fact, the answer is correct to an infinite number of decimal places. But how can you be *sure* it is correct? Remember, if you do set *dx* equal to zero here, it has to be zero at the start of the problem, too. But then the difference quotient reduces to $\frac{f(x+0)-f(x)}{0} = \frac{f(x)-f(x)}{0} = \frac{0}{0}$, which is undefined.

The escape from this logical dilemma is to ask the question, "To how many decimal places do you wish to work?" Given any answer whatsoever, make *dx* smaller than the last decimal place and add a few more zeros for good measure if you want. Of course, if you are working in Maple, you have to tell it to work to that number of decimal places. Your calculations for the difference quotient will then be accurate to that many decimal places.

The theory of limits makes these arguments more precise. You see that the difference quotient itself produces the undefined result 0/0 if *dx* is 0. Yet the difference quotient *approaches* a definite number if *dx* is made smaller and smaller, but not as low as 0. The number that the difference quotient approaches in this way is called the *limit* (if there is one) of the difference quotient as *dx* approaches 0. In symbols, it is expressed as

$$\lim_{x \to 0} f(x) = k \tag{2-13}$$

Using Maple, we express it as

> **Limit(f(x), x=0) = k;**

$$\lim_{x \to 0} f(x) = k \qquad (2\text{-}14)$$

Note that the command *Limit* is capitalized. Use this form when you don't want Maple to work out the answer. (You can always work out the answer later. You'll learn how in the next section.) Notice also that the Maple command syntax is $x = 0$, meaning "as x approaches 0 from either direction." The limit may or may not exist. Maple usually tells you if the limit does not exist.

Calculating the Limit of sin(x)/x

Later on, when we attempt to find the derivative of sin(x), we will encounter an expression like sin(x)/x and we will need to define its value at $x = 0$. The first step in deciding what value to assign is to plot the function. Since we wish to see the function at $x = 0$, we don't want the y axis getting in the way. We'll use the plot option *axes = boxed* to remove the y axis line.

> **plot(sin(x)/x, x = -Pi .. Pi, axes = boxed);**

The graph has a value near 1 when x is near zero.

The graph in Figure 2.10 was produced using Maple Version 3 (V3). Note the small "hole" at $x = 0$. Maple is trying to tell us that there is no value for sin(x)/x at $x = 0$, so it leaves this point out. For some reason, Maple V4 draws a smooth curve that gives no indication that the function is not defined at $x = 0$.

Now that we have a good picture of the behavior of sin(x)/x in the neighborhood of $x = 0$, we can calculate its value for some very small values of x.

> **S := sin(x)/x;**

$$S := \frac{\sin(x)}{x}$$

Let's try $x = 0.001$ and evaluate sin(x)/x.

> **subs(x = 0.001,S);**

$$1000.\ \sin(.001)$$

This wasn't too helpful, because Maple doesn't automatically convert sin(.001) to a decimal number. All we have to do, though, is wrap an *evalf* around the command:

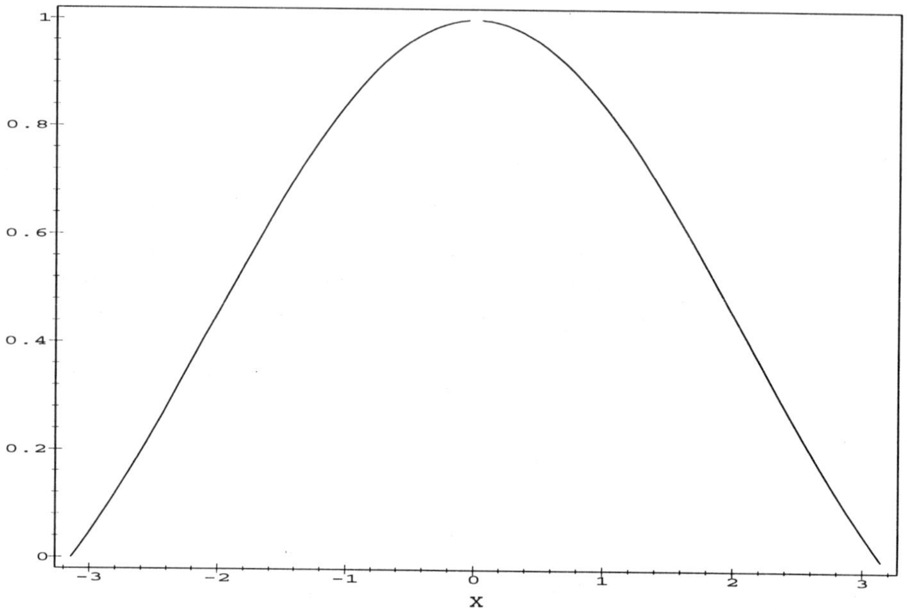

Figure 2.10 Graph of sin(x)/x around $x = 0$

> evalf(subs(x = 0.001,S));

```
.9999998333
```

The result confirms our conclusion based on examining the graph of sin(x)/x. The result differs from 1 in the seventh place.

Your Turn:

Use the diagram in Figure 2.11 to prove that the limit of sin(x)/x is 1. We'll give you some hints as you work along.

1. Sector *DOB* has radius 1. The angle at the vertex of the sector is $\angle DOB = x$. Show that the arc length is also x. Use the formula for the length of a circular arc.

2. Is the straight line distance *DB* greater or less than the arc length x? Is x greater or less than *BC*? Express the ordering of these three lengths as an inequality.

   Answer: ______________________________

3. Express *DA* in terms of x using trigonometry. *Answer*: ______________

4. Express *CB* in terms of x using trigonometry. *Answer*: ______________

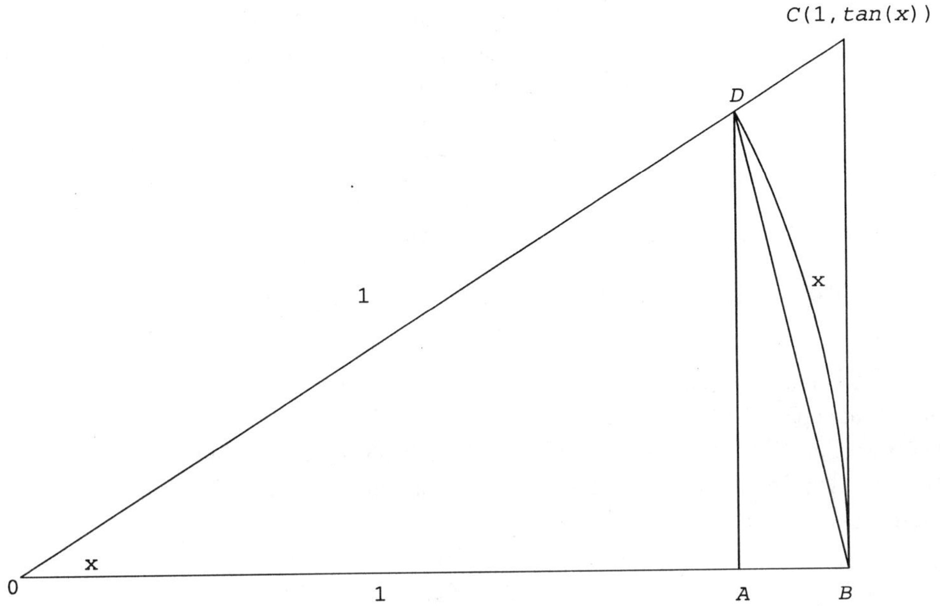

Figure 2.11 Diagram Showing That $\sin(x)/x = 1$

5. Write the lengths *AD*, *x*, and *BC* in order of increasing length. Use inequality signs to express the ordering of the terms. Use step 2 to help you decide where *x* goes in the ordering.

   Answer: ______________________________

6. Substitute in step 5 for *DA* and *CB* using the results in steps 3 and 4. You now have an inequality that contains only *x* and trigonometric functions of *x*.

   Answer: ______________________________

7. Divide each term in the inequality by $\sin(x)$. At this point, you should have an inequality whose terms are 1, $x/\sin(x)$, and $\tan(x)/\sin(x) = 1/\cos(x)$.

   Answer: ______________________________

8. Take reciprocals of each term. Remember to reverse the inequality signs, because if $x > y$, then $1/x < 1/y$.

   Answer: ______________________________

9. Your inequality should have the terms 1, sin(x)/x, and cos(x). What is the limit of cos(x) as x approaches 0? This is easy, because cos(x) is defined as x = 0, so the question is really asking the value of cos(0).

   Answer: ______________________________

10. You have bracketed the quantity sin(x)/x between the number 1 and cos(x), whose limit is 1 as x approaches 0. What does this say about the limit of sin(x)/x?

    Answer: ______________________________

The proof that you have given here by following the outline needs some strengthening to make it acceptable to mathematicians, but it outlines the path to be followed. The result depends in part on the *sandwich* or *squeeze theorem*, which states that if $f(x) \leq g(x) \leq h(x)$ for all x near a, and if both $f(x)$ and $h(x)$ have limit L, then $g(x)$ also has limit L. You can find a proof of this theorem in most calculus textbooks. The end result can be stated in mathematical notation as

$$\lim_{x \to 0} \frac{\sin(x)}{x} = 1 \qquad (2\text{-}15)$$

Attempting to Calculate the Limit of sin(x)/|x|

Equation 2-15 can be used to derive many other trigonometric limits that reduce to 0/0 at the point where the limit is being calculated. Perhaps we can use it to calculate the limit of sin(x)/|x| as x approaches 0. Following the procedure outlined in the previous section, the first thing to do is to plot the graph. In Maple, the absolute value of x is written as *abs*(x).

```
> plot(sin(x)/abs(x), x= -Pi .. Pi);
```

Here we encounter a problem! The graph shows that the function takes a large jump at x = 0 (Figure 2.12). The limit does not exist. A condition for the limit to exist is that the function's value should be close to the value at the limit point for any nearby value of x. Since that is not true for the function sin(x)/|x|, it does not have a limit at x = 0. Let us see what Maple says about the limit of sin(x)/|x| as x approaches 0.

```
> limit(sin(x)/abs(x), x = 0);

undefined
```

The notion of limits can be extended to include *directional limits*. In this case, the limit is computed as x closes in on the limit point from one side only. We can use Maple to see if directional limits exist for sin(x)/|x|.

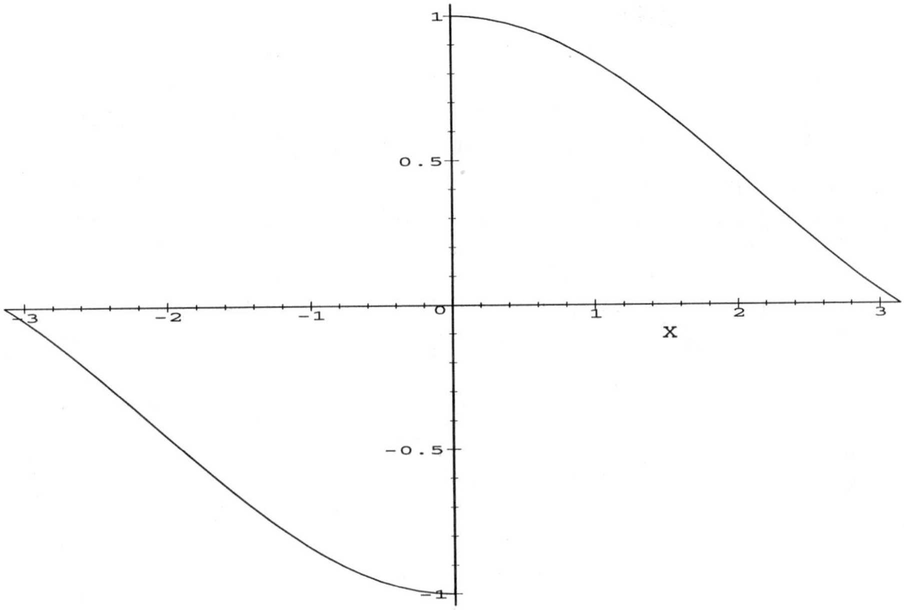

Figure 2.12 Graph of sin(x)/|x| from $-\pi$ to π

> limit(sin(x)/abs(x), x = 0, right);

$$1$$

> limit(sin(x)/abs(x), x = 0, left);

$$-1$$

These answers correspond with the graph.

All that remains is to review the process for finding the limit of a function at a given value for x.

1. If the function exists at x, the limit is simply the value of the function there.

2. If the function evaluates to some undefined value, such as 0/0, use Maple to plot the graph of the function for a range of x values around the point in question. The graph should give you a good idea if the limit exists, and if so, what it is.

3. Use the Maple *limit* command. It is able to calculate most limits of interest in introductory college calculus courses.

The Notion of Continuity

The notion of continuity has been mentioned briefly before in this chapter. The basic idea is that a continuous function is one whose graph you can trace without taking the pencil off the paper. We will restrict the discussion to the Maple command that tests continuity. As in the case of limits, the first approach should be to draw the graph of the function. If the graph has a break, or jump, at some value of x, then the function is not continuous there. It is said that the function is *discontinuous* there, or that x is a *point of discontinuity*.

The Maple command for testing continuity is *iscont*. We can test the continuity of $\sin(x)/x$ and $\sin(x)/|x|$ as follows.

```
> iscont( sin(x)/x, x = -Pi .. Pi);
                                   false
> iscont( sin(x)/abs(x), x = -Pi .. Pi);
                                   false
```

Neither function is continuous at $x = 0$. However, if the definition of $\sin(x)/x$ is extended by *defining* the value of $\sin(x)/x = 1$ at $x = 0$, then the function is continuous.

Paper and Pencil Exercises

PP2–1

Write the Maple command to plot the constant function $f(x) = 10$ from $x = -1$ to $x = 1$ in two different ways. Use (a) an *expression* in the plot command and (b) functional notation for $f(x) = 10$.

(a) *Answer*: ______________________________

(b) *Answer*: ______________________________

PP2–2: The Origins of the Difference Quotient

(a) Find the speed of a car from the 30th and 31st frames if $s(30) = 1{,}150$ feet and $s(30 + 1/15) = 1{,}155$ feet.

Answer: ______________________________

(b) Convert the answer to miles per hour.

Answer: ______________________________

PP2-3: Fractional Change in the Cost of Butter

(a) The price of butter is increased from \$3.40 to \$3.65. What is the fractional change in the cost?

Answer: ______________________________

(b) What is the percent change?

Answer: ______________________________

PP2-4: Fractional Change in Currency Value

(a) During inflationary times, the "goonie," the currency of a small country, sank in value in relation to the U.S. dollar. The goonie was once worth \$1.10 U.S., but after economic hard times beset the country, the goonie was worth \$.72. What is the fractional change in the goonie's worth?

Answer: ______________________________

(b) What is the significance of the minus sign in the answer to part (a)?

Answer: ______________________________

PP2-5

A country reports a fractional increase in its population for the year just ending of 0.015. The population of the country at the beginning of the year was 640,000,000.

(a) How many more people does this fractional increase represent?

Answer: ______________________________

(b) What is the country's population growth for that year, expressed as a percent?

Answer: ______________________________

PP2-6

The formula for converting temperatures from Fahrenheit to Celsius is

$$C = \frac{5}{9}(\mathrm{F} - 32)$$

(a) Find the difference quotient $\frac{\Delta C}{\Delta F} = \frac{C(F + \Delta F) - C(F)}{\Delta F}$. Here, the symbol ΔF has the same meaning as dx. It is a small change in the value of F. Similarly, ΔC is the small change in C resulting from the small change in F.

Answer: ______________________________

(b) What is the interpretation of the differential quotient $\frac{\Delta C}{\Delta F}$? Is it

(i) The size of a Fahrenheit degree in terms of a Celsius degree?

(ii) The size of a Celsius degree in terms of a Fahrenheit degree?

Answer: ______________________________

PP2–7

Find the difference quotient of $y = 1/x$ by algebraic methods.

(a) Write the formula for the difference quotient for this function.

Answer: ______________________________

(b) Simplify the formula by finding the common denominator.

Answer: ______________________________

(c) Set $\Delta x = 0$ (or $dx = 0$, depending on what you are using as the symbol for the small change in x) in your answer to part (b). This is the derivative of $1/x$ with respect to x.

Answer: ______________________________

PP2–8: Maple Syntax

Write the proper commands to accomplish the following tasks.

(a) Plot a graph of the linear function $F = \frac{9}{5}C + 32$, from $C = -40$ to 100 degrees Celsius.

Answer: ______________________________

(b) Define the "cubing function," *cb*, in terms of the variable x. This function cubes its input; for example, $cb(3) = 27$.

Answer: ______________________________

(c) Define the reciprocal function.

Answer: ______________________________

(d) You can define functions of two variables using the syntax, f:= (x, y) -> `expression involving $x$ and $y$\`. Define the symmetric difference quotient and name it *sdq*. Use variables $x1$ and $x2$ instead of x and y in your definition.

Answer: ______________________________

(e) Which command results in a *syntax error* message?

(i) > \`your turn\`;

(ii) > "your turn";

Answer: ______________________________

PP2-9

(a) > plot(9/5*C + 32, C = -40 .. 100);

(b) > cb := x -> x^3;

(c) > x -> 1/x;

(d) > sdq := (x1, x2) -> (f(x2) - f(x1))/(x2 - x1);

PP2-10

Given the logistic equation $x_{n+1} = \frac{1}{2}x_n(1 - x_n)$ and an initial value $x_0 = ½$, mentally calculate

(a) x_1 *Answer*: ______________

(b) x_2 *Answer*: ______________

(c) Estimate x_∞ *Answer*: ______________

PP2-10

State the Maple commands that produce these outputs

(a) $\lim\limits_{x \to 0} e^x = 1$ *Answer*: ______________

(b) $\lim\limits_{x \to 1} 1n(x) = 0$ *Answer*: ______________

(c) $\lim\limits_{x \to 3+} \frac{|x - 3|}{x - 3} = 1$ *Answer*: ______________

(d) $\lim_{x \to 3-} \frac{|x-3|}{x-3} = -1$ *Answer*: ____________________

(e) $\lim_{x \to 1} \frac{\sin(x-1)}{x-1} = 1$ *Answer*: ____________________

(f) Based on your answers to parts (c) and (d), what is $\lim_{x \to 3} \frac{|x-3|}{x-3}$?

Answer: ____

Maple Lab

ML2-1

Verify these general rules for the properties of limits using Maple's *Limit* command. (Really, you are seeing if Maple knows these rules for limits.)

(a) The limit of a constant function:

> assume(c, constant);

> Limit(c, x = a) = value(Limit(c, x = a));

This is *not* the same as General Rule 1 in this chapter. Why?

Answer: ____

(b) The limit of a function multiplied by a constant:

> Limit(c*f(x), x = a) = expand(Limit(c*f(x), x = a));

Is the form of this rule the same as General Rule 2 for the difference quotient?

Answer: ____

ML2-2: The Difference Quotient

Define a general form for the difference quotient as

> dq := f -> (f(x + dx) - f(x))/dx;

$$dq := f \to \frac{f(x + dx) - f(x)}{dx}$$

Note that our definition requires that *f* be a *function*. Thus, if the difference quotient of *f*(*x*) = *x* is required, you would define the Maple function

> X := x -> x;

$$X := x \rightarrow x$$

> dq(X);

$$1$$

Or you could simply put the functional definition in the *dq* command:

> dq(x -> x);

$$1$$

Work out the difference quotients for the following functions. Write the answers by grouping the terms containing *dx* (if there are any) in a way that simplifies the expression. Use the commands *expand* and, if you wish, *collect* and *simplify* to organize your answer in the most understandable form.

(a) The function $f(x) = 3x + 2$

> dq(x -> 3*x + 2);

Answer: ______________________________

(b) The function $f(x) = 20x^2 - 40x + 1{,}234$

> dq(x -> 20*x^2 - 40*x + 1234);

Answer: ______________________________

(c) The function $f(x) = x^3 - 3x^2 + 2x + 1$

> dq(x -> x^3 - 3*x^2 +2*x + 1);

Answer: ______________________________

(d) The function $f(x) = 1/x$

> dq(x -> 1/x);

Answer: ______________________________

(e) The function $f(x) = 1/x^2$

> dq(x -> 1/x^2);

Answer: ______________________________

ML2–3: The Difference Quotient When *dx* Approaches Zero

Find the limit as *dx* approaches 0 in your answers for ML2–2 by using Maple's *limit* command.

(a)

```
> limit(dq(x -> 3*x + 2), dx = 0);
```

Answer: ____________________

(b)

```
> limit(dq(x -> 20*x^2 - 40*x + 1234), dx = 0);
```

Answer: ____________________

(c)

```
> limit(dq(x -> x^3 - 3*x^2 +2*x + 1), dx = 0);
```

Answer: ____________________

(d)

```
> limit(dq(x -> 1/x), dx = 0);
```

Answer: ____________________

(e)

```
> limit(dq(x -> 1/x^2), dx = 0);
```

Answer: ____________________

(f) Compare your answers here to the answers in ML2–2. Do you get these results by letting $dx = 0$ in ML2–2?

Answer: ____________________

ML2–4

Find the difference quotient *dq* of these functions; then apply the limit command (as *dx* approaches 0) to find the formulas for the derivatives of the functions.

(a) f := x -> x^n; (the power function) *Answer*: ____________________

(b) f := x -> sin(x); (the sine function) *Answer*: ____________________

(c) f := x -> cos(x); (the cosine function) *Answer*: ____________________

(d) f := x -> exp(x); (the e^x function) *Answer*: ____________________

(e) f := x -> 10^x; (a power function) *Answer*: ____________________

(f) f := x -> ln(x); (the logarithmic function) *Answer*: ____________________

(g) f := x -> log[10](x); (the $\log_{10}$ function) *Answer*: ____________________

(h) f := x -> sqrt(x); (the square root function) *Answer*: ______________

ML2–5: Why Mathematicians Use Radian Measure

In the last question you were asked to find the limit of the difference quotient for the function $f\colon x \to \sin(x)$. The angle x is in radian measure, since Maple uses only radians for angular measure.

In this question you are to find the limit of the difference quotient of the function $f\colon x \to \sin(180x/\pi)$. You will recognize that you are to find the derivative of $\sin(X)$, where X is in degrees. In your answer, give the formula for the limit and compare it to your answer to ML2–4b. Do you see why mathematicians chose radians in preference to degrees?

(a) The Maple command for the limit is: *Answer*: ______________

(b) Maple's answer for this limit is: *Answer*: ______________

(c) Which formula is simpler, ML2–4b or ML2–5b? *Answer*: ______________

ML2–6

Investigate the limit of $\dfrac{\tan(x)}{x}$ as x approaches zero.

(a) What is the value of $\tan(x)/x$ at $x = 0$? *Answer*: ______________

(b) Graph the function $\tan(x)/x$, x = –Pi/4 .. Pi/4. Does the graph appear to have a limit as x approaches 0?

Answer: ____

(c) Does the graph of $\dfrac{\tan(x)}{x}$ look different from the graph of $\dfrac{\sin(x)}{x}$ in the region $0 < x < 1$? Describe any differences.

Answer: ____

(d) What is the limit of $\dfrac{\sin(x)}{x}$ as x approaches $\pi/2$? *Answer*: ______________

(e) Does the limit of $\tan\dfrac{(x)}{x}$ exist at $x = \pi/2$? *Answer*: ______________

(f) What is the limit of $\frac{5x - 2\tan(x)}{x}$ as x approaches 0?

Answer: ______________________________

ML2-7

Investigate the limit of the function $| 6x - 5 - x^2 |$ at $x = 1$ and at $x = 5$.

(a) Graph the expression, then sketch it here.

Answer: ______________________________

(b) Use Maple's *limit* command to find the limit at $x = 1$ and at $x = 6$.

Answer: ______________________________

(c) Calculate the difference quotient at $x = 1$. First, define the function *f := x -> abs(6*x - 5 - x^2);* then calculate *dq*(*x*). Evaluate the difference quotient with $dx = 0.1$ by using the command *evalf(subs(x = 1, dx = 0.1, dq(x)));*

Answer: ______________________________

(d) Reissue the command in part (c) with smaller values of *dx*, like 0.01 and 0.001. What does the limiting value appear to be?

Answer: ______________________________

(e) Calculate the difference quotient using the symmetric form, *sdq* (define *sdq* as *sdq := (x1, x2) -> (f(x2) - f(x1))/(x2 - x1);*). Start with $x1 = 1.1$ and $x2 = 0.9$, then use $x1 = 1.01$ and $x2 = 0.99$, and so on. What does the limiting value appear to be?

Answer: ______________________________

(f) Use Maple to take the limit of *dq* at $x = 1$. Use the command *limit(dq(1), dx = 0);*

Answer: ______________________________

(g) The function has sharp points at $x = 1$ and $x = 5$. You have to be careful how you take limits at points such as these. If you get different answers depending on the way you choose $x1$ and $x2$ in the symmetric difference quotient, then the limit is undefined. In questions (c) and (d), the difference quotient and the symmetric difference quotient give different answers for the limit. Try *sdq(1.0001, .9);*. What is the result this time?

Answer: ______________________________

ML2–8

If you bank some money in a savings account, your money will grow because the bank gives you interest on your savings. Banks typically compound the interest every month. The formula for compounding the interest over one year is $P_k = P\left(1 + \frac{i}{k}\right)^k$, where k represents the number of interest periods per year. The other parameters are:

P_k: The value of your money at the end of the year

P: Your initial investment

i: The interest rate, expressed as a fraction.

The interest rate is usually stated as a percent and can range from 2% to 10% or more, depending on economic factors. You must convert these percentage figures to their decimal equivalents before using this formula.

Two banks use the same effective interest rate of 8% ($i = 0.08$). Bank A compounds the interest 12 times a year. Bank B decides to become more competitive and advertises that it will compound the interest daily, 365 times a year. A new bank comes to town and gleefully announces that it will compound the interest at every instant, which works out to infinity times a year!

Sally, Jesse, and Rafael each have $1,000 to invest. Sally invests her money in bank A, Jesse invests in bank B, and Rafael invests in bank C.

(a) How much money is in Sally's account at the end of one year?

Answer: ______________________________

(b) How much money is in Jesse's account at the end of one year?

Answer: ______________________________

(c) How much money is in Rafael's account at the end of one year? Use the limit as $k \to \infty$ to compute P_k.

Answer: ______________________________

(d) Bank A charges $2.00 per month in service fees. Banks B and C charge $2.25, claiming that it costs more to do the interest calculations more frequently. All three banks calculate the total service fee owing at the end of the year and apply the charges then, rather than taking the money out of the account every month. Given that the service fees reduce the year-end balance in the accounts, which bank offers the best deal overall?

Answer: ______________________________

ML2-9

(a) Plot the graph of $y = \dfrac{x^4 - 1}{x - 2}$ over the interval $x = -10$ .. 10. Refine the plot picture by choosing an appropriate scale for the y axis. State an appropriate choice for the y axis scale.

Answer: ______________________________

(b) Does the limit for y as $x \to 2$ exist? *Answer*: ______________

(c) Compare the case in part (a) to the function $f{:}x \to \dfrac{x^4 - 1}{x - 1}$. Plot the function. What is the value of the function at $x = 1$, according to the plot?

Answer: ______________________________

(d) What is the value of the function when you evaluate $f(1)$?

Answer: ______________________________

(e) What is the limit of the function as $x \to 1$? *Answer*: ______________

(f) Does the limit of the function f exist as $x \to 2$? *Answer*: ______________

ML2-10

Exercise ML2–7 shows that the limit of a function at a point x may exist, yet the limit of the difference quotient for that same point may not exist. Not only must you be able to calculate the difference quotient, but the result must be independent of the particular choice of the *dx*s. In this problem, you are going to use the equation $y = |x|$ to show that the limit of the difference quotient does not exist at $x = 0$, even though the $|x|$ exists and is continuous at $x = 0$. For the limit of the difference quotient to exist, it must not be dependent on the *particular* sequence of values you use to approach the limit. If one sequence of values gives one result and another sequence gives another, the limit does not exist.

(a) As in ML2–7, use the symmetric form of the difference quotient and take equally spaced points on either side of $x = 0$. The Maple commands might be: *sdq := (x1, x2) -> (f(x2) - f(x1))/(x2 - x1); f := x -> abs(x); sdq(dx, -dx);*, where *dx* is an indeterminate. You do not even have to give *dx* a value. What is the result of the last command, and therefore, what does the limit appear to be?

Answer: ____

(b) Now take points $x1 = 3dx/4$ and $x2 = -dx/4$. The total length of the interval is still *dx*. What does the limit appear to be this time? (If Maple gives you an answer in terms of $|dx|/|x|$, how can you simplify it? What happens if you assume that *dx* is positive?

Answer: ____

(c) Use Maple to calculate the left and right limits at $x = 0$. Are they different? What are they?

Answer: ____

CHAPTER 3

Derivatives: Concepts and Techniques

Objectives for This Chapter

1. Apply the idea of a tangent to a curve in various situations
2. Understand the tangent as the limit of a chord
3. Derive formulas for derivatives, including the following:
   a. The derivative of sums and products
   b. The derivative of a quotient
   c. Related rates
   d. The chain rule
   e. Implicit differentiation

Maple Commands Used in This Chapter

(y@u)(x);	The composition of two functions, *y* and *u*, evaluated at *x*. The result is an *expression*.
alias(e = exp(1));	The *alias* command permits the assignment of a name (*e*) to a value, in this case *exp*(1), the base of the natural logarithms.
allvalues(S1);	Evaluate all values of an expression *S*1 containing roots.
assume(c, constant);	Place restrictions on the variable *c*. In this case, *c* is a constant.
D(f)(x);	Differential operator applied to the function *f*, and then evaluated at point *x*. The result is an *expression*.

D(f);	Differential operator applied to the function *f*. The result is a *function*.
D(sin), D(cos), D(tan), D(sec), D(csc), D(cot);	Derivatives of the trigonometric functions.
D(y@u)(x);	Derivative of the composition of two functions, evaluated at the point *x*. The result is an *expression*.
diff(f(x), x);	Derivative of the *expression* *f*(*x*). The result is an *expression*.
Diff(x^x, x) = diff(x^x, x);	An example of the use of the capitalized form of the *diff* command. *Diff* does not evaluate its argument. Instead, it writes the derivative as it stands. The command *value(Diff(f(x), x);* is equivalent to *diff*.
dq := f -> (f(x+dx) - f(x))/dx;	An example of naming a function. The function—here, the difference quotient—is defined in the same command and named *dq*.
student[makeproc](A, f);	Turn an *expression* (*A*) into a function, with *f* as the dummy variable.
student[makeproc] (A, f, x, dx);	The *expression*, *A*, is made into a function of three variables, *f*, *x*, and *dx*.
eval(subs(f=sqrt, A));	Perform an extra evaluation step on a substitution. Here the substitution of *f* = *sqrt* does not cause a simplification but doing a further evaluation using *eval* does.
evalf(e);	Evaluate in floating point form. Converts the expression *e* into an approximate decimal, with the accuracy specified by the *Digits* variable.
exp(1);	Maple's way of writing the base of the natural logarithms (*e*).
expand(cr1);	Multiply out product terms.
Limit((1 + 1/n)^(n), n = infinity);	The capitalized form of the *limit* command (does not evaluate the limit). Here, the limit occurs as *x* approaches infinity.

limit(sin(x)/x, dx=0);	The limit as x approaches zero. Maple will attempt to fully evaluate this limit.
ln(lhs(eq5)) = expand(ln(rhs(eq5)));	Example of the use of the *lhs* and *rhs* commands. These pick out the left-hand side and the right-hand side, respectively, of an equation.
m[chord]	The *table* notation, used here just to provide subscript notation. The output will be m_{chord}.
normal(D(u/v));	The *normal* command finds a common denominator and writes all terms with this denominator.
op(2, op(1, S));	The command *op*(1, *S*) command picks out the first part of the expression *S*. The outer *op* command picks out the second term of the result of *op*(1, *S*).
plot({x^2, 4*x - 4}, x = 0 .. 4);	The *plot* command, with two plots on the same graph.
plots[display]({p1, p2});	Another method for placing two (or more) plots on the same graph. Both *p*1 and *p*2 must be plot structures (the results of *plot* commands).
simplify(limit(dq(xn), dx = 0));	The *simplify* command. It applies simplification rules to its input (which only sometimes cause simplification).
solve({eq1, eq2}, {x, y});	The *solve* command. Here, solve a system of two equations for two variables.
solve(eq7, diff(y(x), x));	The syntax for isolating the term *dy/dx* in an equation (*eq*7).
solve(subs(x=2, y=4, eq4), {m});	The curly braces around the letter *m* cause the solution to be displayed as {*m* = . . .} rather than as a simple display of the result.
sq := x -> x^2;	An example of the definition of a function. Here, the function *sq* is being defined as the square of *x*.

student[showtangent]
(sqrt(1-x^2), x = 2/3,
x = - 2.. 2, color = black,
scaling = constrained): The command *showtangent* is part of the *student* package. Given an expression, an x value, and a range of x values, it plots the tangent line to the given expression on a graph. The range expression, ($x = -2 .. 2$), sets the scope of the x axis. The last two plot options ensure black lines for the plot and equal scaling on both the x and y axes.

subs(x = 2/3, diff
(sqrt(1 - x^2), x)); The substitute command. Here, it is used to substitute a numeric value for x in an expression containing that variable.

subs(f=sqrt, A); The *subs* command can be used to substitute one variable for another.

The Idea of a Tangent to a Curve

The usual case in a calculus course is that the formula for a function of x is known. We will call this function $y = f(x)$. (Recall that there is another way to describe a function. This alternative mathematical notation is $f{:}x \rightarrow f(x)$.) Let us look at the problem of predicting the change in $f(x)$ as a result of a small increase in x. We know the value of $f(x)$, and we want to predict the value of $f(x + dx)$. The reason for wanting to make this prediction is that it will lead to the most important result of the differential calculus: the *derivative*.

As always, the first step is to plot the function around the value x. Figure 3-1 shows the function in the region of interest.

The function snakes along, growing alternately faster or slower as x increases. At a particular value of x, the height of the vertical line joining the x axis to the function is $f(x)$. This line is shown in the figure. You may be interested in knowing the amount of increase in $f(x)$ as x becomes a little bit larger. If x increases by a small amount, dx, the height of the function is $f(x + dx)$. This situation is shown in Figure 3.2.

We can take a closer look at the region of interest, and we can draw a triangle representing the change in the function f. This is done in Figure 3.3.

The straight line AB in Figure 3.3 is a “fair” approximation to the curve $f(x)$ over this small interval. It is also fairly obvious that the approximation will improve as dx becomes smaller.

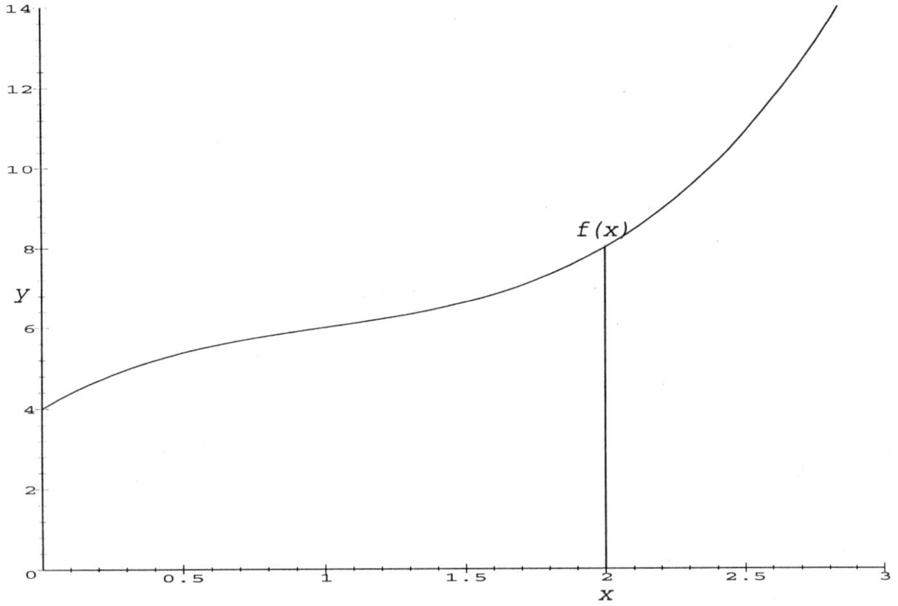

Figure 3.1 The Graph of the Function $y = f(x)$

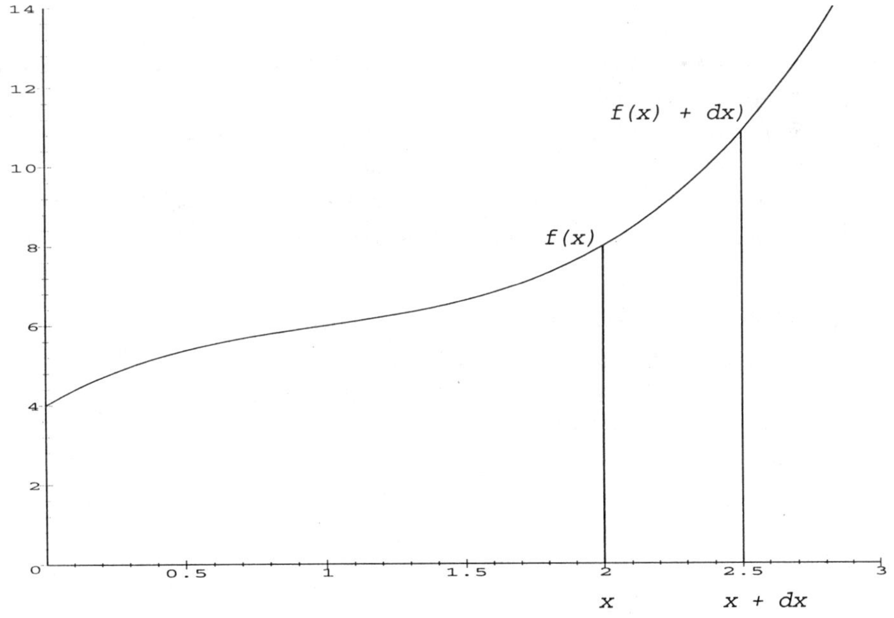

Figure 3.2 Height of the Same Function at Two Nearby Points, x and $x + dx$

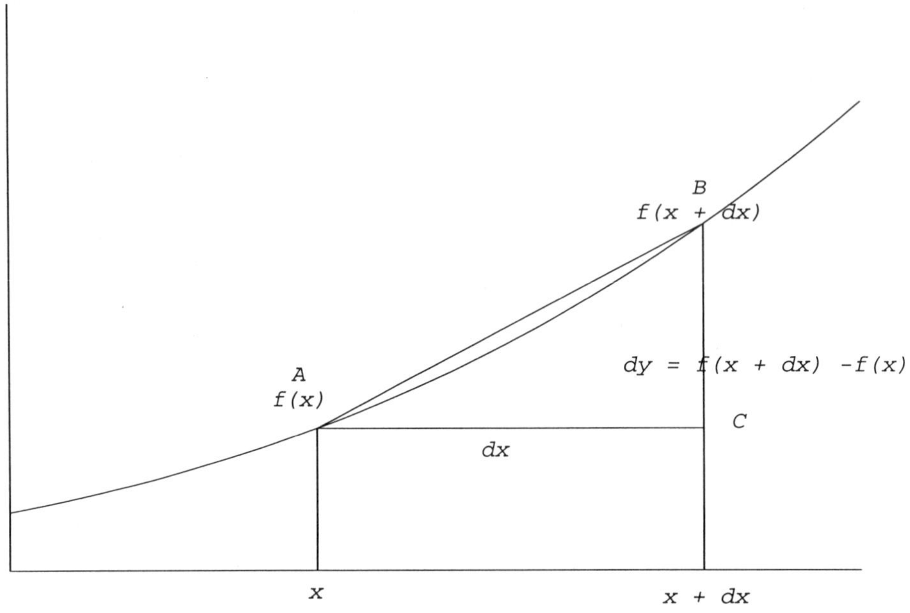

Figure 3.3 The Difference Quotient Triangle for $y = f(x)$

Figure 3.3 contains all the information we need to study changes in the function as x increases to $x + dx$. The straight line joining $f(x)$ to $f(x + dx)$ is called a *chord.* This line cuts the curve at two points, A and B. The length of BC is $f(x + dx) - f(x)$, and the length of line AC is dx. There is another geometrical object that is closely related to the chord. It is called the *tangent.* The defining property of a tangent to a curve is that it has only one point in common with the curve (not two). The tangent seems to lie along the curve at the point where it just touches the curve but does not cut it.

If the curve is defined by a simple function, like $y = x^2$, we can calculate the equation of the line that is tangent to it. It is useful to go through the exercise, then you will see how much work is eliminated by using the techniques of the calculus. Let's see how to get Maple's help in determining the equation of the tangent line to the parabola $y = x^2$. First, define the parabola and the straight line.

> eq1 := y = x^2;

$$eq1 := y = x^2$$

> eq2 := y = m*x + b;

$$eq2 := y = mx + b$$

Find the points of intersection.

> **S := solve({eq1, eq2}, {x, y});**

$$S := \{x = \mathrm{RootOf}(\_Z^2 - m\_Z - b),\ y = m\ \mathrm{RootOf}(\_Z^2 - m\_Z - b) + b\}$$

It's not surprising that the points of intersection, *x* and *y*, are dependent on the slope of the line, *m* and the intercept, *b*. The solution may be somewhat intimidating, though. Maple reports the solution in terms of the root of another quadratic, using the notation *RootOf(_Z*^{2} *– m_Z – b)*. Both *x* and *y*, the points of intersection, can have two values, which Maple combines using the *RootOf* notation. We will find the two values of *x* by solving the *RootOf* equation. It is probably much better to rewrite the quadratic whose roots we want, but we will show you how to extract the *RootOf* expression using Maple's *op* function, which permits you to pick out a part of a larger expression.

> **S1 := op(2, op(1, S));**

$$S1 := \mathrm{RootOf}(\_Z^2 - m\_Z - b)$$

Now that you have the expression you want, you can ask for its values by using the *allvalues* command.

> **allvalues(S1);**

$$\frac{1}{2}m + \frac{1}{2}\sqrt{m^2 + 4b},\ \frac{1}{2}m - \frac{1}{2}\sqrt{m^2 + 4b}$$

We have showed you this method in case you were wondering how to pick out parts of an expression. Also, you may find that Maple reports a result using *RootOf*—which now won't be a complete surprise. If you want to choose the straightforward approach, just copy the quadratic inside the brackets of the *RootOf* expression and give the variable a simpler name.

> **eq3 := z^2 -m*z - b;**

$$eq3 := z^2 - mz - b$$

> **solve(eq3, z);**

$$\frac{1}{2}m + \frac{1}{2}\sqrt{m^2 + 4b},\ \frac{1}{2}m - \frac{1}{2}\sqrt{m^2 + 4b}$$

You get the same result without the need to remember how to use the *op* command. Just be careful when copying that you don't change the expression by mistake! When given the *op* command, Maple does the copying.

We have the result—but what does it mean? These are the *x* values for the points of intersection of the curve $y = x^2$ with the straight line $y = mx + b$. There are two values because the line usually cuts the curve in two places. How can you ensure that the line just touches?

Look at the term inside the square root sign. If $m^2 + 4b = 0$, then both expressions reduce to $m/2$. In this case, there is only one solution for x: it is a repeated root. Solving the equation, $m^2 + 4b = 0$, we have

> solve(m^2 + 4*b = 0, {b});

$$\left\{b = -\frac{1}{4}m^2\right\}$$

We used curly braces around b in the last command so that the output line would be in the form of an equation rather than an expression. We have found that the equation of the tangent line to the parabola is

> eq4 := y = m*x - m^2/4;

$$eq4 := y = mx - \frac{1}{4}m^2$$

We'll test it out by finding the tangent line at $x = 2$. For this value of x, the y value must be 4 if the point is to be on the parabola. If the point is also to be on the straight line, we can solve for m using these values for x and y.

> solve(subs(x=2, y=4, eq4), {m});

$$\{m = 4\},\ (m = 4\}$$

The slope, m, is 4, and the intercept, b, is –4, since $b = -m^2/4$. The equation of the straight line is, therefore, $y = 4x - 4$. The graph of the two equations should show that the tangent line touches, but does not cross, the parabola (Figure 3.4).

> plot({x^2, 4*x - 4}, x = 0 .. 4);

There was a considerable amount of work to finding the equation of the tangent line to a parabola, which is a curve defined by a very simple formula. What might happen if the curve were defined by a more complex formula? By reviewing the steps, you will realize that Maple might report a quite complicated *RootOf* expression or might find no solution at all. It is somewhat startling to realize that by using the technique of the difference quotient, a method can be developed for finding the equation of the tangent line. This method avoids the difficulty of solving for the equation of the tangent line. Instead, you solve for the *slope* of the line. Knowing the slope, you can easily determine the intercept.

The Tangent as the Limit of a Chord

The slope of the chord can be found from triangle *ABC* in Figure 3.3. By definition, the slope is the tangent of angle *BAC*. Since the tangent of this angle is *BC/AC*, and by convention is called m, we have the equation for the slope,

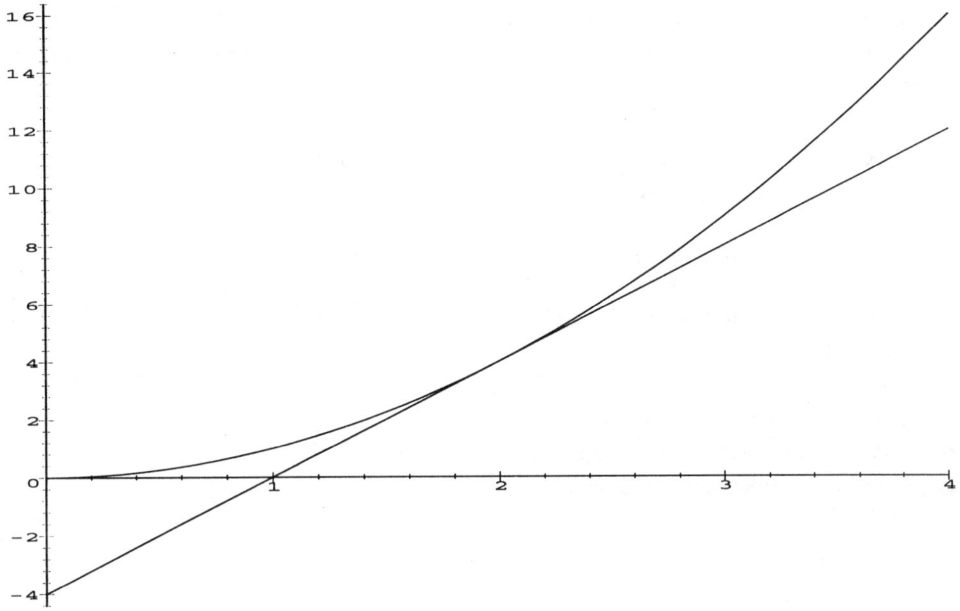

Figure 3.4 The Tangent Line to the Parabola $y = x^2$ at $x = 2$

$$m_{\text{chord}} = \frac{f(x + \Delta x) - f(x)}{\Delta x} \tag{3-1}$$

We can repeat this formula in Maple using the syntax,

> m[chord] = (f(x + dx) - f(x))/dx;

$$m_{\text{chord}} = \frac{f(x + dx) - f(x)}{dx}$$

The Formula Yields the Difference Quotient Once More!

Equation 3-1 is simply the formula for the difference quotient:

> dq := f -> (f(x+dx) - f(x))/dx;

$$dq := f \rightarrow \frac{f(x + dx) - f(x)}{dx}$$

The difference quotient is the formula you use to calculate the slope of a chord, which is one reason why it is so important. But the difference quotient is important for additional reasons. The objective is to find the slope of the tangent line at a given point on a curve defined by a given formula. The way to achieve the objective is to consider the limit of a sequence of chord lines as the other point of the chord approaches the required point. Example 3-1 illustrates the process.

Example 3-1: A Sequence of Chords Passing through $x = 2$ and the Curve $y = x^2$

Graph a few chord lines, all of which pass through the point $x = 2, y = 4$.

Solution: This example shows the construction of the chord lines, and also how to plot a number of lines on the same plot, and how to individually select the start and end-points of the lines. These techniques are useful in any situation where you want to display several curves on the same plot. In the next example, you'll be shown how to label the curves; in other words, how to add text to a plot. For the current example, the first thing we need to do is construct the equations of the chord lines.

The Equation of a Chord Line

We need the equations for the lines passing through the point $x = 2, y = 4$, on the parabola. We will use slopes of $m = 4.5, 5, 5.5$, and 6 for the chord lines. The equation of a line in point-slope form is

> y = m*x + b:

$$y = mx + b$$

Since the line passes through the parabola at $x = 2, y = 4$, it must satisfy

> 4 = m*2 + b;

$$4 = 2m + b$$

We are free to choose any slope m we wish, but once the slope has been chosen, the intercept must satisfy the last equation shown above.

> solve(4 = m*2 + b, {b});

$$\{b = 4 - 2m\}$$

Putting together all the information, we define the parabola and the chord lines:

> ex1 := x^2;

$$ex1 := x^2$$

> ex2 := 4.5*x - 5;

$$ex2 := 4.5x - 5$$

> ex3 := 5*x - 6;

$$ex3 := 5x - 6$$

> ex4 := 5.5*x - 7;

$$ex4 := 5.5x - 7$$

> ex5 := 6*x - 8;

$$ex5 := 6x - 8$$

Now examine the plot command. It's more complicated than previous commands because the straight lines are described in parametric form. The expressions for the straight lines are enclosed in square brackets. Three items are listed: the x coordinate (x), the y coordinate (expressed in terms of x through *ex2*, and so on), and a range value. The range value sets the start and end-points of the line.

> plot({ ex1, [x,ex2, x=2..2.5], [x, ex3, x = 2..3], [x, ex4, x = 2 .. 3.5], [x, ex5, x = 2 .. 4]}, x = 1.5 .. 4, y = 0 .. 16, color=black);

Study the plot command and identify which parametric descriptor corresponds to each chord. Check that the slopes seem correct and the ranges of the chords on the graph correspond to those in the command.

Your Turn: Do you observe four chord lines in Figure 3.5? The chord with the slope $m = 4.5$ is difficult to see because it fits the curve so well.

(a) Extend this line by changing its range value to $x = 2 .. 3$. Is the chord line clearly visible now?

Answer: ______________________________

(b) You know that this chord line cuts the parabola at $x = 2$, $y = 4$. Find the other point of intersection *exactly*.

Hint: Use the command

> solve({y = x^2, y = 9/2*x - 5}, {x, y});

One solution will be $x = 2$, $y = 4$. State the other solution.

Answer: ______________________________

(c) The second intersection point in part (b) lies to the right of the point (2, 4). Does the second intersection point of the chord line $y = 4.01, x - 4.02$, lie to the right of (2, 4)? Solve for the intersection point, give its coordinates, and state whether it lies to the left or right of the point (2, 4).

Answer: ______________________________

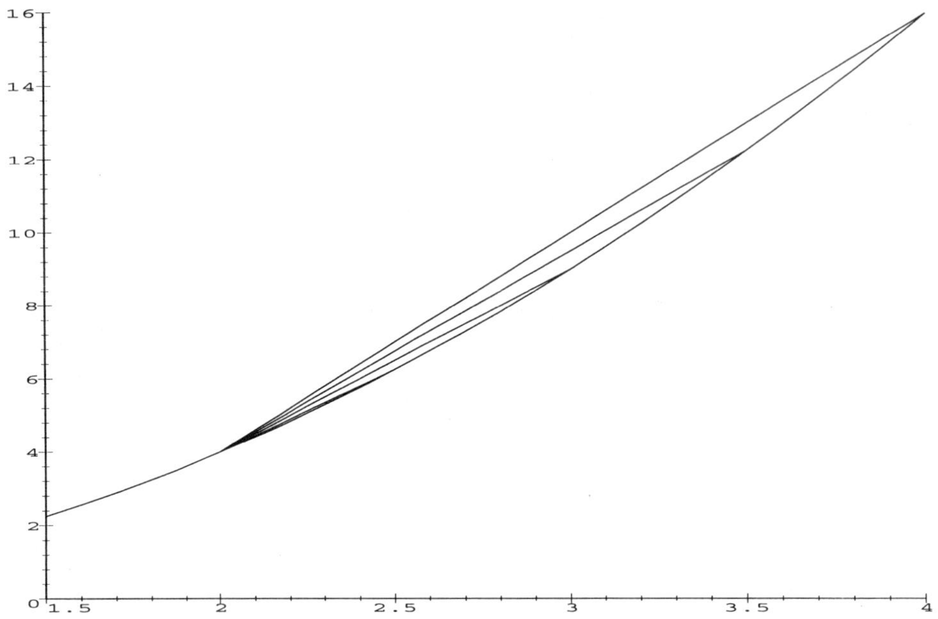

Figure 3.5 Chord Lines Passing through $x = 2, y = 4$, on the Parabola $y = x^2$

(d) Try a chord line with a slope less than 4. Use the equation $y = 3.99x - 3.98$. Solve for the intersection points, state them, and state whether the other point lies to the right or to the left of (2, 4).

Answer: ____

The Equation of a Tangent Line

As you choose different values for the slopes of the chord lines in Example 3-1, you see that the second intersection point lies to the right of point (2, 4) when $m > 4$ and to the left of (2, 4) when $m < 4$. What happens when $m = 4$? The equation of the line with slope 4 passing through the point (2, 4) is $y = 4x - 4$. Since the point (2, 4) is on the parabola as well, this point is on the parabola and also on the straight line. If we attempt to solve for the second intersection of this supposed chord line, we find a repeated root in the solution.

```
> solve({ y = 4*x - 4, y = x^2}, {x, y});
```

$$\{y = 4, x = 2\}$$

The point (2, 4) is the *only* solution. We call this line the *tangent line*. It has only one point in common with the curve.

The objective of the differential calculus is to find efficient methods for finding tangents to curves. In the case of the parabola, we were able to derive the tangent line by solving for its point of intersection, with the requirement that it have only one point in common with the parabola. What we are looking for now is a simpler method of finding the tangent line to a curve whose equation may be much more complicated than a simple parabola. The method involves finding the limit of the difference quotient.

How the Difference Quotient Is Defined in Maple

We'll show you yet another way of defining the difference quotient in Maple. Since this book is an extended tutorial on Maple, you may find these methods useful in other situations, too. The method shown here corresponds to a more or less natural progression of ideas. If you were presented with the need to define the difference quotient, your first thought might be to make it a named expression.

> A := (f(x+dx) - f(x))/dx;

$$A := \frac{f(x + dx) - f(x)}{dx}$$

This is the formula you want, but it is in the form of an *expression* rather than being defined as a *function*. Here, *A* is the *name* of an *expression*. You can turn any expression into a function using the *student[makeproc]* command, which gives you a great deal of control over the type of function you create without forcing you to bother with the details of syntax. Some examples follow.

Example 3-2

Define the difference quotient using the *student[makeproc]* command.

Solution: As a result of the previous command, the difference quotient is a named expression. We would like to have the option of using an expression or a function. We could simply define the function *dq* as we have done previously, but we can also choose to have Maple produce the function for us. Thus, we ask Maple to make up a function using a given expression and telling it which variable in the expression to use as the functional parameter.

Case 1: Create the function that we defined in Chapter 2.

> **dq := student[makeproc](A, f);**

$$dq := f \rightarrow \frac{f(x + dx) - f(x)}{dx}$$

You see from this example that the makeproc command can take two parameters, the expression, *A*, and the function name, *f*. You can work with the expression *A* or the function *dq*. Note the different ways of using the two forms by choosing the square root function, *sqrt*, as an example.

> **subs(f=sqrt, A);**

$$\frac{\mathrm{sqrt}(x + dx) - \mathrm{sqrt}(x)}{dx}$$

> **dq(sqrt);**

$$\frac{\sqrt{x + dx} - \sqrt{x}}{dx}$$

The two methods produce the same result, but they look different. You can make them look the same by using the *eval* command.

> **eval(subs(f=sqrt, A));**

$$\frac{\sqrt{x + dx} - \sqrt{x}}{dx}$$

Either notation has its advantages and disadvantages, so it is useful to know how to use both forms.

Case 2: Create a difference quotient function with more parameters. You might want to make every part of the difference quotient a parameter. The command for accomplishing this is:

> **dq1 := student[makeproc](A, f, x, dx);**

$$dq1 := (f, x, dx) \rightarrow \frac{f(x + dx) - f(x)}{dx}$$

Now all three items, *f*, *x*, and *dx*, are parameters. Observe what happens when we ask for the difference quotient of sin(*a*) with interval *da*.

> **dq1(sin, a, da);**

$$\frac{\sin(a + da) - \sin(a)}{da}$$

The function *dq1* has three variable parameters, all of which must be specified when you write the command. The next command shows what happens if one of the parameters is left out:

> dq1(sin, a);

```
Error, (in dq1) dq1 uses a 3rd argument,
        dx, which is missing
```

Your Turn: Use the function *dq* to find the difference quotient of x^n.

(a) If you want to take the difference quotient of the *n*th power of the variable *x*, would you use the following command? Give the result of the command in your answer.

> dq(x^n);

Answer: ____________________

(b) How would you modify the command in (a) to produce the desired results? *Hint:* You need to give the command *dq* a *function* as a parameter. (Remember that x^n is an expression rather than a function, while *x* -> *x*^*n* is a function.)

Answer: The command is __________, and the output of the command is __________.

The Limit of the Difference Quotient

We are closing in on our objective, which is finding a method of calculating the formulas for tangents. Let's review (refer to Figures 3.3 and 3.4).

1. The tangent to a curve *f*(*x*) at the point *x* is required.

2. Construct a chord passing through *x*. Construct the chord so that it cuts the curve again to the right of *x*. Call the *x* coordinate of that point x_1.

3. Choose a point on the curve *f*(*x*) that is closer to *x* than x_1. Construct a new chord.

4. Continue the process of constructing chords. The two points where they cut the curve *f*(*x*) will successively grow closer together.

5. In the limiting case, there will be a line that cuts (touches) the curve *f*(*x*) only once. This line is called the *tangent* to the curve. The *slope* of the tangent is what is required.

The chord's slope is given by the difference quotient. The limit is found using Maple's *limit* command. There are two forms of the command. The *inert* form begins with a capital L. It does not attempt to work out the answer. It merely shows the form of the command.

> Limit(dq(f), dx=0);

$$\lim_{dx \to 0} \frac{f(x + dx) - f(x)}{dx}$$

This formula has a special name: the *derivative of the function* f *at the point* x. Often, the derivative is called *dy/dx* and is written in one of the following two ways:

> dy/dx = Limit(dq(f), dx=0);

$$\frac{dy}{dx} = \lim_{dx \to 0} \frac{f(x + dx) - f(x)}{dx}$$

df(x)/dx = Limit(dq(f), dx=0);

$$\frac{df(x)}{dx} = \lim_{dx \to 0} \frac{f(x + dx) - f(x)}{dx}$$

One could work with the expression A instead:

> dy/dx = Limit(A, dx = 0);

$$\frac{dy}{dx} = \lim_{dx \to 0} \frac{f(x + dx) - f(x)}{dx}$$

The difference between the two notations appears when an attempt is made to actually evaluate the limits. Let us evaluate the derivative of our old favorite, $y = x^2$. The squaring function is defined as $f{:}x \to x^2$, but it is important to realize that the variable x in this expression is a *dummy variable*, meaning that it can be replaced by any other symbol without changing the meaning of the function. For example, the function $f{:}p \to p^2$ works just as well. Look at the two commands that follow. The answer is given in terms of x, even though we use the dummy variable p as the function's parameter. More importantly, *dq* uses the function as a parameter, while the *subs* command must be used to substitute a function for f in the expression A.

> dy/dx = limit(dq(p -> p^2), dx=0);

$$\frac{dy}{dx} = 2x$$

> dy/dx = limit(subs(f = (p -> p^2), A),dx = 0);

$$\frac{dy}{dx} = 2x$$

Either form produces the same result, but the first form is slightly easier to use.

Formulas for Derivatives

We can proceed to listing the formulas for the derivatives of common functions.

The Derivative of a Constant

> assume(c, constant);

> dc(x)/dx = limit(dq(x -> c), x = 0);

$$\frac{dc(x)}{dx} = 0$$

You have already seen that a constant plots as a horizontal straight line; consequently, its derivative (the slope of the line) must be zero.

The Derivative of a Linear Function

A linear function is defined as

> assume(m, constant); assume(b, constant);

> L := x -> m*x +b

$$L := x \rightarrow mx + b$$

Both *m* and *b* have now had restrictions placed on them. Maple will write a variable *m* having assumptions as ~*m*. This is done to remind you that the variable is not totally free. In this case, *m* is not a variable; it is a constant.

> dL(x)/dx = limit(dq(L), dx = 0);

$$\frac{dL(x)}{dx} = m\sim$$

A straight line has a constant slope. Since *m* is the slope of the line $y = mx + b$, the derivative must be *m*.

The Derivative of the Squaring Function

The squaring function is not defined in Maple (even though the square root function is defined). Thus, we will define it here.

> sq := x -> x^2;

$$sq := x \rightarrow x^2$$

> d(x^2)/dx = limit(dq(sq), dx = 0);

$$\frac{d(x^2)}{dx} = 2x$$

The derivative of x^2 is $2x$.

The Derivative of x^n

Without placing any assumptions on *n*, the derivative of x^n is

```
> xn := x -> x^n;
```

$$xn := x \to x^n$$

```
> d(x^n)/dx = limit(dq(xn), dx = 0);
```

$$\frac{d(x^n)}{dx} = \frac{x^n n}{x}$$

This result is usually given in calculus textbooks as

```
> d(x^n)/dx = simplify(limit(dq(xn), dx = 0));
```

$$\frac{d(x^n)}{dx} = x^{(n-1)} n$$

The Derivative of the Square Root Function

```
> d(sqrt(x))/dx = limit(dq(sqrt), dx = 0);
```

$$\frac{d(\sqrt{x})}{dx} = \frac{1}{2}\frac{1}{\sqrt{x}}$$

Note that this is just a specific example of the previous general rule. No assumptions were made about *n* there, and it is valid for fractional powers of *x* (like $x^{1/2}$ or any other fractional power).

Derivatives of Trigonometric Functions

The derivatives of trigonometric functions are

```
> dsin(x)/dx = limit(dq(sin), dx = 0);
```

$$\frac{d\sin(x)}{dx} = \cos(x)$$

```
> dcos(x)/dx = limit(dq(cos), dx = 0);
```

$$\frac{d\cos(x)}{dx} = -\sin(x)$$

```
> dtan(x)/dx = limit(dq(tan), dx = 0);
```

$$\frac{d\tan(x)}{dx} = 1 + \tan(x^2)$$

This last formula differs from the one usually given for the derivative of tan(x) in most textbooks:

$$\frac{d\tan(x)}{dx} = \sec(x)^2$$

You can use elementary trigonometry to show that $1 + \tan^2(x) = \sec^2(x)$. And, as you know by now, Maple writes $\tan(x)^2$, where most texts write $\tan^2(x)$. The notations mean exactly the same thing.

Maple Commands for the Derivative

Up to now, we have asked Maple to calculate the limit of the difference quotient as our way of showing you the well-known formulas for the derivatives of the standard functions. You can (and should) check these formulas with those given in any standard calculus text. You'll have an opportunity to develop some of the standard formulas algebraically in the exercises at the end of this chapter.

You could continue to ask Maple to compute the limit of the difference quotient as a way of finding the derivative of a function of an expression, but there is an easier way. In fact, there are two easier ways. One is for expressions and the other is for functions. We'll demonstrate how to find the derivative of an expression first. The commands for finding the derivative of an expression are constructed as follows.

Case 1: The Derivative of an Expression. If you want to find the derivative of the *expression* $f(x)$, with respect to the variable x, the command is

> diff(f(x), x);

$$\frac{\partial}{\partial x} f(x) \tag{3-2}$$

Maple writes the answer using the partial derivative sign. It is one of Maple's idiosyncracies, and you'll have to get used to it. If you like to use rules, this is one. In Maple, if y is a function of x, the derivative of y with respect to x is written as the right-hand side in Equation 3-2, while the same derivative in most textbooks is written as the left-hand side.

$$\frac{dy}{dx} = \frac{\partial}{\partial x} y(x) \tag{3-3}$$

Note that no matter whether standard or "curly" ds are used, dy/dx reminds you of the sides of the triangle in Figure 3.3. The derivative is the limit of the sides of the triangles as the chord's two points approach one another.

To permit you to become accustomed to this notation, we'll show you its application to some of the derivatives computed so far.

(a) Linear variation

```
> Y := m*x + b;
```

$$Y := mx + b$$

```
> diff(Y, x);
```

$$m$$

(b) The quadratic

```
> Y := x^2;
```

$$Y := x^2$$

```
> diff(Y, x);
```

$$2x$$

(c) The Power Law

```
> Y := x^n;
```

$$Y := x^n$$

```
> diff(Y, x);
```

$$\frac{x^n n}{n}$$

The *same* command was used to calculate the different derivatives. This was possible because the name *Y* was defined differently each time. The important thing to note is that in each of the definitions, *Y* is an *expression*. It bears repeating that *diff* works on expressions, not functions.

If you attempt to find the derivative of a *function* using *diff*, you obtain an incorrect result. Here's what happens if you attempt to find the derivative of the sin function by issuing the command

```
> diff(sin, x);
```

$$0$$

The function name *sin* does not contain the variable *x*, so Maple does not know that *sin* is meant to be a function of *x*. If you want to find the derivative of the sin function, one way is to convert it to an expression. Simply write it as *sin*(*x*).

> diff(sin(x), x);

$$\cos(x)$$

Now everything is correct. Be careful to understand the difference between the *name* of a function, like *sin*, and an *expression*, like *sin*(*x*). These examples may help make it clear.

> Y := sin;

$$Y := \sin$$

> Y := sin(x);

$$Y := \sin(x)$$

In the first of the two commands, *Y* is the name of a *function*, whereas in the second, *Y* is the name of an *expression*. In both cases *Y* is a *name*, but in these two commands, it stands for two different types of mathematical object.

Case 2. The Derivative of a Function. There are times when it is desirable to find the mathematical *form* of the derivative without mentioning a variable for the function. People will often say, "The derivative of sin is cos," yet they won't mention an angle. The notation for this case is the *D* operator. Again, we'll show some examples of its use.

(a) *D* applied to the linear function.

> f := x -> m*x + b;

$$f := x \rightarrow mx + b$$

> D(f);

$$x \rightarrow m$$

The result is also a *function*. This is the important distinction. The result of the command *diff* is an *expression*, just like its inputs. The result of applying the *D* (differential operator) to a function is also a *function*.

When using the *D* operator, be *very careful* how you write your command. Make sure *D* is applied to functions. The difficulty arises when you want to get an answer as an expression. This is done by asking Maple to apply the *D* operator to the function *f* and then evaluating the result at point *x*. The result will be an expression. Here is the proper command:

> D(f)(x);

$$m$$

Things have worked out as they should. Now check the brackets in the next command.

> **D(f(x));**

$$D(m)\,x + mD(x) + D(b)$$

This is not at all what was wanted! You must realize that *f*(*x*) is an expression. The *D* operator is applied to an expression in this case, and the result is certainly not simple—or simple to understand. We won't try to explain here how to reduce Maple's result to *m*, the answer we were looking for. Just note that this is *not* the correct way of asking Maple to apply the *D* operator.

(b) *D* applied to the trigonometric functions

> **D(sin), D(cos), D(tan), D(sec), D(csc), D(cot);**

$$\cos,\ -\sin,\ 1 + \tan^2,\ \sec\tan,\ -\csc\cot,\ -1 - \cot^2$$

The result of the command just above is technically known as a *sequence.* The individual commands are separated by commas. Each command is executed by Maple, and the results are displayed one after another, again separated by commas. This is a good way of displaying several results on one line. Decode Maple's output by looking up any of the trigonometric derivatives that you don't remember. You'll find that they correspond to the derivatives shown here.

(c) *D* applied to the squaring and square root functions: Since there is no built-in squaring function in Maple, one must be defined. We'll call it *sq*, as we did before.

> **sq := x -> x^2;**

$$sq := x \to x^2$$

> **D(sq);**

$$x \to 2x$$

Since the square root function is defined in Maple, *D* can be directly applied to it.

> **D(sqrt);**

$$\frac{1}{2}\,\frac{1}{\text{sqrt}}$$

The result is somewhat surprising, but it is the correct answer. To see that it is correct, apply the result to the variable *x*.

> **D(sqrt)(x);**

$$\frac{1}{2}\,\frac{1}{\sqrt{x}}$$

This is indeed the result we derived earlier.

Derivatives of Exponential and Logarithmic Functions

The last broad class of functions for which we need formulas for the derivatives are the exponential and logarithmic functions.

(a) The derivative of the exponential: You'll find the notation e^x for the exponential in most textbooks and, in fact, most mathematical literature This might lead you to believe that you can write it this way in Maple.

> e^x;

$$e^x$$

This looks correct, but it doesn't work! Mathematicians reserve the letter *e* to mean the number 2.7182818. . ., but Maple doesn't assign any special value to *e*. To demonstrate that this is the case, try to convert *e* to an approximate decimal using *evalf.*

> evalf(e);

$$e$$

Maple does not convert *e* to a decimal value, indicating that *e* has no special meaning in Maple. The way of writing *e* in Maple is *exp*(1).

> exp(1);

$$e$$

Maple reports the result as *e*, but this *e* is *different*! Its decimal value is

> evalf(exp(1));

```
2.718281828
```

This is the *e* we want. It is the base of the natural logarithms, and its derivative has the remarkable property that

> D(exp);

```
exp
```

The derivative of the exponential function is the exponential function. This bears repeating, so we'll use the *diff* command on *exp*(*x*).

> diff(exp(x), x);

$$e^x$$

Another thing that bears repeating is the fact that Maple knows nothing about the letter *e* itself. If you ask Maple to take the derivative of e^x by issuing the command

> diff(e^x, x);

$$e^x \ln(e)$$

you get another answer that is hard to interpret. You know that ln(e) = $\log_e(e) = 1$. Why did Maple leave the answer in terms of ln(e)? The answer is that Maple treats e as a variable, not the base of the natural logarithms.

You could ask Maple to define e as the base of the natural logarithms. Using the basic definition of e,

> e = Limit((1 + 1/n)^(n), n = infinity);

$$e = \lim_{n \to \infty} \left(1 + \frac{1}{n}\right)^n$$

The last equation says that e is equal to the given limit. Changing the command just a little bit, let us define the *name e* to be this limit.

> e := limit((1 + 1/n)^(n), n = infinity);

$$e := e$$

Now that the name e has a meaning in Maple, you can use it in any exponential expression. Here are some examples:

> evalf(e), ln(e), diff(e^x, x);

$$2.718281828,\ 1,\ (e)^x$$

Maple will remember that e has been defined until you close the current session. (When you restart Maple, e will not be defined until you provide the definition.) Another way of defining e is by using the *alias* command

> alias(e = exp(1));

$$1,\ e$$

Just remember that these definitions last only as long as the current session. When you exit Maple, the definitions will be forgotten and e will become an undefined letter value once more.

(b) The derivative of the logarithm: You may be used to using the name *ln* for the natural logarithms and the name *log* for logarithms to base 10. In most textbooks on advanced mathematics, both names stand for the natural logarithm. The way to specify the *common logarithm* is to use the notation log_{10}. This is Maple's way as well. Check that this is the case by evaluating the logarithm of 10 using both the *ln* and the *log* notation.

> **evalf(log(10) = ln(10));**

```
2.302585093 = 2.302585093
```

The previous command has tested the equality for one number only, and that to only ten digits of accuracy. There is another way to ask Maple to check for sameness. It is the *testeq* command. Be aware of this quote from the Maple Help page on *testeq*: "The result false is always correct; the result true may be incorrect with very low probability."

> **testeq(log(10) = ln(10));**

```
true
```

Maple thinks the result is very probably true. Actually, we know it is true because Maple's Help page for ?*log* states that the command log(x) is the same as the command ln(x).

The derivative of ln(x) is

> **diff(ln(x), x);**

$$\frac{1}{x}$$

Both the exponential and logarithmic functions have very simple derivatives. The fact that these derivatives are so simple is the reason why *e* and *ln* were defined in the first place.

The Derivative of Sums and Products

All the derivatives that have been presented up to now have involved only a single function. As more complex expressions arise, it is important to be able to use a "divide and conquer" technique to work out the complete derivative. Some of these general rules are presented in this section and the next.

Rule D1

The derivative of a sum of functions is the sum of the derivatives. If *f* and *g* are two functions, Maple automatically expands the result, showing that it is applying Rule D1.

> **D(u + v);**

$$D(u) + D(v)$$

Rule D2

The derivative of a product of two functions, *f* and *g*, is the sum of *f* times the derivative of *g* and *g* times the derivative of *f*.

> **D(u*v);**

$$D(u)\,v + u\,D(v)$$

Let's look at this result again, this time using *diff*

> **diff(u(x)*v(x), x);**

$$\left(\frac{\partial}{\partial x}u(x)\right)v(x) + u(x)\left(\frac{\partial}{\partial x}v(x)\right)$$

Using the *D* operator, the answer is more succinct, but both answers mean the same thing.

The Derivative of a Quotient

Rule D3

The derivative of a quotient of two functions, *f* and *g*, is given by both the formulas

> **D(u/v), normal(D(u/v));**

$$\frac{D(u)}{v} - \frac{u\,D(v)}{v^2},\ \frac{D(u)\,v - u\,D(v)}{v^2}$$

In the previous command, we used the *normal* command to bring both terms over a common denominator. This is the way this rule is most often stated.

The Chain Rule

We have seen how to compute the derivative of $3 - 4x$ and $x^2 + x$. How can we use the known results to find the derivative of $(3 - 4x)^2 + (3 - 4x)$? The derivative of an expression like this one can be calculated by expanding the quadratic term and then collecting terms in x^2 and x. No new rule is needed for a case like this, but what if the expansion is impossible to carry out? We know the derivative of $\sin(x)$ and of x^2, but how would we calculate the derivative of $\sin(x^2)$? The *chain rule* is a powerful technique for calculating derivatives of this type.

The situation is usually described by defining *y* as a function of *u* and then defining *u* as a function of *x*. Remember that the variable in a function definition is a *dummy variable.* Its name is unimportant; it serves to help us think about the problem in terms of two differently named variables. The definitions of the two functions are

> y := u -> u^2 + u; u := x -> 3 - 4*x;

$$y := u \to u^2 + u$$

$$u := x \to 3 - 4x$$

These two definitions seem to say that y is a function of u and u is a function of x. In actuality, only the functional *form* is important, not the name of the variable used in the definition, which is used only to help our thought processes.

Let us check that the composition of y and u applied to x results in the expression we want.

> (y@u)(x);

$$(3 - 4x)^2 + 3 - 4x$$

This is indeed the expression with which we began the section on the chain rule. Now we want to find the derivative of this expression with respect to x.

> D(y@u)(x);

$$-28 + 32x$$

We can check this result by calculating the derivative of $(3 - 4x)^2 + (3 - 4x)$. Give the expression the name *cr1*:

> cr1 := (3 - 4*x)^2 + (3 - 4*x);

$$cr1 := (3 - 4x)^2 + 3 - 4x$$

> cr2 := expand(cr1);

$$cr2 := 12 - 28x + 16x^2$$

You can determine the derivative of *cr2* with respect to x by inspection. It is $-28 + 32x$, which checks with our result using the differential operator D. Of course, you could ask Maple for the derivative, without the need to expand the expression.

> diff(cr1, x);

$$-28 + 32x$$

The process of taking the derivative of the composition of two functions by taking the derivative of each with respect to its own variable is called the *chain rule*:

$$\text{If } y = f(u) \text{ and } u = g(x), \text{ then } \frac{dy}{dx} = \frac{dy}{du}\frac{du}{dx}$$

Example 3-3

Find the derivative of $\sin(x^2)$ using the chain rule.

Solution: When you attempt to apply the chain rule, it's important to know which functions are being composed. In this example, the functions are $y = \sin(u)$ and $u = x^2$. Since $dy/du = \cos(u)$, and $du/dx = 2x$,

$$\frac{dy}{dx} = \frac{dy}{du}\frac{du}{dx} = \cos(u) \cdot 2x = \cos(x^2) \cdot 2x$$

Note that the term $cos(u)$ appears in the answer and that we used the fact that $u = x^2$ to write the final answer in terms of x alone.

Let us check the result directly using Maple.

```
> dy/dx = diff( sin(x^2), x);
```

$$\frac{dy}{dx} = 2\ \cos(x^2)\, x$$

The chain rule makes calculations simpler, but if Maple is doing the work, there is no need for it. Even if a problem is given in terms of two functions, you can simply define them as we have done here and apply the differential operator to the composition of the two functions. Maple takes care of the details—you make sure of the solution. With Maple, it becomes less important to be able to apply the chain rule.

Related Rates

The concept of related rates arises in problems concerning the time rate of change of one variable with respect to another. As far as the calculus is concerned, time is no different than any other variable. Therefore, any problem in related rates is solvable using the chain rule. An example of a related rates problem can be found in ML3–11.

Implicit Differentiation

Even though now you have Maple as an assistant, it is still important to understand the technique of *implicit differentiation*. An equation in x and y where it is not possible to solve for y is called an *implicit function*. Actually, the more correct term in most cases is *implicit relation*. A function is allowed only one value for each choice of y. Oftentimes, however, more than one value of x is possible in implicit relations. A good example is the equation for a circle. The general point on the circumference of a circle of radius 1, centered at the origin, is $x^2 + y^2 = 1$. This equation is a *relation*, because any value of x between, but not including, -1 and 1 has two possible values for y. Sketch the circle and you will see why. For any given value of x, $-1 < x < 1$; a vertical line through x cuts the circle at y. Thus, there are two solutions for any given x in the domain $-1 < x < 1$.

Example 3-4

Find the equations of the tangent lines to the circle $x^2 + y^2 = 1$ at the point $x = 2/3$.

Solution: We'll solve this problem in two ways. First, the straightforward way. The equation for the circle can be solved explicitly for y in terms of two functions, $y = \sqrt{1 - x^2}$ and $y = -\sqrt{1 - x^2}$. The graph for this example shows you how to draw a tangent to a function at a given point using the *showtangent* function. Since *showtangent* is part of the student package, it is accessed by the command

> **p1 := student[showtangent](sqrt(1-x^2), x = 2/3, x = - 2.. 2, color = black, scaling = constrained):**

> **p2 := student[showtangent](-sqrt(1-x^2), x = 2/3, x = - 2.. 2, color = black, scaling = constrained):**

The parameters to the command are the function, the point where the tangent is required ($x = 2/3$), the range of values for the x axis (x = -2 .. 2), the color specification (color = black), and finally, a parameter that makes sure that a circle looks like a circle. The parameter *scaling = constrained* tells Maple to use the same scale on both the x and y axes. Figures will be shown undistorted on this kind of a graph. The two graphs are given names, and the commands are ended with a colon. This prevents Maple from giving immediate output. (The output is a plot data structure. If you want to see what one looks like, use a semicolon—but be prepared for lots of output!)

Now that *p*1 and *p*2 have been defined, they can be plotted on the same graph using *plots*[*display*] (Figure 3.6).

> **plots[display]({p1, p2});**

The tangent lines have a certain symmetry. They are mirror images, the mirror being the x axis. The value of the slope for positive y is

> **subs(x = 2/3, diff(sqrt(1 - x^2), x));**

$$-\frac{2}{15}\sqrt{5}\sqrt{9}$$

The value of the slope of the tangent at negative y is

> **subs(x = 2/3, diff(-sqrt(1 - x^2), x));**

$$\frac{2}{15}\sqrt{5}\sqrt{9}$$

As expected, the slopes are negatives of one another.

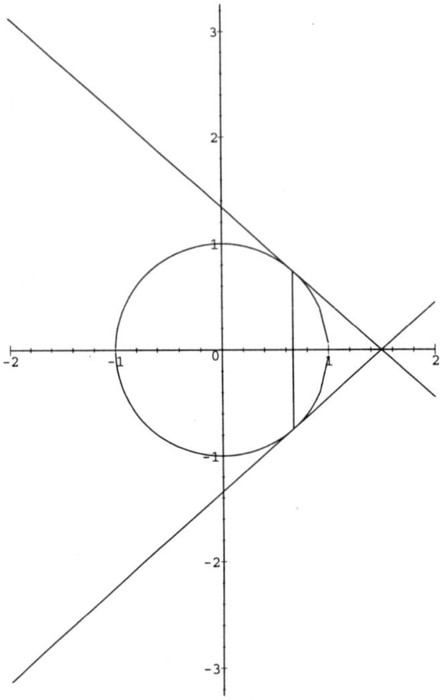

Figure 3.6 Tangents to $x^2 + y^2 = 1$ at $x = 2/3$.

The slopes for the tangent lines have now been determined, and it remains to construct the equations for the tangent lines. Since we know the x coordinate of the tangent point ($x = 2/3$), the y value is found from the equation of the circle. Then the equation for the tangent line is constructed using the point-slope form

> y = sqrt(1 - (2/3)^2), y = -sqrt(1 - (2/3)^2);

$$y = \frac{1}{3}\sqrt{5}, \quad y = -\frac{1}{3}\sqrt{5}$$

> y - sqrt(5)/3 = -2*sqrt(5)*sqrt(9)/15*(x 2/3), y - (-sqrt(5)/3) = 2*sqrt(5)*sqrt(9)/15*(x 2/3),

$$y - \frac{1}{3}\sqrt{5} = -\frac{2}{5}\sqrt{5}\left(x - \frac{2}{3}\right), \quad y + \frac{1}{3}\sqrt{5} = \frac{2}{5}\sqrt{5}\left(x - \frac{2}{3}\right)$$

Note that Maple has simplified the formula for the slope in addition to displaying the tangent equations.

Your Turn

(a) Find the equations of the tangent lines to the circle $x^2 + y^2 = 1$ at the point $x = 1/3$.

Answer: ______________________________

(b) Find the equations of the tangent lines if the slopes are both $\sqrt{3/2}$.

Answer: ______________________________

Example 3-5: The Derivative of x^x

Find the derivative of x^x using Maple, then verify the result using implicit differentiation.

Solution: This used to be a challenging problem before the appearance of computer algebra systems like Maple. It is still a good example of the use of implicit differentiation to solve for a complicated derivative. No special knowledge is required to find the derivative using Maple. We'll write the solution out in full by writing the command using both the capitalized and lower-case versions of the *diff* command.

> Diff(x^x, x) = diff(x^x, x);

$$\frac{\partial}{\partial x}x^x = x^x(\ln(x) + 1)$$

Now that you know Maple's solution, you can verify it by using known derivatives and implicit differentiation. The technique is as follows.

(a) Express the problem as an equation. The equation is $y(x) = x^x$, and the derivative we want is $dy(x)/dx$.

> eq5 := y(x) = x^x;

(b) So far, nothing useful has resulted. This is the step that makes the difference: take the logarithm of both sides of the equation in step (a). We'll need to use *expand* on the right-hand side to force Maple to write $\ln(x^x)$ as $x \ln(x)$.

> eq6 := ln(lhs(eq5)) = expand(ln(rhs(eq5)));

$$eq6 := \ln(y(x)) = x\ln(x)$$

(c) Take the derivative with respect to x of each side of *eq*6. Since the left-hand side is given in terms of a function of x and not x itself, the chain rule is used. Maple applies the chain rule automatically, but you should do it mentally as you follow along in this example. Since $d/dy(\ln(y) = 1/y$, the left-hand side is $\frac{1}{y}\frac{dy}{dx}$ by the application of the chain rule. The right-hand side contains a product of expressions in x, which means that the product rule (Rule D2) can be applied. Again, Maple applies this rule automatically, and you should verify the result by hand or as a mental exercise.

> eq7 := diff(lhs(eq6), x) = diff(x*ln(x), x);

$$eq7 := \frac{\frac{\partial}{\partial x} y(x)}{y(x)} = \ln(x) + 1$$

Equation *eq*7 can be solved for *dy/dx*. We ask for *dy/dx* just as we would for any other variable.

> eq8 := solve(eq7, diff(y(x), x));

$$eq8 := -(\ln(x) - 1)\, y(x)$$

For the sake of completeness, we'll substitute for $y(x)$.

> eq9 := subs(y(x) = x^x, e8);

$$eq9 := -(-\ln(x) - 1)\, x^x$$

Compare *eq*9 with Maple's solution at the beginning of this example. The results are the same except for the way the terms are written in *eq*9. Maple seems to have a preference for minus signs. You can remove them by expanding the last result, if you wish.

> expand(eq9);

$$x^x \ln(x) + x^x$$

Your turn: Find the derivatives of $y(x) = x^{(x^x)}$ and $y(x) = (x)^{x^x}$ directly using Maple and also by implicit differentiation.

(a) Is $\left(x^x\right)^x = x^{(x^x)}$? *Answer*: ____________________

(b) What is the derivative of $\left(x^x\right)^x$? *Answer*: ____________________

(c) What is the derivative of $x^{(x^x)}$? *Answer*: ____________________

Paper and Pencil Exercises

PP3–1

Given the graph of the function $y = x^2 - 8x + 19$ in Figure 3.7, draw the tangent line using a ruler or straightedge. Extend the tangent line from the point of tangency to the x axis.

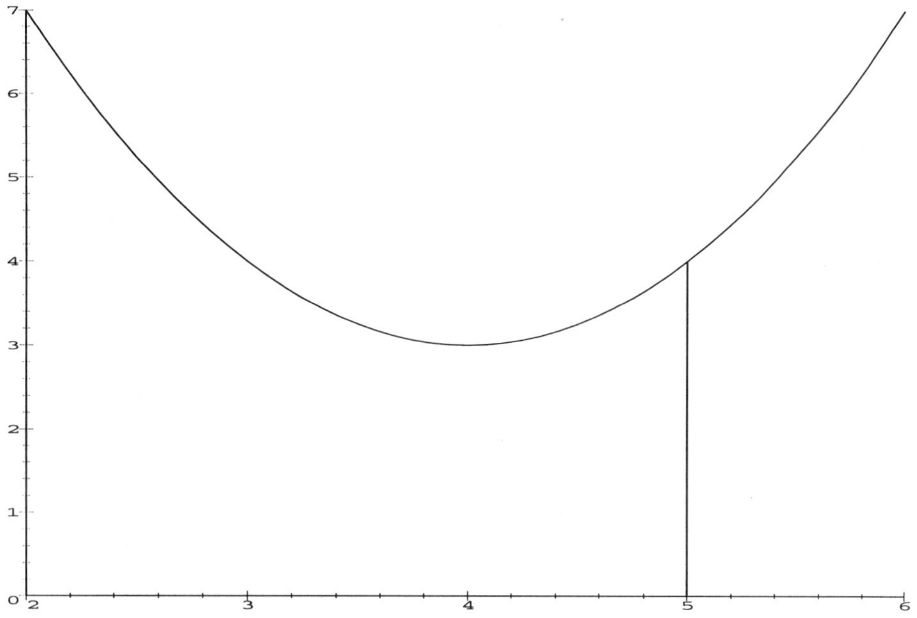

Figure 3.7 Finding the Tangent Line Graphically

(a) Where does the tangent line cut the x axis?

Answer: ______________________________

(b) Consider the triangle formed by the tangent line, the x axis, and the line $f(x)$.

(i) Measure the length of the base of this triangle. Call the result b.

Answer: ______________________________

(ii) Measure the altitude of the triangle, that is, measure $f(5)$.

Answer: ______________________________

(c) Find the slope of the tangent line by calculating $f(x)/b$. Call the result m.

Answer: ______________________________

(d) Compare your answer to the derivative of the function, evaluated at $x = 5$.

Answer: ______________________________

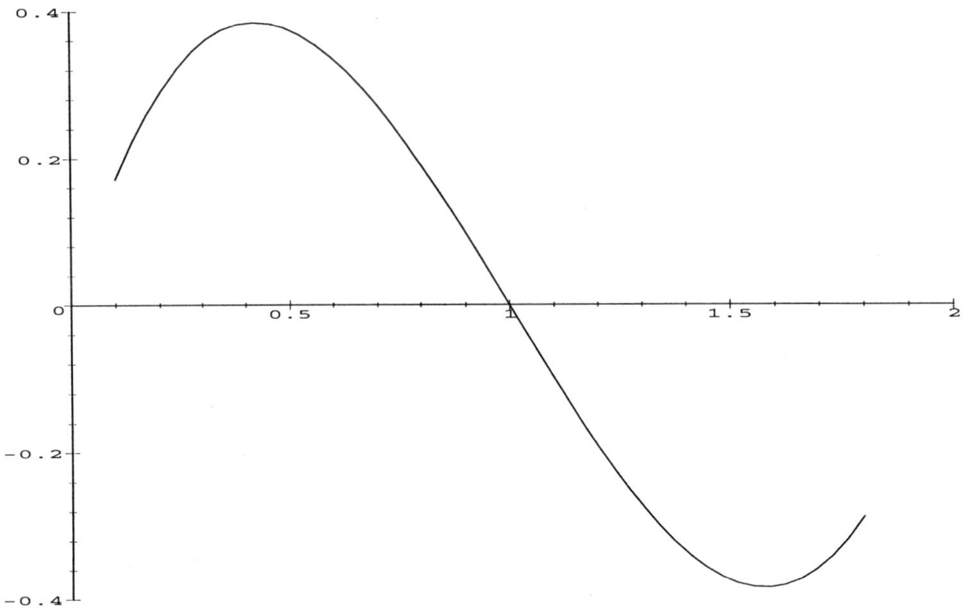

Figure 3.8 Finding Maximum, Minimum, and Zero Points Graphically

PP3–2

Given the function in Figure 3.8:

(a) Find the x coordinate of the maximum. *Answer*: ______________________

(b) Find the x coordinate of the minimum. *Answer*: ______________________

(c) Find the x coordinate where the function is 0. *Answer*: ______________________

(d) Sketch the tangent lines at the three points: the maximum point, the minimum point, and the point where the curve passes through $y = 0$.

Answer: ____

(e) Is the slope of the tangent line where $f(x) = 0$ also 0? If not, measure the slope by constructing a triangle based on the tangent line.

Answer: ____

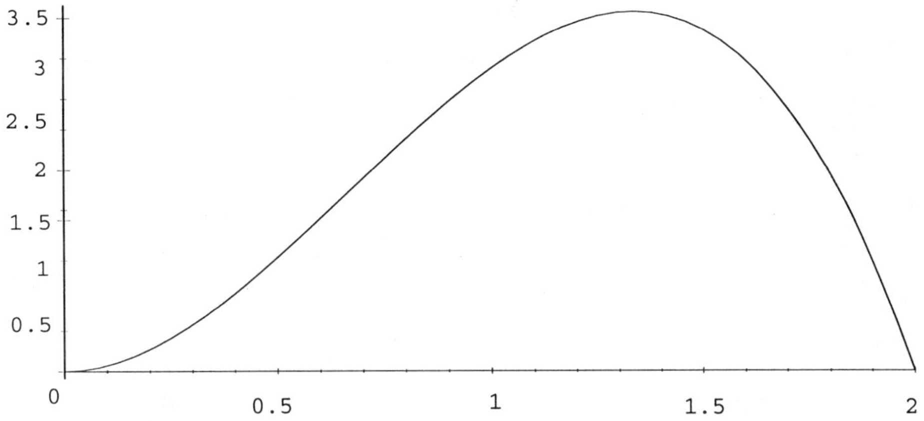

Figure 3.9 Graph for Problem PP3–3

PP3–3

Given the function in Figure 3.9, find the equation of the tangent line at $x = 1$ graphically.

Answer: ____

PP3–4

Given $y = 3x^2 + 2$, find

(a) The slope at $x = 2$. *Answer*: ________________

(b) The equation of this tangent line at $x = 2$. *Answer*: ________________

PP3–5

Estimate the derivative of $f\colon x \to 4x - x^2$ at $x = 1$, using the formulas for both the difference quotient (*dq := x → (f(x + dx) – f(x))/dx;*) and the symmetric difference quotient (*sdq := (x1, x2) → (f(x2) – f(x1))/(x2 – x1);*).

Hint: use $x1 = 1 + dx/2$ and $x2 = 1 - dx/2$ to bracket the point $x = 1$ evenly.

(a) Choose $dx = 0.01$ and enter the difference quotients and the symmetric difference quotients in Table 3.1.

Table 3.1

Table for PP3–5

Difference Quotient (DQ)		Symmetric DQ	
$f(1)$		$f(1.005)$	
$f(1.01)$		$f(0.995)$	
Δx	0.01	Δx	0.01
dq		sdq	

(b) Find the derivative at $x = 1$ symbolically. *Answer*: ____________

(c) Compare the two estimates, *dq* and *sdq*, to the derivative. Which is the better form to use?

Answer: ____________

PP3–6

Given $y = e^x$, find the slope of the chord at $x = 1$:

(a) Using *dq*, with $dx = 0.01$. *Answer*: ____________

(b) Using *sdq*, with $dx = 0.01$. *Answer*: ____________

(c) Compare your answers in (a) and (b) with the symbolic derivative of e^x at $x = 1$. Which formula for the difference quotient, *dq* or *sdq*, gives the more accurate result for the derivative?

Answer: ____________

(d) Apply the formula for *dq* (*not sdq*) with $dx = 0.00001$. State the result and compare it to (b).

Answer: ____________

(e) Based on the above results, which formula, *dq* or *sdq*, should you use in numerical work?

Answer: ____________

PP3–7

Apply the chain rule to find the derivatives with respect to *x* of these expressions. Do these problems by hand and check your answers using Maple if you wish.

(a) $4(3x + 1)^2 + 2(3x + 1)$ *Answer*: ____________

(b) $\dfrac{1}{3x^2 + 2x + 1}$ *Answer*: ____________

(c) $\sqrt{5x - 3}$ *Answer*: ____________

(d) $\dfrac{1}{\sqrt{1 - x^2}}$ *Answer*: ____________

(e) $\sin(kx)$ *Answer*: ____________

(f) $e^{(x^2)}$ *Answer*: ____________

(g) $e^{\sin(x)}$ *Answer*: ____________

(h) $\sin(e^x)$ *Answer*: ____________

(i) $\ln\left(\dfrac{1}{x}\right)$ *Answer*: ____________

(j) $ln(ln)(x))$ *Answer*: ____________

PP3–8

State the derivative with respect to x of the following expressions. In all cases, first attempt to do the problems mentally.

(a) $ax^2 + bx + c$ *Answer*: ____________

(b) $x^3 + px + q$ *Answer*: ____________

(c) $9x^4 - 12x^3 + 4x^2 - 8x$ *Answer*: ____________

(d) $\sin(x) + \cos(x)$ *Answer*: ____________

(e) $ln(x) + \dfrac{1}{x}$ *Answer*: ____________

(f) $\dfrac{3x + 2}{2x + 3}$ *Answer*: ____________

PP3–9

Find dy/dx by implicit differentiation.

(a) $\dfrac{x^2}{3^2} + \dfrac{y^2}{2^2} = 1$ *Answer*: ____________

(b) $x^2 - y^2 = 1$ *Answer*: ______________________

(c) $\sin(xy) = 1$ *Answer*: ______________________

(d) $ln(xy) = 1$ *Answer*: ______________________

(e) $e^{(xy)}$ *Answer*: ______________________

PP3–10

The two lines in each of Figures 3.10 through 3.12 represent the function and its derivative. Which is which?

Answer: ______________________

Answer: ______________________

Answer: ______________________

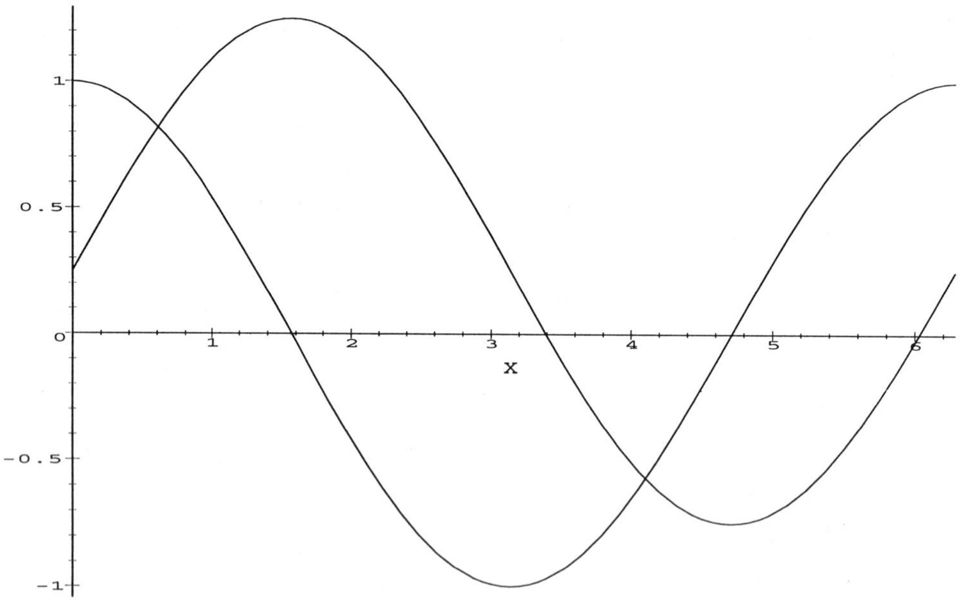

Figure 3.10 Diagram for PP3–10(a)

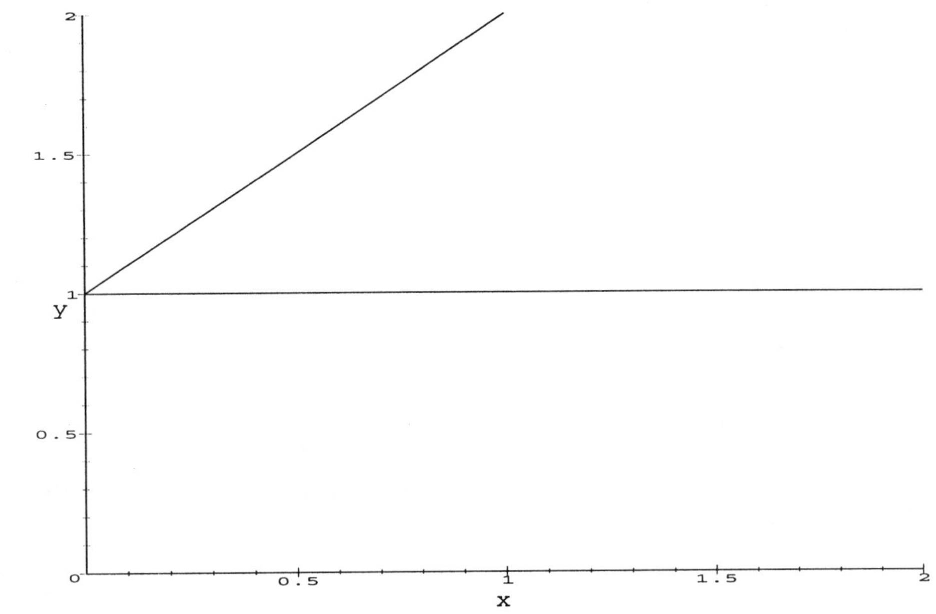

Figure 3.11 Diagram for PP3–10(b)

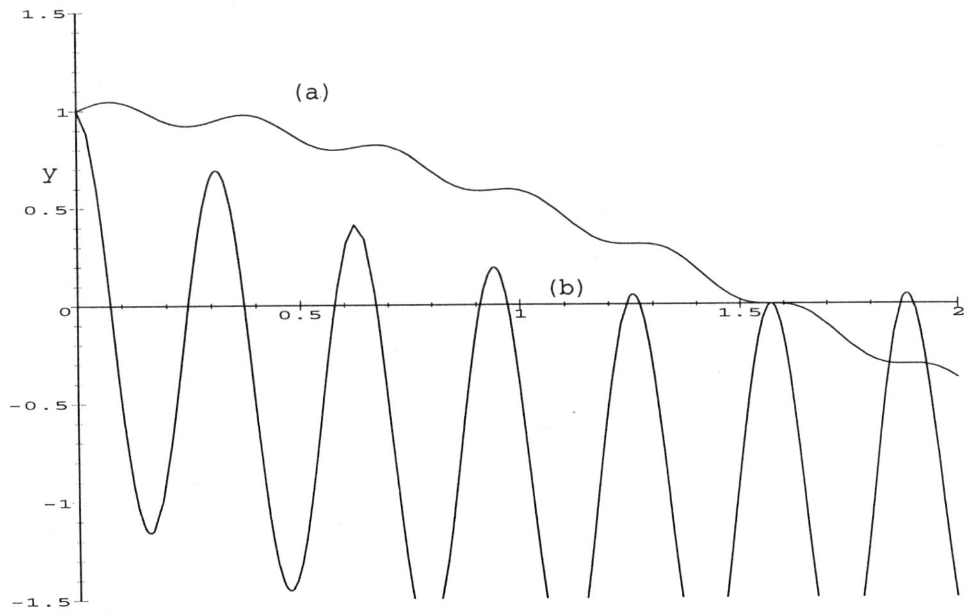

Figure 3.12 Diagram fro PP3–10(c)

Maple Lab

ML3–1

Define the symmetric difference quotient as

```
> sdq := (f, x, dx) -> (f(x + dx/2) - f(x - dx/2))/dx;
```

$$sdq := (f,\ x,\ dx) \rightarrow \frac{f\left(x + \frac{1}{2}dx\right) - f\left(x - \frac{1}{2}dx\right)}{dx}$$

This definition of the symmetric difference quotient is different than the one used before, but this time you pass the function *f* to *sdq*, rather than defining it separately. Also, the definition ensures that equal intervals on either side of *x* are taken.

(a) Apply *sdq* to the cubic function. Note that the first argument to *sdq* is given as a function, and not simply as x^3.

```
> sdq(x -> x^3, x, dx);
```

$$\frac{\left(x + \frac{1}{2}dx\right)^3 - \left(x - \frac{1}{2}dx\right)^3}{dx}$$

Now take the limit as $dx \to 0$.

```
> limit(sdq(x -> x^3, x, dx), dx = 0);
```

$$3x^2$$

Compare the result with the derivative of the cubic function.

```
> D(x -> x^3)(x);
```

What is your conclusion?

Answer: ______________________________

(b) Use this definition to find the derivative of the function f: $x \to |x - 1|$ at the point $x = 1$. Issue these commands and note the results.

```
> sdq(x -> abs(x - 1), x, dx);
```
Answer: ____________________

```
> limit(sdq(x -> abs(x-1), x, dx), dx = 1);
```
Answer: ____________________

(c) The answers seem to be terribly complicated, but Maple is trying to tell you that it does not know any information about *x*, and it is trying to be completely general. Give Maple some help by instructing it that you want to evaluate the answer at $x = 1$.

> subs(x=1, A);

Answer: ______________________________

Since this doesn't seem to be quite enough, ask Maple to evaluate the answer once again. Or, you could simply "wrap" an *eval* around the previous command.

> eval(subs(x=1, A));

Answer: ______________________________

(d) You have arrived at a numerical answer, but is it correct? Remember that it is not enough that the limit exists for a particular choice of decreasing intervals *dx*. The limit must be the same for all possible ways of defining decreasing intervals. If the curve has a sharp point at the value of *x* under consideration, different ways of defining the way that decreases in the interval *dx* will produce different results. Define *sdq* as

> sdq := (f, x, dx) -> (f(x + dx/10) - f(x - dx/(9/10)))/dx;

This keeps the total length of the interval the same, namely, *dx*. This time, though, most of the interval lies to the left of *x*. Find the numerical value of the *sdq* using the command

> eval(subs(x=1, A));

Answer: ______________________________

ML3-2

The symmetric difference quotient defined in ML3–1 can be used to calculate the slope of a chord that goes through the points whose *x* coordinates are $x - dx/2$ and $x + dx/2$. Its slope should approximate the derivative of the function *f* at the point *x*. Evaluate the symmetric difference quotient for the data given in each part of this question and compare it to the derivative, evaluated at the same point. Your answer should be in three parts, (i) the result of applying *sdq* to the function, (ii) the value of the derivative at the same point, and (iii) the number of digits of accuracy in *sdq*.

As you are doing this exercise, remember to check your results by plotting the graph of the function. Some results may be misleading unless you check whether the derivative really exists.

Example: Apply this procedure to the function f: $x \rightarrow \cos(x)$ at $x = 0$ and $dx = 0.01$.

> sdq(cos, 0, 0.01);

$$0$$

Since the derivative of cos is –sin and –sin(0) = 0, this result is accurate to at least ten decimal places (since Maple is evaluating its results to ten places here). The graph shows that cos has a maximum at $x = 0$, so the derivative must be 0 at $x = 0$. Not all results will be this accurate when you use *sdq*.

Answer: (i) 0.0061; (ii) 0; (iii) *sdq* is accurate to two decimal places.

(a) f:= *sin*; at $x = 1$ and $dx = 0.01$.

Answer: (i) ________ (ii) ________ (iii) ________

(b) f:= *exp*; at $x = 1$ and $dx = 0.001$.

Answer: (i) ________ (ii) ________ (iii) ________

(c) f:= *ln*; at $x = 1$ and $dx = 0.001$.

Answer: (i) ________ (ii) ________ (iii) ________

(d) f:= *x -> x*(x^2-1)/(x^3-x^2-x+1);* at $x = 0$ and $dx = 0.01$.

Answer: (i) ________ (ii) ________ (iii) ________

(e) f:= *x -> 1/x;* at $x = 0$ and $dx = 0.01$.

Answer: (i) ________ (ii) ________ (iii) ________

(f) f:= *sin;* at $x = 1$ and $dx = 0.01$.

Answer: (i) ________ (ii) ________ (iii) ________

ML3–3

Most times the derivatives of interest arise in problems where the formulas contain numerical constants whose values are not simple whole numbers. Use the method of ML3–2 to evaluate the symmetric difference quotient in the following cases. Note that the interval *dx* matches the rest of the problem. In some cases it can be greater than one and still yield accurate results.

(a) The function $f:=R \to \dfrac{6.71R}{2310. + R^2}$, $R = 2x\ 10^3$ and $dR = 1$.

Answer: (i) ________ (ii) ________ (iii) ________

(b) The function $f:=T \to 32T^2\left(59.5 - \dfrac{T}{2.5}\right)$, $T = 3.55$ and $dT = 0.1$.

Answer: (i) ________ (ii) ________ (iii) ________

(c) The function $f:=t \to -16.1t^2 + 70.5t - 10.11$, $t = 1.5$ and $dt = 0.01$.

Answer: (i) ________ (ii) ________ (iii) ________

ML3–4

The previous three problems demonstrated, among other things, that the symmetric difference quotient is a good approximation to the derivative. If the function in question is given by a formula, the derivative can be found easily using Maple. Show that this is so by using Maple to find formulas for the derivatives of these common functions.

(a) f := x -> x^n; *Answer*: ____________

(b) f := x -> sin(x) *Answer*: ____________

(c) f := x -> cos(x) *Answer*: ____________

(d) f := x -> tan(x) *Answer*: ____________

(e) f := x -> cot(x) *Answer*: ____________

(f) f := x -> sec(x) *Answer*: ____________

(g) f := x -> csc(x) *Answer*: ____________

(h) f := x -> ln(x) *Answer*: ____________

(i) f := x -> log[10](x) *Answer*: ____________

(j) f := x -> exp(x) *Answer*: ____________

(k) f := x -> 10^x *Answer*: ____________

ML3–5

Given $f := 2x + \sin(x)$, plot the function and its derivative over the interval $x = 0 .. 2$. Sketch the curves here and label the derivative.

(a) State how you can tell which is the curve for the derivative and which is the graph of the original function.

Answer: ______________________________

(b) What is the maximum value of the derivative over the given interval?

Answer: ______________________________

(c) From the graph, what is the minimum value of the derivative over the interval?

Answer: ______________________________

ML3–6

Use Maple to find the derivative of the general polynomial.

(a) Define the general polynomial as

> px := Sum(a[i]*x^i, i = 0 .. n);

Then take the derivative using

> diff(px, x); *Answer*: ______________

(b) Write the general polynomial of ten terms. Use the command, *expand(sum(a[i]*x^i, i = 0 .. 10));*. Find the derivative of each term without using Maple. Compare your work to the general result in part (a) by working out the first few terms. Is the pattern simply that of the derivative of kx^n?

Answer: ______________________________

ML3–7

Find the derivative of f: $x \rightarrow \tan(2x + 3)$ by using the *D* operator and by using the *diff* command.

(a) To use the operator command, you might think that you can find the derivative by issuing the command

> D(tan)(2*x + 3);

Write the output of this command here. *Answer*: ______________

(b) However, the output obtained in (a) is not what the problem is asking. Take the derivative using *diff*.

> diff(tan(2*x +3), x); *Answer*: ______________

(c) Compare (a) to (b). They differ by a factor. What is that factor?

Answer: ______________________________

(d) What is the correct form of the command if you wish to use the D operator? Here's how to do it: you regard tan($2x + 3$) as the composition of u and v, where

> u := x -> tan(x);

> v := x -> 2*x +3;

To verify that the procedure yields the correct answer, write the results of all the following commands:

(i) u@v; *Answer*: ______________

(ii) (u@v)(x); *Answer*: ______________

(iii) D(u@v); *Answer*: ______________

(iv) D(u@v)(x); *Answer*: ______________

(v) D(tan@(x -> 2*x + 3)); *Answer*: ______________

(vi) D(tan@(x -> 2*x + 3))(x);

Answer: ______________________________

(e) If you wish to find the derivative of sin(exp($x^2 + \sqrt{x}$) with respect to x, which of the following commands produces the correct result?

(i) D(sin)(exp)(x^2 + sqrt(x));

(ii) D(sin@exp)(x^2 + sqrt(x));

(iii) D(sin@exp@(x -> x^2 + sqrt(x)));

(iv) D(sin@exp@(x -> x^2 + sqrt(x)))(x);

Answer: ______________________________

(f) Check your answer by using the *diff* command

> diff(sin(exp(x^2 + sqrt(x))), x);

Answer: ______________________________

ML3-8

Find the derivatives dy/dx of the following expressions by both methods, D and *diff*. Write the answer only once, after checking that both answers are the same.

(a) $y = \dfrac{\sqrt{3x^2 + 10}}{25x}$ *Answer*: ______________

(b) $y = 2x^3(9x^4 + 1)^2$ *Answer*: ______________

(c) $y = \dfrac{3}{4}\dfrac{1}{(3 - 4x^2)^{\frac{1}{4}}}$ *Answer*: ______________

(d) $y = \dfrac{2 - x^2}{2 - x^3}$ *Answer*: ______________

(e) $\dfrac{\dfrac{1}{x}}{\dfrac{1}{x + 1}}$ *Answer*: ______________

(f) $y = \dfrac{1 + x + x^2 + x^3}{x - \frac{1}{3}x^3 + \frac{1}{5}x^5}$ *Answer*: ______________

(g) $y = 10\sin(3x)$ *Answer*: ______________

(h) y = A*sin(k*x) *Answer*: ______________

(i) $y = A\sin(kx - \omega t)$ *Answer*: ______________

(j) y = A*sin(exp(k*x)) *Answer*: ______________

ML3-9

The second derivative can be found using the notation (D@@2), in the case of the differential operator, or *diff*(*f*(*x*), *x, x*), in the case of the *diff* command. The *diff* command can also be written as *diff*(*f*(*x*), *x*$2). These commands can be generalized to (D(@@n) and *diff*(*f*(*x*), *x*$*n*) in cases where the *n*th derivative is required. Find the second derivative, d^2y/dx^2, of the following expressions.

(a) $y = 16x^2 + 10x + 25$ *Answer*: ______________

(b) $y = \sin(x)$ *Answer*: ______________

(c) $y = e^x$ *Answer*: ______________

(d) $y = \sin(x) - \cos(x)$ *Answer*: ______________

(e) $y = x^x$ *Answer*: ______________

ML3–10

Use the technique of implicit differentiation to find *dy/dx* for these examples.

(a) $xy^2 - yx^2 = 1$ *Answer*: ____________________

(b) $xy^2 + xy - yx^2 = 1$ *Answer*: ____________________

(c) Sketch the plot of $xy^2 - yx^2 = 1$ using

> plots[implicitplot](x*y^2-y*x^2=1, x = -10 .. 10, y = -10 .. 10, numpoints = 10000);

The command will take some time to complete because of the large number of points to be plotted.

(i) Is the plot a function or a relation? *Answer*: ____________________

(ii) Find the values of y at $x = 10$ *Answer*: ____________________

(iii) Substitute the values of x and y into your formula for the derivative. State the exact answer.

Answer: ____

(iv) Reduce the answers to an approximate decimal.

Answer: ____

(d) Sketch the graph of $xy^2 + xy - yx^2 = 1$.

(e) Compare the curves in (d) to $xy^2 - yx^2 = 0$. Can you find infinitely many points on this curve where $dy/dx = 1$?

Answer: ____

ML3–11

An aircraft is flying on a straight flight path that is taken as the x axis in this problem. An observer who is located on the negative y axis is looking upward from point $P(0, -5)$, where the coordinate distances are measured in miles. The aircraft passes overhead at a speed of 300 miles per hour. At any time, t, the position of the plane is given by $x(t) = 300t$, where t is in hours.

(a) Find an expression for the distance between the plane and the observer. Name this expression for the distance, L. Write the expression in terms of x, the plane's position at any time.

Answer: ______________________________

(b) Find the derivative of the distance with respect to x between the observer and the plane. Name the result *Dlx*.

Answer: ______________________________

(c) Find the derivative of the distance with respect to time by multiplying *Dlx* by $dx/dt =$ 300. You are applying the chain rule in the form $dL/dt = (dL/dx)(dx/dt)$. This is also a related rate.

Answer: ______________________________

(d) At what rate is the distance between the observer and the plane changing when $t = 0$? Find dL/dt at $t = 0$.

Answer: ______________________________

(e) At what rate is the distance between the observer and the plane changing when $t = \infty$? That is, find dL/dt when t is large.

Answer: ______________________________

(f) If $dL/dt = 0$, does that mean that the plane is standing still? Explain.

Answer: ______________________________

CHAPTER 4

Applications of Derivatives

Objectives for This Chapter

Use the ideas of the differential calculus to solve problems in:

1. Linear variation
2. Distance versus time
3. Force versus extension in a spring
4. Resistance versus current
5. Voltage versus charge
6. Energy versus temperature
7. Economics: Interest and principal
8. Quadratic variation
9. Distance versus time
10. Energy versus velocity
11. Power versus current
12. Rate of change of area with radius in a circle
13. Higher-order variation
14. Inverse variation
15. Trigonometric variation
16. Exponential variation
17. Maxima and minima

Maple Commands Used in This Chapter

C := 1e-13;	Maple's *e* notation for powers of ten.
D(s)(1);	The differential operator applied to the function *s* and evaluated at the point 1.
evalf(subs(k = 1093.75, x0 = 0.05, D(F)(0)));	The *evalf* and *subs* commands, applied to the result of applying the differential operator to the function *F*, evaluated at 0.
exp(V/0.027);	Maple's notation for the exponential function.
f := x -> if x > 1 then 4 - x^2 else x + 4 fi;	The *if* command as part of the definition of a piecewise defined function.
fsolve(D(f)(x) = 0);	Solve an equation numerically.
plot(s, 0 .. 6);	The form of the *plot* command when *s* is a function instead of an expression.
s := t -> 30*t -1/2*9.8* t^2; s(1);	Naming *s* as a function of *t*, and evaluating it at *t* = 1.
solve(e4, {t});	The *solve* command. This form of the command causes Maple to write the answer in the form of an equation in *t*.
solve(subs(F = 35, x = 3.2/100, F=k*x), {k});	An example of the use of the substitute (*subs*) command.
subs(x = sols[3], D(f)(x)) ;	The command *sols*[3] picks out the third expression in the sequence called *sols*.
v[t = 3*seconds]	The *table* notation, used here as a convenient way of forming a subscript.

Linear Variation

The topic of linear variation hardly needs the calculus, but it serves as a place to start in your investigations of the applications that contain the notion of change. In the first case, consider any of the definitions in physics that attempt to capture some intrinsic property of matter. The definition of the property should be a number that is capable of being put in a handbook in a way that makes it as useful as possible. Let us take the specific example of that property of matter called the *specific heat*. The specific heat of a substance is a measure of how much energy it takes to change the temperature.

There are several possibilities for constructing a working definition. Perhaps an experimenter found that it took 2,000 joules of energy to heat a lump of mineral from 20 degrees Celsius to 25 degrees Celsius. As a result of the experimental measurements, the experimenter could claim that the specific heat of the mineral was 2,000 joules because that is the amount of energy needed to change the temperature. Is this number, namely 2,000 joules, worthy of being put in a handbook? The answer is no, because the number relates to a narrowly defined situation. What if two equal lumps of the same mineral was used? Since the new lump takes 2000 joules to heat it through the same temperature range, you realize that the number that goes in the handbook should be the amount of heat, in joules, that it takes to heat a unit mass of the substance. With a number like this in the handbook, you could easily determine the energy it would take to heat any mass of substance: multiply the mass by the number in the handbook, which is the specific heat of a unit mass. The argument applies to temperature as well. If it takes a certain amount of joules to heat a substance one Celsius degree, you can assume that it takes double that amount to heat the same substance through a two-degree temperature change. (Realize that this really is an assumption—there is no reason why this *has* to happen.)

By making the two assumptions about mass and temperature range, you have *linearized* the problem of determining how much heat is required to heat a substance. Stated another way, you have used proportional thinking. Most people use proportional thinking and employ linearization when making a stab at a completely new problem. Here is what has been accomplished: the definition of specific heat is the amount of heat required to heat a unit mass of substance by one Celsius degree. In symbols, the specific heat of a substance is defined as

$$c = \frac{\Delta Q}{\Delta T}\frac{1}{m} \tag{4-1}$$

where c is the specific heat, m is the mass of the substance in kilograms, ΔQ is the quantity of heat required, and ΔT is the temperature change. Note that this is a *definition*. For the definition to be useful, the value of c that goes into the handbook must work adequately over a range of temperatures. There is no guarantee at the outset that the *same* value of c will work for a change of 0°C to 1°C and a change of 99.5°C to 100.5°C. In fact, if the material is water, the amount of heat required to change the temperature from 99.5°C to 100.5°C is enormous compared to the amount of heat needed to change the temperature from 0°C to 1°C. The reason is that water boils at 100°C (at sea level), and a great deal of heat is required to turn water from a liquid into a vapor. This energy is in addition to the relatively small amount needed to change the temperature.

Let us not dwell on the exceptions to the rule. Instead, consider it established that for water between the temperatures of 10°C and 50°C, the specific heat is $c = 1$ calorie (cal) per gram (gm°C).

Example 4-1

Calculate the number of joules of heat energy required to raise the temperature of a tank containing 3,000 kilograms of water (a small swimming pool) from 16°C (unpleasantly cold) to 28°C (comfortably warm).

Solution. Before applying Equation 4-1, we must convert the specific heat of water to numbers consistent with kilograms (kg) and joules (J). Since 1 cal = 4.1818 J and 1 kg = 1,000 gm, the specific heat is

$$c = 1\frac{\text{cal}}{\text{gm}\cdot{}^\circ\text{C}} \times 4.18\frac{\text{J}}{\text{cal}} \times 1{,}000\frac{\text{gm}}{\text{kg}} = 4{,}186.8\frac{\text{J}}{\text{kg}\cdot{}^\circ\text{C}}$$

We substitute the given numbers after solving for ΔQ, and we round the answer because we presume that the volume of the tank is only roughly 3,000 kg:

$$\Delta Q = mc\Delta T \qquad (4\text{-}2)$$

$$\Delta Q = 3{,}000 \cdot 4{,}186.8 \cdot (28 - 16)$$

$$= 1.5 \cdot 10^8 \text{ J}$$

The example shows the usefulness of defining physical quantities like the specific heat of substances in such a way that they can be used for any mass of material and any allowable temperature range.

From the point of view of the calculus, observe that the definition of the specific heat contains the quantity $\Delta Q/\Delta T$. In the limit, this is the derivative, dQ/dT. If this quantity were known for a broad range of temperatures, the amount of heat required to change the temperature of a substance, even though it varies with temperature, could be calculated. You'll see how to do this in Chapter 6, on applications of integration.

Your Turn: Materials scientists know that the *resistivity* of a material is a property that allows them to calculate the resistance of a wire. For instance, copper has a resistivity of 0.003 ohms per meter. The resistance of the copper wire is found using the formula $R = \rho A/L$, where R is the resistance in ohms, ρ is the resistivity in ohms per meter, A is the cross-sectional area in meters, and L is the length of the wire. Note that the geometry of the wire does not appear in the definition of resistivity. The resistance of any particular wire is calculated by using its particular cross-sectional area and length, along with the resistivity value for the material.

What is the resistance of 40 kilometers of transmission-line copper wire of cross-sectional area 0.004 square meters?

Answer: ______________________________

Distance versus Time

The detailed study of motion became possible only after the discovery of calculus, which is necessary to understand many concepts relating to motion. The first of these concepts is the definition of *instantaneous velocity* or *instantaneous speed.* The common term for this concept is *speed.* We will not make a distinction between speed and velocity here, so the terms can be used interchangeably. This quantity is derived from a simpler one, called *average velocity* or *average speed.* For example, imagine that you take a trip by car. You travel from your home to the beach, a distance of 50 miles. The trip takes you 2 hours. By definition, the average speed for this trip is the ratio of the distance traveled to the time it takes. If we use the symbols $\bar{v}$ for the average speed, Δx for the distance traveled, and Δt for the elapsed time, then the formula for average speed is

$$\bar{v} = \frac{\Delta x}{\Delta t} \tag{4-3}$$

We use the symbols Δx and Δt for the distance and time variables to remind us that it is the time interval for a corresponding change in distance that is required in the formula. Stated another way, Δt is a time difference, such as the reading on a clock when the trip ended minus the time on the same clock when the trip started. For example, the average speed of the car for the trip to the beach is 50/2 = 25 miles per hour. What is the meaning of this result? You could have taken the trip to the beach by driving at 65 miles per hour most of the way but stopping for a lunch break at a fast food restaurant. Even though your speed while the car was moving was 65 miles per hour, during the time when you were having lunch, the car was stopped. The meaning of the result can be understood by imagining another car taking the same trip, leaving at the same time, and arriving at the same time, but traveling at the same speed throughout. Obviously, a car traveling at a constant rate of 25 miles per hour for 2 hours would cover a total distance of 50 miles. This car would arrive at the beach at the same time as you, who drove at 65 miles per hour but stopped for lunch during the trip.

Thus, the definition of average speed gives information about how fast an object must go, but it provides only a constant rate of speed, if the object is to cover a given distance in a given time. The average speed of any object, even though its speed varies, can be found by the use of Equation 4-3. All that is important is the total distance traveled and the total time it took.

There are times when the average speed is not applicable to the situation. For example, if you are stopped for speeding because you were going 65 miles per hour in a 45-mile-per-hour zone as you entered the beach road, it won't help your case to state that your average speed during the last couple of hours was 25 miles per hour! In the case of obeying the posted speed limit, it is the *instantaneous velocity* (or *instantaneous speed*) that governs the situation. But how is this quantity, instantaneous speed, defined? It is the limit of the ratio $\Delta x/\Delta t$ as t approaches zero. The symbol for instantaneous speed is v, so as a formula, the definition of instantaneous speed is

$$v = \lim_{t \to 0} \frac{\Delta x}{\Delta t} = \frac{dx}{dt} \tag{4-4}$$

The definition shows that v is a derivative. It is the derivative of x with respect to t. Let us examine the situation by means of a graph. A particularly uneven walk/jog is shown in Figure 4.1. The elapsed time is shown on the horizontal axis, and the y axis gives the distance traveled.

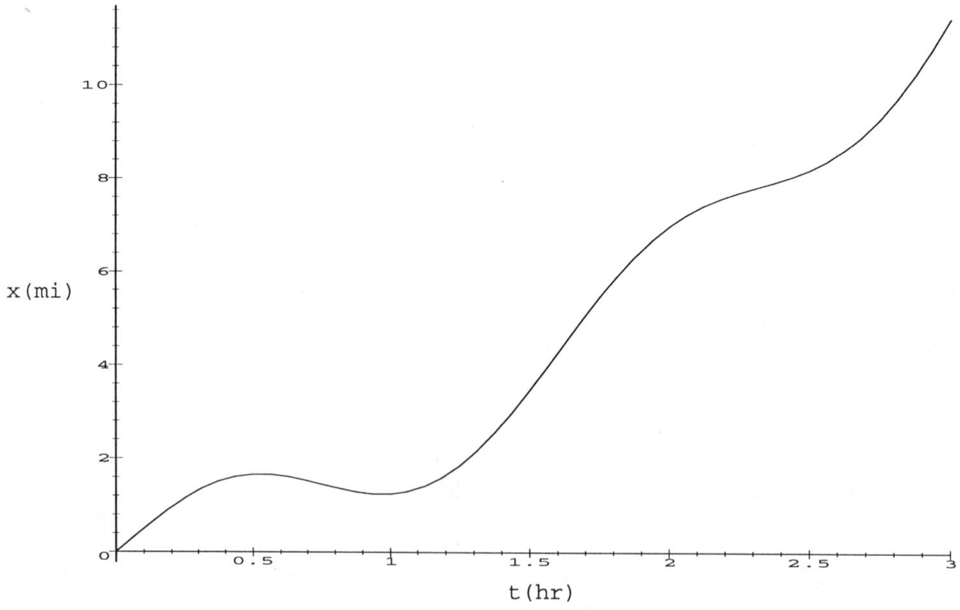

Figure 4.1 The Distance-Time Graph of a Walk/Jog

Example 4-2: Finding Some Average Speeds

(a) Approximately what is the average speed for the first half hour?

(b) What is the average speed during the second half hour?

(c) What is the average speed during the first hour?

Solution:

(a) The distance traveled during the first half hour is approximately 1.6 miles. Therefore, by Equation 4-4, the average speed is 1.6/0.5 = 3.2 miles per hour. During this time, the jogger is warming up by taking a slow walk.

(b) During the second half hour, the distance traveled is *negative*. At the beginning of the half hour, the person is at the $x = 1.6$-mile point, and at the end of the time period, he or she is near the 1.2-mile point. Thus, the distance traveled is 1.2 – 1.6 = –0.4 mile. The average speed is –0.4/0.5 = –0.8 mile per hour. (Perhaps the jogger decided to turn around and go home.)

(c) During the first hour, the definition of the average speed tells us to use 1.2 miles as the total distance traveled, since this is now the distance away from the starting point, and to use 1 hour as the elapsed time. The average speed for this trip is 1.2/1 = 1.2 miles per hour. Note that the average speed for the full hour is *not* the sum of the two average speeds.

Your Turn. What is the average speed during the period from 0 to 2 hours?

Answer: ______________________________

The jogger decides not to go home. Instead he (or she) turns around and begins to jog and jogs for about a mile, between points $x = 1.2$ and $x = 2.2$ miles. We will try and determine the jogger's instantaneous speed at $t = 1.5$ hours. The distance-time curve looks linear for this time interval. We will show how linear it is by passing a straight line through the $t = 1.5$-hour point. This is shown in Figure 4.2.

The tangent line approximates the curve very well for a small time interval around $t = 1.5$. The slope of this line is dx/dt; therefore, it is the instantaneous speed at $t = 1.5$. You can work out the slope by measuring the lengths of the base and the altitude of the triangle in Figure 4.2. The base is 1.5 – 1.05 = 0.45 hours, and the height is 3.5 miles. The instantaneous speed is 3.5/0.45≈7.8 miles per hour. The jogger is still not going very fast. (Perhaps the problem is a sore ankle.)

We found the instantaneous velocity by taking large distance and time measurements. Is this acceptable? The answer is yes, because we measured the slope of the tangent line. Since all

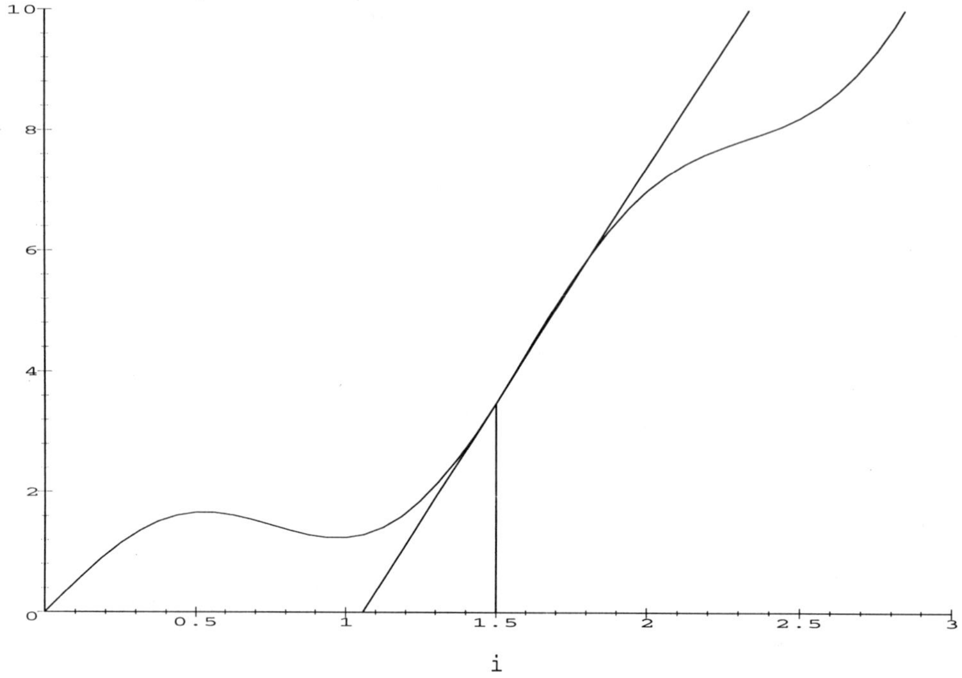

Figure 4.2 The Tangent to the Distance-Time Graph at $t = 1.5$ Hours

triangles built on this line have the same slope, we can use any of them that we choose. We could have used an even bigger triangle to make our measurements easier and more accurate. If you are making measurements from a graph, its a good idea to use the biggest figure available.

If we want more accuracy than the graph can give us, we need a formula that relates the distance to the time. We need to make x a *function* of time. This is not a simple task in the case of the jogger's little outing. Even had the jogger been following a plan, there are too many variables to make for an easy derivation of a formula for distance as a function of time. We can look to other situations where formulas are readily available. The classic case is that of a falling object.

Example 4-3: The Speed of a Falling Object

Physics tells us that the distance-time relation for an object falling under the influence of gravity is

$$s = v_o t - \frac{1}{2}at^2 \tag{4-5}$$

where s is the distance traveled, v_o is the initial speed, a is the acceleration due to gravity, and t is the elapsed time. In the International System of Units (SI), s is measured in meters, a = 9.8 meter (m)/second(s)^2, and t is measured in seconds. The quantity v_o is the initial velocity, that is, the speed of the object when $t = 0$.

(a) If a ball is thrown straight up with an initial velocity of 30 meters per second, how high will it be after 1 second?

(b) How high will it be after 5.1 seconds?

(c) What is the velocity at 1 second; at 5.1 seconds?

Solution: We will set the problem up so that it can be done in Maple. We can express the distance-time relation as a function.

> s := t -> 30*t -1/2*9.8*t^2;

$$s := t \rightarrow 30t - 4.900000000t^2$$

(a) Since s is a function of t, the distance at $t = 1$ is $s(1)$.

> s(1);

$$25.10000000$$

(b) Similarly, the distance at $t = 5.1$ seconds is $s(5.1)$.

> s(5.1);

$$25.5510000$$

The ball rose to a height of 25.1 meters after 1 second. How is it that at 5.1 seconds it is only at 25.55 meters? If you don't see the answer immediately, it will become clearer when we calculate the derivative of the function and evaluate it at these two times. Since the derivative is *dx/dt*, it represents the velocity of the ball at these times.

(c) The velocities are

> D(s)(1);

$$20.20000000$$

> D(s)(5.1);

$$-19.98000000$$

At 1 second, the velocity is positive, and at 5.1 seconds the velocity is negative. At $t = 1$, the ball is going *up,* whereas at $t = 5.1$, the ball is going *down.* At some time between 1 second and 5.1 seconds, the ball reaches its maximum height. Then it begins to fall back to the earth.

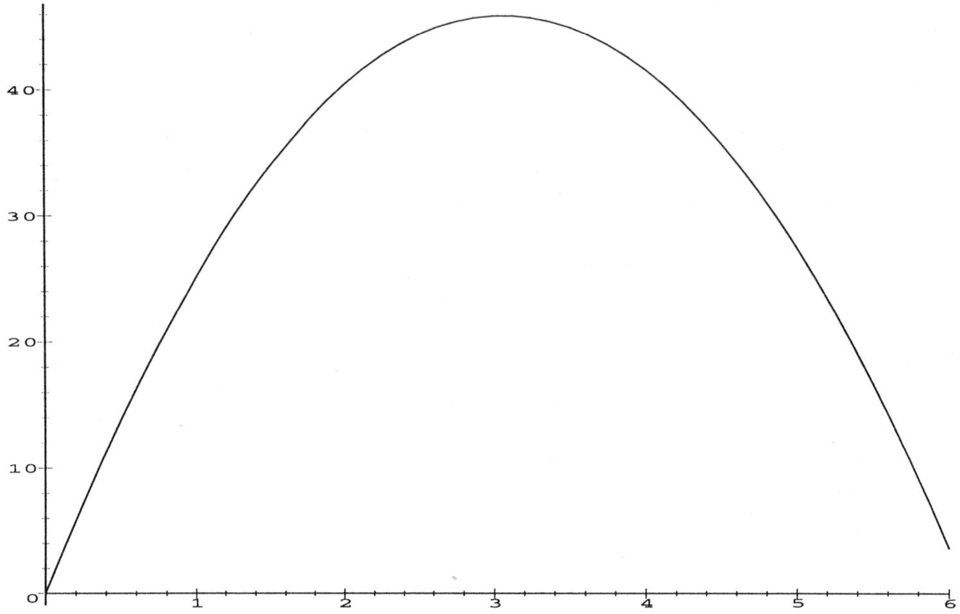

Figure 4.3 Distance-Time Graph for Example 4-3

At $t = 5.1$ seconds, it is close to the ground again. That is why s is nearly the same then as at $t = 1$ second.

To get an even better picture of the problem, draw the graph of the distance-time relationship.

```
> plot(s, 0 .. 6);
```

From the graph, you can see that the ball returns to earth shortly after the 6-second mark and that it reaches a maximum height of 46 meters at a time of 3 seconds after being thrown (Figure 4.3).

The example shows how to apply the calculus to the problem of finding the speed of an object, given its distance, as a function of time. It's also possible to find the maximum height by making the following important observation: at the top of its flight, the ball's instantaneous speed must be zero. Note the shape of the distance-time graph in Figure 4.3. The slope of the tangent line is zero at the maximum point. In symbols, $dx/dt = 0$. To find the time at which the height is a maximum, solve the equation

```
> e4 := D(s)(t) = 0;
```

$$e4 := 30 - 9.800000000t = 0$$

The equation is derived by setting the derivative of the distance function equal to zero. The derivative is evaluated at the representative time, *t*. The solution of the equation is

```
> solve(e4, {t});
```

$$\{t = 3.061224490\}$$

The analytic solution indicates that the ball reaches maximum height slightly after $t = 3$ seconds. This agrees well with our graphical solution, where we estimated the maximum time at 3 seconds. To find the distance, substitute this time in the function *s*.

```
> s(3.06122);
```

$$45.91836735$$

You don't have to type in all ten digits. The data in the problem was given to two digits, so the answer should be reported to the same accuracy. The height attained is approximately 46 meters.

We can find the total time of flight as well. You can argue that the time of ascent should equal the time of descent, or you can solve for the time when the total distance from the starting point is zero. Here is the solution found by setting the distance equal to zero.

```
> solve( s(t) = 0, {t});
```

$$\{t = 0\}, \{t = 6.122448980\}$$

There are two solutions. The first, $t = 0$, says that the height of the ball is zero initially. The other solution is the one we are looking for. It tells us that the ball falls to earth 6.1 seconds into its flight. This value agrees with Figure 4.3, where $s = 0$ at $t = 0$ and also at a time shortly after 6 seconds.

The velocity can be found at any time by evaluating the derivative of *s* with respect to *t* at that time. At $t = 3$ seconds, the velocity of the ball is

```
> v[t = 3*seconds] = D(s)(2)*`meters per second`;
```

$$v_{t\,=\,3\ seconds} = 10.40000000\ meters\ per\ second$$

Here again we have used the square brackets after the variable *v* simply to produce nice-looking output. The output is read as follows. The velocity at $t = 3$ seconds is 10.4 meters per second. Both *seconds* and *meters per second* are treated as free variables by Maple. Back quotes (`) must enclose the words *meters per second*, since this is a variable as far as Maple is concerned and it contains spaces. The back quotes inform Maple that the phrase is to be treated as one object. You don't have to write the solution using such elaborate techniques, but you may be interested in using them in a formal report. The example is included here just to show you how it is done.

Force versus Extension in a Spring

A spring produces a return force when it is stretched. An *ideal* spring produces a force that is strictly proportional to the stretch. The formula for the return force is

$$F = kx \qquad (4\text{-}6)$$

where F is the return force in newtons; x is the stretch, or *displacement*, of the spring in meters; and k is a constant of proportionality, called the *spring constant*.

Example 4-4: Calculating the Spring Constant

Find the spring constant, k, if a force of 35 newtons extends the spring by 3.2 centimeters.

Solution: Since this is a linear problem, the spring constant is truly a constant. It is calculated by solving Equation 4-6 for k and then substituting the given values for F and x. Here is a Maple "one-liner" that accomplishes the solution. Read the commands from the inside out to see how it accomplishes the solution.

```
> solve(subs(F = 35, x =3.2/100, F=k*x), {k});
```

$$\{k = 1093.750000\}$$

Your turn

(a) What are the units of k?

Answer: ____

(b) Why did we divide the x value by 100 in the Maple command?

Answer: ____

(c) How much force, in newtons, is required to stretch the spring a distance of 4.5 centimeters?

Answer: ____

Example 4-5: When the Spring "Constant" Is No Longer a Constant

Real springs cannot obey the simple law of Equation 4-6. One reason is that no restrictions were placed on the displacement, x, whereas a real spring cannot be extended indefinitely.

One possible behavior of a real spring is that the return force flattens out at larger displacements. Let the more realistic equation for the spring in the interval $0 < x < 1$ be

```
> F := x -> k*x*(exp(-(x-x0)));
```

$$F := x \to kxe^{(-x + x0)}$$

where k is the same as in Example 4-5 and $x0$ is 0.05. We can define an effective spring constant for small displacements. It is no longer k, but it can be defined as dF/dx. For small displacements around the extension x, dF/dx will be the effective force constant. If it is multiplied by the small displacement, Δx, it will produce the incremental return force, ΔF.

(a) What is the effective spring constant at $x = 0$?

(b) What is the effective spring constant at $x = 0.8$?

Solution:

(a)

```
> evalf(subs(k = 1093.75, x0 = 0.05, D(F)(0)));
```

1149.827761

(b)

```
> evalf(subs(k = 1093.75, x0 = 0.05, D(F)(0.8)));
```

103.3301834

At $x = 0$, the spring constant is almost the same as in the previous example. However, when the extension x becomes large, the effective spring constant becomes much less. This means that it takes less *extra force* to extend the spring a small amount at 0.8 meters than it does at 0 meters. The effective spring constant has become much weaker.

Your Turn

(a) Find the effective spring constant at $x = 1$ meter.

Answer: ______________________________

(b) What happens to the spring constant for $x > 1$ meter?

Answer: ______________________________

Resistance versus Current

In the 1820s, Georg Ohm startled the scientific world by suggesting that the electrical current through a material is proportional to the applied voltage. He had discovered that the rule was true for a broad class of materials over a wide range of voltages and currents. Today, this rule is known as *Ohm's law*. Much more is known about materials science today, and there are many examples of departures from Ohm's law. We will look at a case where the law applies, and we will use the calculus to investigate the relationship between current and voltage when the law no longer applies.

Example 4-6: The Resistor

The resistor is one of the most common components in electrical and electronic technology. Today, these devices are designed and constructed so that they will obey Ohm's law for their planned operating ranges. If V is the applied voltage in volts and i is the current through the device in amperes, then Ohm's law is given by the linear relation

$$V = Ri \tag{4-7}$$

The constant of proportionality is R; it represents the *resistance*, in ohms. The component itself is called a resistor. Normally, when people talk about a resistor, they mean one that obeys Ohm's law.

(a) Find the voltage across a resistor when the current through it is 116 milliamperes if an applied voltage of 25 volts causes a current of 53.2 milliamperes (1 milliampere = 10^{-3} amperes).

(b) A resistor has a resistance of 22 kilohms (1 kilohm = 1,000 ohms). Find the current through it if the applied voltage is 12 volts.

Solution:

(a) The resistance can be found with the information that an applied voltage of 25 volts causes a current of 53.2 milliamperes. Using the formula for Ohm's law and solving for R, $R = V/i = 25/53.2 = 0.47$ kilohms, or 470 ohms. Note that in the calculation we used a current in milliamperes to obtain a resistance in kilohms. You should redo the calculation by converting the current to amperes to obtain a resistance in ohms. You will get the same result, thus demonstrating the correctness of the shortcut.

Apply Ohm's law once more to solve for the voltage across a resistor when the current through it is 116 milliamperes. $V = 470 \cdot 116 \cdot 10^{-3} = 55$ V.

(b) Again using Ohm's law, $12 = 22\,i$, $i = 12/22 = 0.55$ milliamperes.

Your Turn. You are asked to choose a resistor so that a current of 25 milliamperes will flow through it when the voltage across it is 110 volts. What resistance value will do the job?

Answer: ____

Example 4-7: The Nonohmic Resistor

Electrical devices that do not obey Ohm's law are called nonohmic. One of the most common of these is the solid-state device called the *diode*. The relationship between the current and voltage in a diode is given by the relation

$$i = I_S(e^{V/V_O} - 1) \tag{4-8}$$

where I_s is a small number, perhaps 10^{-13}, and V_O is 0.025.

The diode can be placed at an operating point by applying a fixed voltage across its terminals. This so-called direct current (DC) voltage causes a current to flow. The amount of current is determined by Equation 4-8. A small alternating current (AC) voltage is also applied to the device. The total voltage across the terminals of the diode is therefore the sum of the constant DC voltage and the small AC voltage. Since the AC voltage is small, the total voltage remains close to the DC level. There is a distinction between the DC resistance and the AC resistance. The DC resistance is, by definition, V/i, whereas the AC resistance is defined as dV/di. In other words, it is the ratio of the voltage change to the current change. We will see that this quantity can be very different than the DC resistance in a diode.

(a) Plot the diode characteristic curve for the range $V = 0$ .. 1. Use a y range of $i = 0$ .. 10e-3 (10 milliamperes).

(b) Find the DC resistance of the diode when the applied voltage is 0.5 V.

(c) Find the DC resistance of the diode when the applied voltage is 0.6 V and 0.7 V.

(d) Find the AC resistance of the diode when the applied voltage is 0.5 V.

(e) Find the AC resistance of the diode when the applied voltage is 0.6 V and 0.7 V.

Solution:

(a) Using the given values for I_S and V_O, the plot command is

```
> plot(1e-13*(exp(V/0.027) - 1), V = 0 .. 1, i = 0 .. 10e-3);
```

The diode current remains very small until the voltage across the device exceeds 0.6 volts. Then the diode "turns on" and the current rises rapidly with increasing voltage (Figure 4.4).

(b) The DC resistance of the diode when the applied voltage is 0.5 V is found by applying Ohm's law. The current at $V = 0.5$ V is

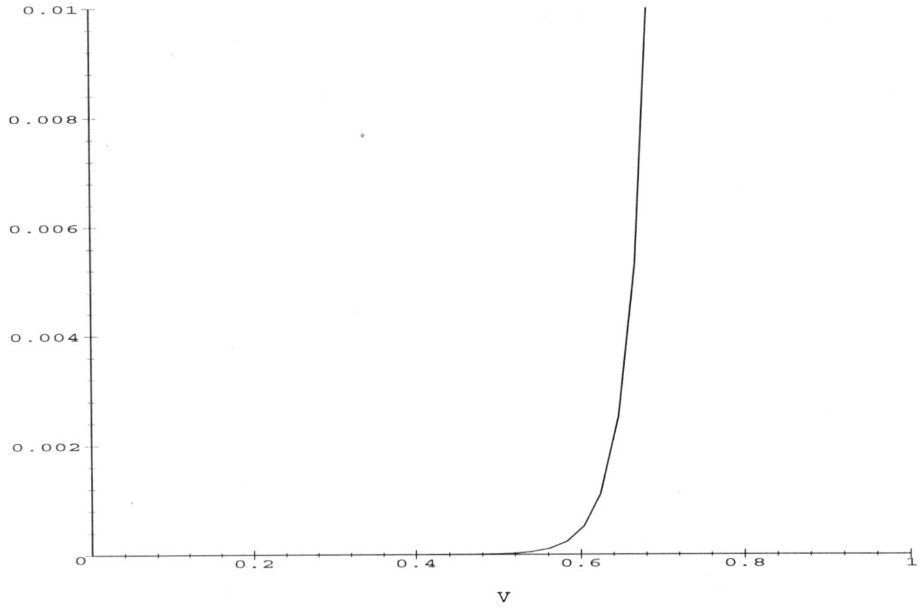

Figure 4.4 The Diode Characteristic Curve

> **evalf(0.5/subs(V = 0.5, 1e-13*(exp(V/0.027) - 1)));**

```
45339.80019
```

The resistance is over 45,000 ohms. This is a high resistance, which demonstrates the fact that the diode has not yet turned on.

(c) The DC resistance of the diode when the applied voltage is 0.6 V and 0.7 V is found by copying and pasting the command in (b) and then changing the substitution value for *V*.

> **evalf(0.5/subs(V = 0.6, 1e-13*(exp(V/0.027) - 1)));**

```
1116.815721
```

> **evalf(0.5/subs(V = 0.7, 1e-13*(exp(V/0.027) - 1)));**

```
27.50954674
```

The resistance drops rapidly as the applied voltage increases. Clearly, the device is nonohmic.

(d) The AC resistance of the diode when the applied voltage is 0.5 V is found by taking the derivative. We want to find dV/di, but it is easier to find di/dV. Since $dV/di = (di/dV)^{-1}$, it really doesn't matter which derivative we take. Calling the current *C*, we define the expression

> C := 1e-13*(exp(V/0.027) - 1);

$$C := .1\ 10^{-12} e^{(37.03703704\ V)} - .1\ 10^{-12}$$

We will construct a Maple command to evaluate *dV/di*. The first step is to take the derivative *di/dV*. Next, we find the reciprocal of the derivative, since we really want *dV/di*. Then we substitute the value $V = 0.5$ into the result. We are almost finished, but if we stop here, Maple will not fully evaluate the expression, which contains an exponential. Lastly, we use *evalf* to reduce everything to a decimal result.

> dV/di = evalf(subs(V = 0.5, (diff(C, V))^(-1)));

$$\frac{dV}{di} = 2448.349188$$

The AC resistance is about 2,500 ohms, much less than the DC resistance. Remember that this is the resistance for small changes around a 0.5-V operating point.

(e) Find the AC resistance of the diode when the applied voltage is 0.6 V and 0.7 V.

The command has already been set up in part (d). All that has to be done here is to copy it and change the operating point value.

> dV/di = evalf(subs(V = 0.6, (diff(C, V))^(-1)));

$$\frac{dV}{di} = 60.30804891$$

> dV/di = evalf(subs(V = 0.7, (diff(C, V))^(-1)));

$$\frac{dV}{di} = 1.485515524$$

In all cases, the AC resistance is less than the corresponding DC resistance. The reason is easily seen by plotting the tangent to the curve, which represents the AC condition, and joining a line from the origin to the operating point, which represents the DC condition. This is shown in Figure 4.5.

The slope of the DC line is very much less than the slope of the AC line. However, the resistance is the reciprocal of the slope, so the AC resistance is much less than the DC resistance.

Maxima and Minima

The calculus is often used to find the maximum and minimum values of a function. As long as the function is continuous, you can find the derivative of the function over its defined

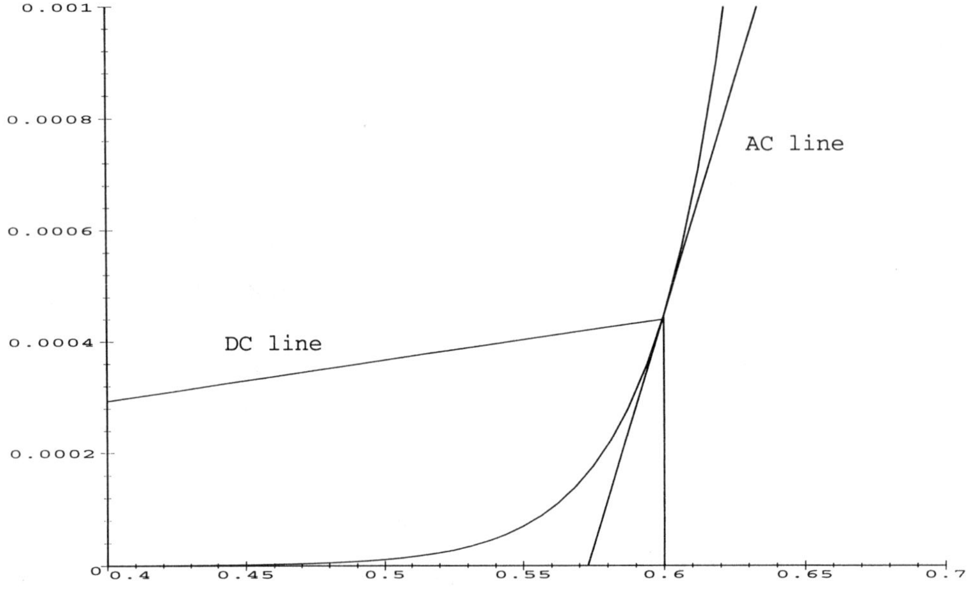

Figure 4.5 DC and AC Reciprocal Resistance Lines for the Operating Point $V = 0.6$ Volt

range. A maximum or minimum point of the function is signaled by the fact that the derivative must be 0 there. Consider the graph of a function shown in Figure 4.6. For the domain $0 < x < 1.6$, the function has one maximum value and two places where a local minimum exists. A local minimum is a point where the value of the function is less than nearby points on either side.

The tangent lines at the maximum and minimum points are shown in Figure 4.6. They are horizontal lines, and their slope is 0. The fact that the slope is zero at a local maximum or minimum can be used to find such points if the function's formula is known. Here are the steps to follow. We will assume that the independent variable is x.

1. Plot the function and check for maximum and minimum points. You can estimate the maximum and minimum values using the graph.

2. Find the derivative of the function.

3. Equate the derivative to zero and solve for x.

4. Check that the value, or values, of x are within the domain defined by the problem.

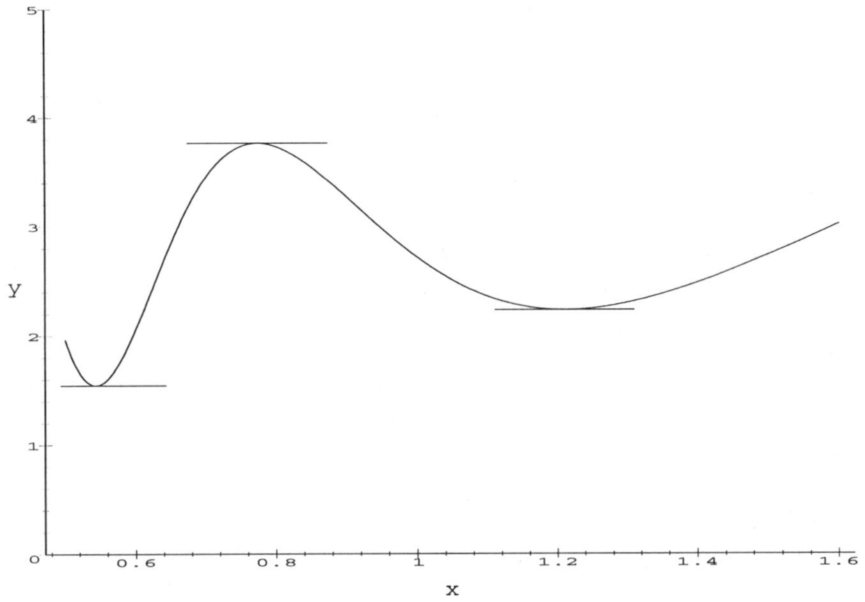

Figure 4.6 A Function with One Local Maximum and Two Local Minima

5. Make sure that you have found all the maximum and minimum points. Sometimes the *solve* command cannot find a solution and you must use *fsolve*. The *fsolve* command often needs help to find all solutions. Give *fsolve* help by supplying it with an interval within which to look for a solution. For example, the command

   >fsolve(f(x) = 0, x, x = 0..2);

   causes *fsolve* to search for solutions only in the interval $0 \le x \le 2$.

6. Check your solution by back-substitution.

Example 4-8: Finding Maxima and Minima

Find the maximum and minimum points of the function $y = 2x^4 - 3x^3 - 24x^2 + 23x + 92$.

Solution: Define the function to Maple.

> f := x -> 2*x^4 -3*x^3 - 24*x^2 +23*x + 92;

$$f := x \to 2x^4 - 3x^3 - 24x^2 + 23x + 92$$

1. Plot the function to check for maxima and minima.

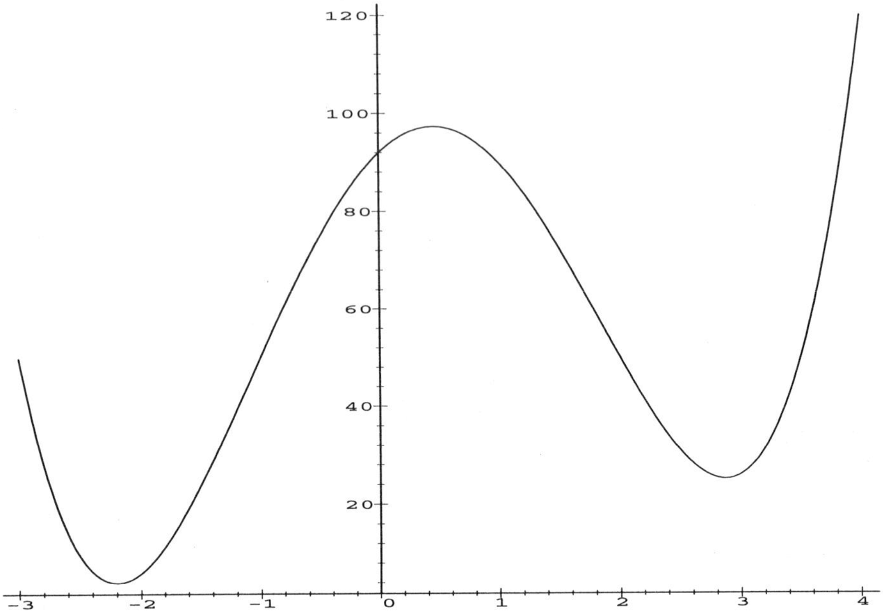

Figure 4.7 The Graph of $y = 2x^4 - 3x^3 - 24x^2 + 23x + 92$, Showing Maxima and Minima

> **plot(f, -3 .. 4);**

The graph shows that there are two local minima, at $x \approx -2$ and $x \approx 3$. There is one local maximum, at $x \approx 1$ (Figure 4.7).

2. The derivative is

> **D(f);**

$$x \rightarrow 8x^3 - 9x^2 - 48x + 23$$

We have left the derivative in functional form. In the next step, we will use the expression form as we attempt to solve the equation.

3. We try the solve command on the equation *dy/dx* = 0.

> **solve(D(f)(x) = 0, x);**

$$\%1^{1/3} + \frac{137}{64}\frac{1}{\%1^{1/3}} + \frac{3}{8},$$

$$-\frac{1}{2}\%1^{1/3} - \frac{137}{128}\frac{1}{\%1^{1/3}} + \frac{3}{8} + \frac{1}{2}I\sqrt{3}\left(\%1^{1/3} - \frac{137}{64}\frac{1}{\%1^{1/3}}\right),$$

$$-\frac{1}{2}\%1^{1/3} - \frac{137}{138}\frac{1}{\%1^{1/3}} + \frac{3}{8} - \frac{1}{2}I\sqrt{3}\left(\%1^{1/3} - \frac{137}{64}\frac{1}{\%1^{1/3}}\right)$$

$$\%1 := -\frac{133}{512} + \frac{1}{64}I\sqrt{39901}$$

Ouch! The result is ugly in two ways. First, the expression is very complicated, and second, the quantity *I* makes an appearance. Maple uses *I* as the square root of negative one; in other words, it is the imaginary quantity. Its appearance here is most unwelcome, because we know from the graph that the solutions for the places of zero slope are real, not imaginary.

Some help can be obtained by converting the answer to decimal form.

> **evalf(solve(D(f)(x) = 0, x));**

$$2.867679726 + .1\ 10^{-9}\ I,\ -2.198662672 - .1\ 10^{-9}\ I,\ .4559829460 - .1\ 10^{-9}\ I$$

The imaginary part of the solution is so small that you can conclude that it should be zero. Try setting *Digits := 30* and issue the command again. You will see that the imaginary term is either 10^{-29} or that it disappears altogether. Thus, the conclusion is that the imaginary parts in the solution are to be ignored and the solutions really are

$$2.867679726,\ -2.198662672,\ .4559829460$$

Unless the functions you are dealing with are no more complicated than quadratics, it is probably better to use the *fsolve* function, which avoids pitfalls such as the example above. Here is the solution using *fsolve.*

> **sols := fsolve(D(f)(x) = 0);**

$$sols := -2.198662672,\ .4559829459,\ 2.867679726$$

Except for the order in which the solutions are displayed, the result is almost exactly the same as that obtained using *solve*. You have to be sure that *fsolve* finds all the solutions, though. You may have to give it some help by telling it the domain in which to search.

4–5. Steps 4 and 5 are not required in this example, because we have allowed for any domain and we have already shown the use of *solve* and *fsolve.*

6. Back-substitution is the most important part of any problem. At this point, we have three solutions, all of them given to ten-digit accuracy. However, those ten digits are useless unless they are correct! To prove that they are, we substitute the *x* values in the expression for the derivative. We'll put all the commands on the same line.

> subs(x = sols[1], D(f)(x)), subs(x = sols[2], D(f)(x)), subs(x = sols[3], D(f)(x)) ;

$$0,\ 0,\ .1\ 10^{-6}$$

The first two solutions verify exactly, but the third is correct to five decimal places. Is it really the solution, or should we be looking for some other number? Once again you could set *Digits* to a higher number, like 30. You would have to execute the *fsolve* command again to obtain the solutions to the higher accuracy. Then you could check the solution with the command shown here. You will find that the discrepancy now comes in the 16th place, confirming that the maximum and the minimum have been found.

Your Turn. Find where the derivative of the function $y = 42 + 4x - 3x^2$ is 0. At this point, is the function a maximum or a minimum?

Answer: ____

Functions for Which Calculus Does Not Work—Discontinuous Functions

The method of setting the derivative equal to zero and solving only works for continuous functions. For example, examine the graph in Figure 4.8. It has a maximum value, but this value cannot be found by setting the derivative equal to zero.

The function in Figure 4.8 has a step discontinuity at $x = 1$. It also has a maximum value at $x = 1$. This is easy to calculate from the definition of the function, which is

> f := x -> if x > 1 then 4 - x^2 else x + 4 fi;

```
f := proc(x) options operator, arrow; if 1 < x then 4-x^2
                else x+4 fi end
```

The function is defined by two formulas. One is valid as long as x is less than or equal to 1, and the other is valid for x greater than 1. The function is defined and single-valued at all values of x, but it has a step discontinuity at $x = 1$. From the graph in Figure 4.8, it is apparent that the maximum value is $f(1) = 5$, which is easily verified by evaluation.

> f(1);

$$5$$

However, the method of finding the maximum value by setting the derivative equal to zero fails.

> solve(D(f)(x) = 0, x);

```
Error, (in unknown) cannot evaluate boolean
```

The graph of the derivative shows that it is nowhere near zero in the domain specified (Figure 4.9).

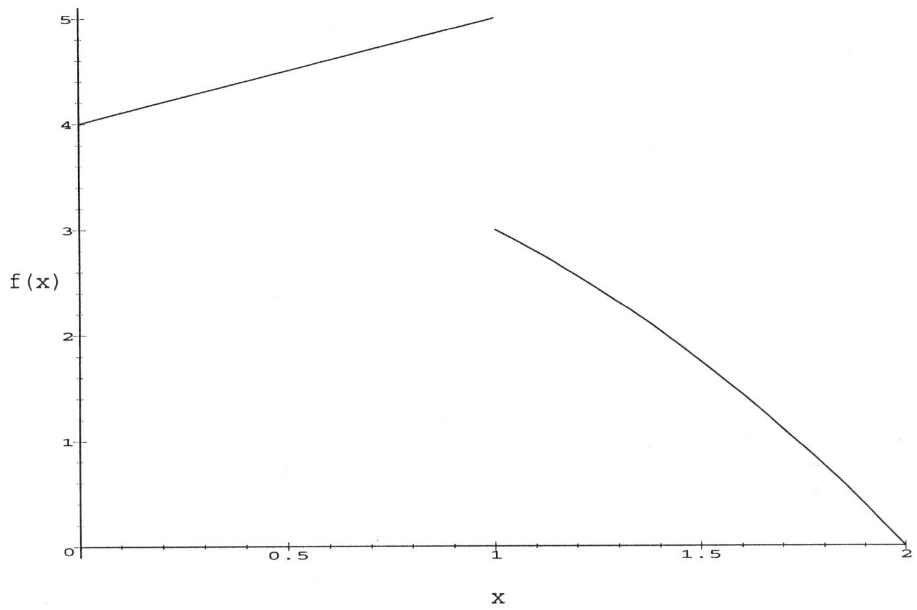

Figure 4.8 Plot of a Discontinuous Function

The derivative of this discontinuous function is also discontinuous. Part of the graph is positive, and part is negative. If it were a continuous function, there would be a value of x in the interval where the function would attain the value 0, but because the function is discontinuous, this does not have to happen.

We could change the definition just slightly to cause the function to have no maximum. Here is the definition:

> f := x -> if x >= 1 then 4 - x^2 else x + 4 fi:

> f(1);

3

Now the value of the function at $x = 1$ is only 3. The value of the function at any value less than 1 comes arbitrarily close to 5, so one would expect that the maximum of the function would be 5, but this does not happen.

> f(.99999);

4.99999

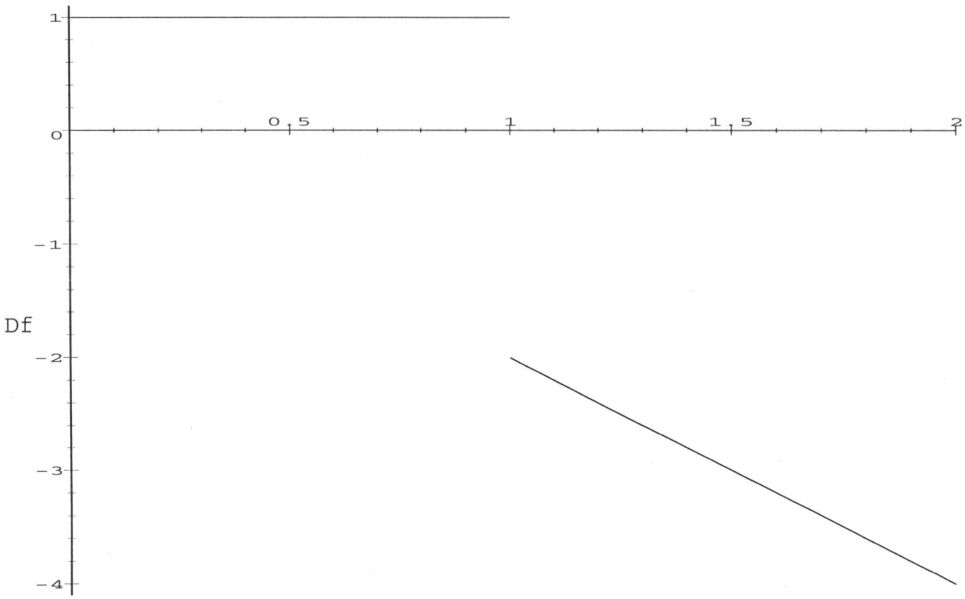

Figure 4.9 Graph of the Derivative of the Discontinuous Function for Example 4-8

Conclusion: The method of setting the derivative equal to zero works for continuous functions but may fail for discontinuous functions. As always, plot the graph before deciding on your next move.

Paper and Pencil Exercises

PP4–1: Voltage versus Charge

In an electronic device called a capacitor, the voltage, V, across the device is proportional to the charge on its plates. The formula is $q = CV$. In a different type of electronic device, there is a property akin to capacitance. This hypothetical device has a capacitance that is a function of the voltage, given by the equation $C = kV^2$. Find the relationship between q and V, and then find dq/dV.

Answer: ____

PP4–2: Energy versus Temperature

The formula for the amount of energy needed to change the temperature is given by Equation 4-1. If, in a certain substance, it is known that $Q = m(c_0 + c_1T^3)T$, relate the specific heat, c, to constants c_0 and c_1 at $T = 100°$.

Answer: ____

PP4–3: Economics: Interest and Principal

An investor places \$10,000 in a mutual fund whose interest rate varies over time according to $i = 0.07 + 0.01t$, where i is the rate of interest and t is time in years. The formula for the growth of the investment is $P(t) = P(0)(1 + i)^t$. What is the rate of change of the investment, $dP(t)/dt$, when $t = 1$ year?

Answer: ____

PP4-4: Quadratic Variation

The energy (E) stored in the capacitor of problem PP4–1 is $E = \frac{1}{2}CV^2$. Find $\frac{1dE}{CdV}$ at $V = 2V$.

Answer: ____

PP4–5: Distance versus Time

The distance as a function of time for a UFO sighted in 1956 was determined to be $d = 2.0t + .4t^2 + .3t^3$ (t in seconds, d in kilometers). How fast was the UFO going at $t = 3$ s, according to the formula?

Answer: ____

PP4–6:

Take the derivative again. What was the acceleration, at $t = 3$ s, of the UFO in problem PP4–5?

Answer: ____

PP4–7: Energy versus Velocity

The kinetic energy of a moving body is $E = \frac{1}{2}mv^2$ (E in joules, m in kilograms, and v in meters per second). What is the rate of change of energy with respect to time if $dv/dt = 5t$?

Answer: ____

PP4–8: Power versus Current

The power dissipated by a resistor is $P = Ri^2$, where i is the current in amperes, R is the resistance in ohms, and P is the power in watts. Use Ohm's law to find dP/dV.

Answer: ____

PP4–9: Rate of Change of Area with Radius in a Circle

(a) Find the rate of change of the circumference of a circle with respect to its radius.

Answer: ____

(b) Find the rate of change of the area of a circle with respect to its radius.

Answer: ____

PP4–10: Higher-Order Variation

Calculate the first and second derivatives of $y = x^{5/3}$ at $x = 0$.

Answer: ____

PP4–11: Inverse Variation

The pressure-volume relationship for an enclosed gas at constant temperature is $P = k/V$, where P is in pascals and V is in cubic meters. Find dP/dt and state the physical units for k.

Answer: ____

PP4–12: Trigonometric Variation

Without Maple's help, find $\frac{d(\sin^2\theta)}{d\theta} + \frac{d(\cos^2\theta)}{d\theta}$ and interpret the result.

Answer: ____

PP4–13: Exponential Variation

Given $R(t) = R_0(1 - e^{-kt})$, find $dR(t)/dt$ at $t = 1/k$.

Answer: ______________________________

PP4–14: Rationalizing the Numerator

Given $f(x) = \sqrt{x}$, find df/dx by applying the formula for the difference quotient and rationalizing the denominator before letting $\Delta x \to 0$.

Answer: ______________________________

Maple Lab

ML4–1

Measurements taken at a test track were analyzed by an engineer, who found, by fitting the data to a cubic polynomial, that the car's position could be accurately described by $s = -\frac{1}{5}t^3 + 10t^2 + 20t$, where s is the distance in feet from the starting point and t is the elapsed time in seconds. The engineer verified that the formula gave an accurate description of the car's position from $t = 0$ to $t = 30$ seconds.

The driver of the car was told to use maximum acceleration initially and then to smoothly decrease the acceleration. About halfway through the test, the driver was to begin braking, gently at first and then harder, in an attempt to decelerate smoothly.

(a) Find the car's speed as a function of time. Name the speed v. (The car's speed is given by $v = ds/dt$.)

Answer: ______________________________

(b) Find the car's acceleration as a function of time. Name the acceleration a. (The car's acceleration is given by $a = dv/dt$, or $a = d^2s/dt^2$.)

Answer: ______________________________

(c) Plot the distance, velocity, and acceleration on the same graph for the interval $t = 0$ .. 30. (Note that you may need to scale the velocity and the acceleration curves. Your command should be, *plot({s, m*v, n*a}, t=0..35);*, where m and n are numbers that produce the best-looking graph.)

Answer: ______________________________

(d) At what time is the velocity at the maximum?

(i) Estimate the maximum velocity from the graph.

Answer: ____

(ii) Compute the maximum velocity by solving $dv/dt = 0$.

Answer: ____

(e) At what time is the acceleration at the maximum? (Note that you *cannot* use the formula $dv/dt = 0$ because the maximum acceleration is at one of the end-points of the time interval.)

Answer: ____

(i) Read the maximum acceleration from the graph.

Answer: ____

(ii) Calculate the maximum acceleration by substitution in the formula.

Answer: ____

ML4–2

An investment analyst predicts that the future price of a company's stock will be $p = \frac{100t}{3t^2 + 25t + 100}$, where p is the stock price in dollars per share and t is the time in days. Today at midnight is $t = 0$. Past experience has shown that this analyst's predictions are usually good for a period of 3 months.

(a) Plot the price of the stock predicted by the formula. Sketch the graph here.

Answer: ____

(b) What is the long-term prospect for the company as predicted by this model?

Answer: ____

(c) If the formula gives an accurate prediction, when, according to the graph, is the best time to sell the stock?

Answer: ____

(d) Find the maximum of the expression p by solving $dp/dt = 0$ for t.

Answer: ____

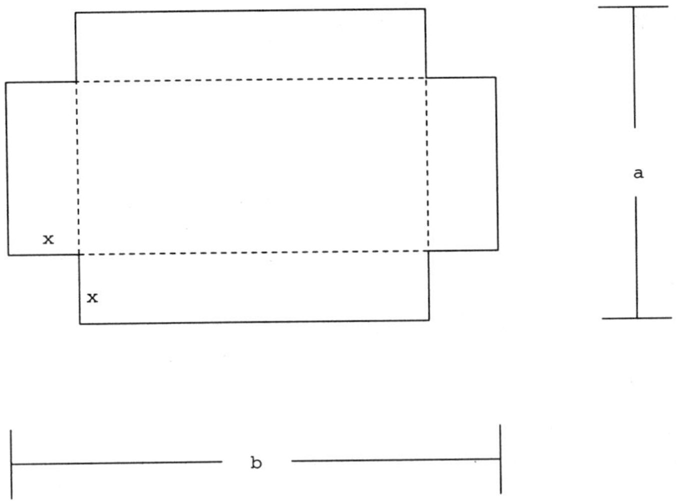

Figure 4.10 Fold on the Dotted Lines to Make a Box

ML4–3

A classic exercise in optimization is to find the maximum volume of a box, given some restrictions on the dimensions. One example of this type of problem arises when a box is to be made from a flat sheet of material by cutting out the corners and folding up the sides.

A cardboard box is made from a flat rectangular sheet. The box is formed by cutting squares from all four corners, then folding the sides up to form the box. Figure 4.10 shows the situation. The dotted lines are the fold lines.

(a) If the sides of the cardboard sheet are $a = 15$ centimeters and $b = 25$ centimeters, derive a formula for the volume of the box in terms of x. Name the expression V.

Answer: ______________________________

(b) Plot the volume as x ranges from 0, the obvious minimum value, to the maximum possible value of x.

What is the maximum possible value of x?

Answer: ______________________________

(ii) Sketch the graph here:

(c) Solve for the value of x that maximizes the volume.

(i) Using the *solve* command, x is:

Answer: ______________________________

(ii) Which answer from the *solve* command did you discard, and why?

Answer: ______________________________

(d) Generalize the problem to a sheet whose sides are the indeterminates a and b. Write the volume as a function of the three variables, a, b, and x.

(i) Define (and name) V as a function of x. The function will contain parameters a and b.

Answer: ______________________________

(ii) Can you plot this function?

Answer: ______________________________

(iii) Find the value of x that maximizes the volume. Give the symbolic solution.

Answer: ______________________________

ML4–4

You are building an A-frame cottage. The back of the cottage, which faces a lake, has the triangular shape shown in Figure 4.11.

You would like to install a large picture window in the shape of a rectangle, through which you want to let in as much light as possible. This means that the area of the picture window must be a maximum.

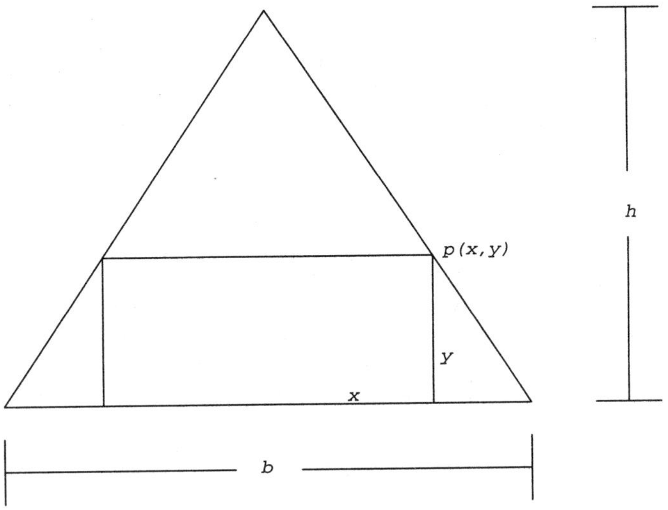

Figure 4.11 A Large, Rectangular Picture Window in an A-frame Cottage

Suppose that $b = 25$ feet and $h = 20$ feet.

(a) Find the y coordinate of point P in terms of x. (You must find the equation for the line that represents the right-hand sloping roof of the cottage. Choose the origin at the midpoint of the base of the cottage.)

(i) The slope of this line is:

Answer: ________________________________

(ii) The y intercept of the line is:

Answer: ________________________________

(iii) The slope-intercept equation of the straight line is:

Answer: ________________________________

(b) Express the area of the picture window in terms of x. Remember to double the area of the rectangle found by multiplying together the coordinates of point P.

(c) Plot the area as a function of x. What is the domain of x for the plot?

Answer: ________________________________

(d) Find the maximum area analytically.

Answer: ______________________________

ML4–5

Old MacDonald's farm lies on the north side of a riverbank. His son, a student of mathematics, has determined that the shape of the riverbed bounding the southern extent of his property is closely approximated by the curve

$$y + \frac{1}{100}x\left(\frac{1}{100}x - 5\right)\left(\frac{1}{100}x - 20\right) + 800$$

Old MacDonald's property extends for 2 miles north of the riverbank. His property line along the river goes from $x = 0$ to $x = 2{,}500$, where x is in feet. The eastern and western property lines go directly north-south. That is, the western boundary is the line $x = 0$, and the eastern boundary line is $x = 2{,}500$ feet. The eastern boundary starts at approximately $y = 900$ feet, and the western boundary line starts at $y = 3{,}309$ feet.

Farmer MacDonald has 4,000 feet of fencing. For reasons known only to himself, he wants to run the fencing in a straight line from the local maximum of the riverbank, near $x = 0$, to an as-yet undetermined point, and from there in another straight line to the riverbank once more, this time meeting the bank at right angles.

(a) Plot the shape of the riverbank to get an idea of the shape of the southern part of Old MacDonald's farm.

Answer: ______________________________

(b) The farmer decides to choose point $P(1{,}500, 2{,}500)$ as the top corner of the fence. He runs the fencing from the local maximum, near $x = 0$, to point P, and from there over to the riverbank, near $x = 2{,}500$, where the fence meets the riverbank at right angles.

(i) Find the point of the local maximum by solving $dy/dx = 0$. State both the x and y coordinates.

Answer: ______________________________

(ii) Find the general expression for the slope of the normal to the riverbank, in terms of x. (*Remember:* The slope of the normal line is $-1/(dy/dx)$).

Answer: ______________________________

(c) Form the expression for the slope of the line joining P to the general point (x, y) on the riverbank, in terms of x.

Answer: ______

(d) The line Farmer MacDonald is looking for is the one going through P and perpendicular to the riverbank. Plot expressions (bii) and (c) on the same graph and find the point of intersection near $x = 2{,}500$. The x and y values at this point determine the point on the riverbank where a normal line will pass through point P. State these values.

Answer: ______

(e) Use the Pythagorean relationship to find the total length of the fencing.

Answer: ______

(f) Has Farmer MacDonald enough fencing to complete his project?

Answer: ______

ML4-6

You want to design a curve passing through the points (-3, 0), (0, 0), and (4, 0). The curve should have a local maximum of 2 somewhere in the domain $-3 < x < 0$ and should have a local minimum in the interval $0 < x < 4$. Find a possible cubic with these properties.

(a) If a cubic is to have zeros at –3, 0, and 2, the cubic expression must have factors x, $(x + 3)$, and $(x - 2)$. If it is to possess a local maximum of 2, the cubic can be stretched vertically by an adjustable scaling factor, k. Putting it all together, define the cubic in terms of the parameter k and name it $y1$.

Answer: ______

(b) Take the derivative of the expression in (a) and solve for the x coordinate of the local maximum.

Answer: ______

(c) At the local maximum, the value of the expression is 2. Substitute the value of x that you found in (d) into $y1$ to find the corresponding y component. Your answer will be an expression containing k. Equate the expression to 2 and isolate k.

Answer: ______

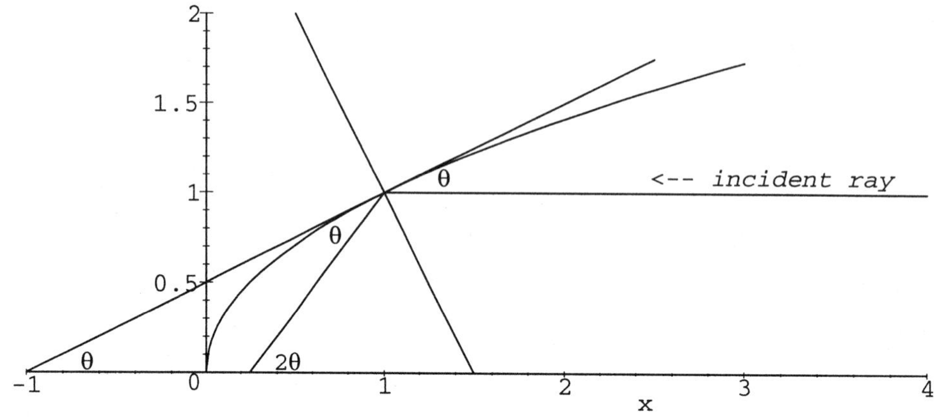

Figure 4.12 Diagram for ML4–7

(d) Substitute this value for k in $y1$ and plot the result for the range $x = -3 .. 2$. Sketch your curve here. Does it have all the required characteristics?

Answer: ____

ML4–7

A parabolic reflector is given by the equation $y = \sqrt{x}$. Both x and y are measured in feet. A pencil-thin horizontal ray of laser light is incident from the right. Since the ray is horizontal, it maintains a constant height, $y = c$, above the x axis (Figure 4.12).

(a) Find the coordinates of the point where the ray of light strikes the parabolic reflector's mirrored surface.

Answer: ____

(b) Find the slope of the tangent to the parabola at the point where the light ray strikes the surface. Call this angle θ.

Answer: ____

(c) When the beam of light is reflected by the mirrored surface of the parabola, it strikes the x axis at a point called the focus. The law of reflection states that the angle that the reflected beam makes with the tangent is the same as the angle the incident ray makes with the tangent. Show by elementary geometry that the slope of the reflected ray is 2θ.

Answer: ____

(d) Find the slope of the reflected ray. *Note:* $\tan(2x) = 2\tan(x)/(1 - \tan^2(x))$

Answer: ____________________

(e) Find the x intercept of the reflected ray.

Answer: ____________________

(f) Does the x intercept depend on c? If it does not, you can conclude that any ray that is parallel to the x axis will pass through the same x intercept. For this reason, the point is called the *focus*.

Answer: ____________________

(g) Stretch the parabola by a factor k. The equation becomes $y = k\sqrt{x}$. What change does this make to the focus; that is, where is the x intercept now?

Answer: ____________________

(h) Stretch the parabola by multiplying x by k. The equation becomes $y = \sqrt{kx}$.

(i) Repeat the steps you used to solve for the focal point in step (g).

Answer: ____________________

(ii) Is there a way of arriving at the answer without repeating all the steps in the Maple solution?

Answer: ____________________

ML4-8

Suppose you want to make a trough out of a piece of galvanized sheet metal. The sheet is rectangular in shape, with dimensions of 5 feet by 15 feet. You want to bend the metal in the shape of a trough as shown in Figure 4.13.

(a) Assume that $x = 1$ foot. This problem asks you to find the angle θ that maximizes the cross-sectional area of the trough, thereby allowing it to hold the greatest amount of water when the end pieces are welded on. Knowing the angle, you can determine the height of the trough and the breadth of the top.

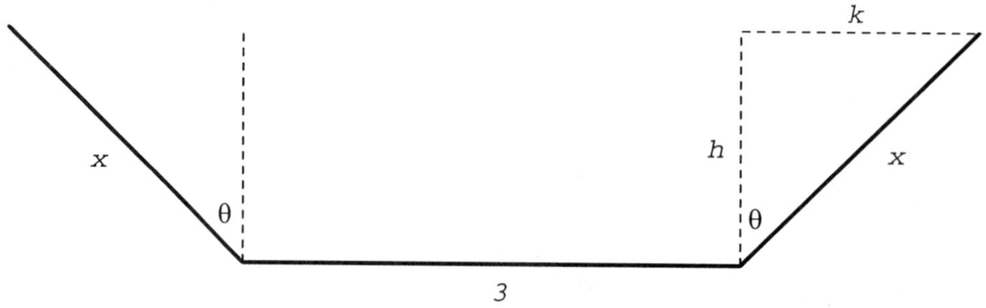

Figure 4.13 Side View of the Trough in Problem ML4–8

Sketch the diagram and label the dimensions. Leave the angle θ as an unknown.

(ii) Derive a formula for the cross-sectional area A based on the angle θ.

Answer: ______________________________

(iii) Find $dA/d\theta$.

Answer: ______________________________

(iv) Transform the equation in (iii) into one containing sin(θ).

Answer: ______________________________

(v) Let $y = \sin(\theta)$. The substitution transforms the equation in (iv) to an equation in y. Write the transformed equation here.

Answer: ______________________________

(vi) Find the value of y that maximizes (v).

Answer: ______________________________

(vii) To what angle θ does (vi) correspond?

Answer: ______________________________

(viii) Since Maple works in radians, your answer to part (vii) is in radians. Convert θ to degrees.

Answer: ______________________

(b) Plot the area function in part (aii)

(i) Find the maximum angle graphically.

Answer: ______________________

(ii) Convert θ from radians to degrees.

Answer: ______________________

ML4–9

In problem ML4–8, the base of the trough is 3 feet. In this problem, however, the base of the trough is a variable and it can be any length between 0 and 5 feet.

(a) Write the formula for the cross-sectional area in this case. Name the result A.

Answer: ______________________

(b) The area A depends on two variables, x and θ. If $x = 0$, the flat sheet has no sides. If $x = 2.5$, the trough is in the shape of a V. Any value of x in between will form a trough. You would expect that the trough of maximum capacity would simultaneously satisfy $dA/dx = 0$ and $dA/d\theta = 0$. Define *>Ax := diff(A, x) = 0;* and *>At := diff(A, theta) = 0;*. Write the derivatives here.

Answer: ______________________

Answer: ______________________

(c) Solve these equations simultaneously for both x and θ. You will get many choices for θ. From step (b), you can discard any solution where $x = 0$ or $x > 2.5$. You can discard solutions where $\theta < 0$ because you know that the maximum volume will be produced only if the angle is positive. When you rule out all these possibilities, you are left with the strange-looking result, *arctan(1,RootOf(_Z^2-3))*. This expression can be reduced to a decimal number by first using the *allvalues* command, *allvalues(theta = arctan(1,RootOf(_Z^2-3)));* and then applying *evalf*.

Answer: ______________________

Answer: ______________________

(d) Plot the function in three dimensions using > *plot3d(A, x = 0 .. 2.5, theta = 0 .. Pi/4, orientation=[90,90], axes=normal,style=patchcontour);* and > *plot3d(A, x = 0 .. 2.5, theta = 0 .. Pi/4, orientation=[0,90], axes=normal,style=patchcontour);*. Observe that the graph reaches a maximum for both x and θ. You can try other orientations to see the shape of the surface better, but the orientations given here allow you to measure the maximum values of x and θ.

(i) Estimate the maximum value of x.

Answer: ______________________________

(ii) Estimate the maximum value of θ.

Answer: ______________________________

(iii) Do your measurements agree with the calculations in part (c)?

Answer: ______________________________

ML4–10

A classic problem in minimization is that of a lifeguard who can run at 6 miles per hour along a sandy beach and can swim at 2 miles per hour. She spots a swimmer in trouble. The swimmer is 0.2 miles down the beach and 0.15 miles from shore. The lifeguard's observation post is 0.1 miles from the shoreline. Figure 4.14 shows the situation.

Point X is on the shoreline, somewhere between the observation post and the swimmer. The lifeguard is a rapid calculator. She quickly determines the location of point X so that the time it will take her to reach the swimmer is a minimum. You will recheck her calculations in this problem. First, choose a coordinate system. Let the point on the shoreline nearest the observation point be the origin (0, 0). Choose the shoreline as the x axis. Thus, the lifeguard's observation post lies on the y axis and has the coordinates $A(0, -0.1)$. The location of the swimmer is $B(0.2, 0.15)$.

(a) Let the coordinate of point X be $(x, 0)$. Form the expression for the straight-line distance from the lifeguard's observation post, A, to her entry point in the water, X.

Answer: ______________________________

(b) Form the expression for the straight-line distance from X to the swimmer at B.

Answer: ______________________________

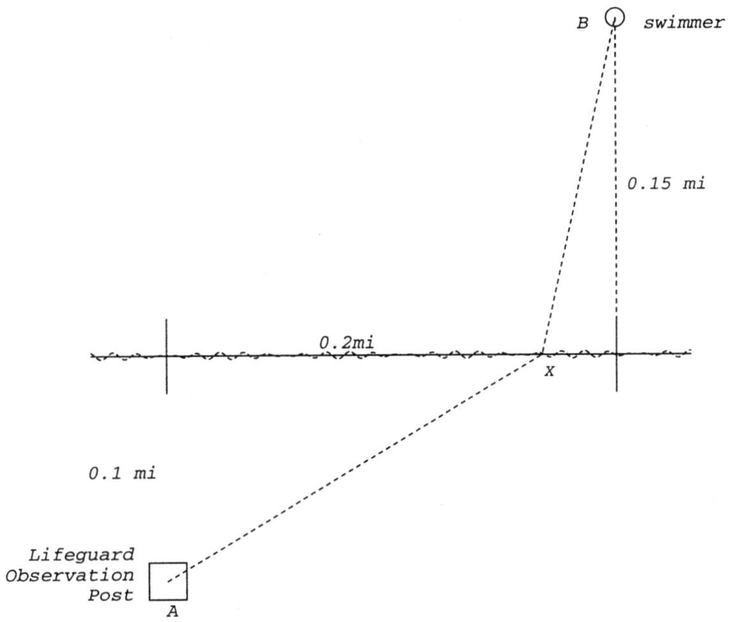

Figure 4.14 Diagram for ML4–10

(c) Knowing the distances and the lifeguard's running and swimming speeds, form the expressions, containing the parameter x, for the time it takes her to reach point X and then for the time in the water to reach the swimmer. Name the expressions $t1$ and $t2$.

Answer: ____

(d) Plot the expression for $t1 + t2$, after choosing a reasonable domain for x.

Answer: ____

(e) Does the plot exhibit a minimum? If so, estimate the value of x that minimizes the time.

Answer: ____

(f) You may run into trouble if you use the *solve* command to find the value of x. It is better to use *fsolve* in this problem. Since you know that x must be between 0 and 0.2, use the command, > *fsolve(diff(t1+t2, x) = 0, x=0..0.2);* to find the minimum time as an approximate decimal.

Answer: ____

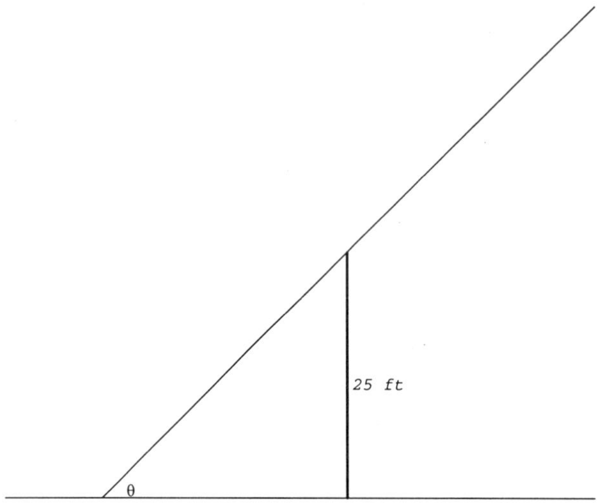

Figure 4.15 Diagram for ML4–11

(g) Compare the two ways of finding the minimum.

(i) Which is easier?

Answer: ______________________________

(ii) Which is more accurate?

Answer: ______________________________

ML4–11

Suppose you live in the south, where the sun is directly overhead at noon on a summer day. Since the earth rotates once a day, the sun's angle changes at 0.25 degrees per minute. A pole 25 feet high casts a shadow, except when the sun is directly above. The angle of the sun is measured in such a way that $\theta = 0°$ at dawn, when the sun is in the east and $\theta = \pi$ when the sun sets in the west (Figure 4.15).

(a) Find a formula for the length of the shadow cast by the pole in terms of the sun's angle, θ. Choose a coordinate system such that the base of the pole is at the origin (0,0). The tip of the shadow is at point x, 0, and the other end of the shadow will always be at point 0, 0.

Answer: ______________________________

(b) Find the rate at which the shadow cast by the sun is changing as a function of θ.

Answer: ________________________________

(c) How fast is the shadow changing when the angle is 15°, one hour after sunrise?

Answer: ________________________________

(d) How fast is the shadow lengthening, in feet per minute, when the sun is at an angle $\theta = 165°$, one hour before sunset?

Answer: ________________________________

(e) How fast is the shadow changing at noon, when the sun is directly overhead?

Answer: ________________________________

(f) In part (e), the sun is directly overhead, so the length of the shadow is zero. Does this imply that the rate of change of the length of the shadow should be zero? Explain.

Answer: ________________________________

____

____

ML4–12

There is a way of approximating the root of an equation numerically by calculating the tangent line. It is called Newton's method.

Step 1. Graph the function to get an idea where the roots are.

Step 2. Choose some value of x near one of the roots. Calculate $f(x)$ and $df(x)/dx$.

Step 3. Find the x intercept of the tangent line. Use the point-slope formula for the straight line to construct the equation for the line. Since $y = 0$ at the x intercept and since the line goes through $(x, f(x))$, the point-slope formula $y - y_o = m(x - x_o)$ becomes $-f(x) = (df(x)/dx)(x_i - x)$. Solving for the x intercept, x_i, $x_i = x - f(x)/f'(x)$, where $f' = df/dx$.

Step 4. Since the intercept x_i should be closer to the root, use the value of x_i as a new starting value and perform step 3 again to get an even better estimate. The process in step 3 can be automated using Maple.

Example: Find the root of $y = x^3 - 2x^2 - x + 2$ near $x = 0.3$.

Solution:

1. The plot of the expression is shown in Figure 4.16.

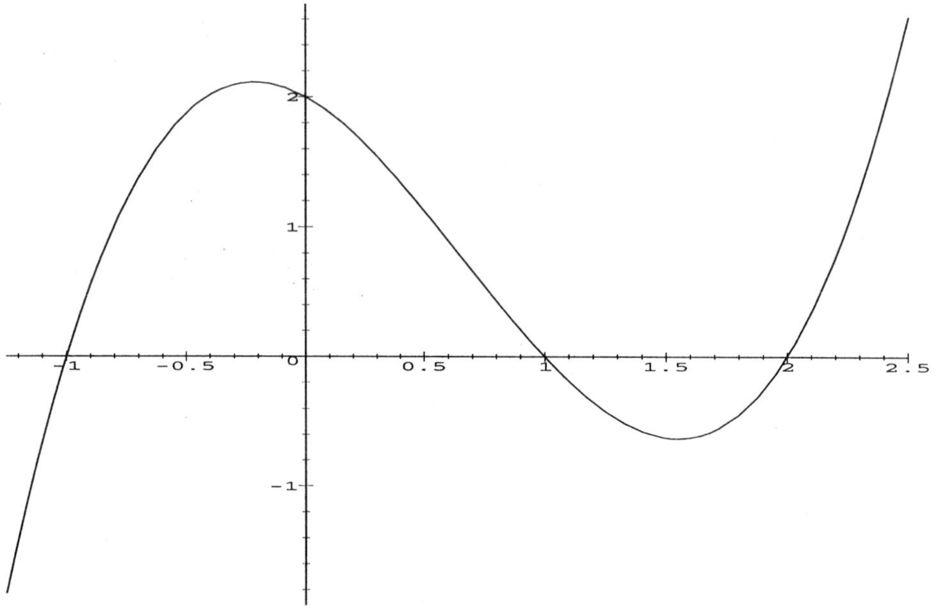

Figure 4.16 Plot of the Expression

2. Define > *f* := *x^3 - 2x^2 - x + 2;*. The derivative is > *D(f)(x)*. Let us say that the root you are looking for is $x = 1$ and you begin the process at $x = 0.3$. The tangent line and the x intercept are shown in Figure 4.17.

Note that the x intercept is very close to 1 even though the initial estimate is 0.3, which is not very close to 1 at all. You can see that you must get close enough to the root that the tangent line at your initial guess has a slope that will get you even closer. If you happened to pick a point near the local maximum or minimum, the tangent line will be nearly parallel to the x axis and the intercept will be far out on the positive or negative x axis. Eventually, you will find one or other of the exterior roots, but not the one in the middle.

3. Assign your initial estimate for x and then find the x intercept: > *x := x - f(x)/D(f)(x);*. Note that x receives a new value through this assignment, which makes it easy to calculate the intercept again.

```
> f:= x -> x^3-2*x^2-x+2;
```

$$f := x \rightarrow x^3 - 2x^2 - x + 2$$

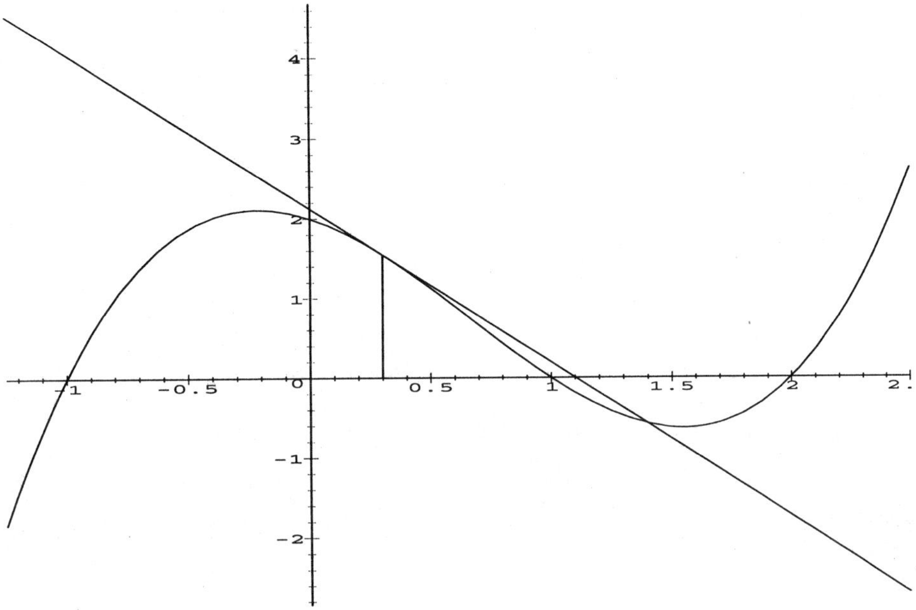

Figure 4.17 The Tangent Line and the x Intercept

> **x := 0.3:**

> **x := x - f(x)/D(f)(x);**

```
x := 1.101554404
```

4. Place the cursor on the command and hit Enter, thereby executing the command again. After a few iterations, the result should be the root, accurate to ten digits. As the command is reexecuted, the sequence of values generated is

```
.9929737484, .9999758310, .9999999999
```

Thus, the answer is accurate to ten digits after only four repetitions of the command.

Use Newton's method to find all the real roots of the following equations.

(a) $x^4 - 4.3x^3 + 1.1x + 25 = 0$ *Answer*: ____________________

(b) $2.34x - 12.222 + 2\sin(x) = 0$ *Answer*: ____________________

(c) $x^2 + \cos(10x) = 0$ *Answer*: ____________________

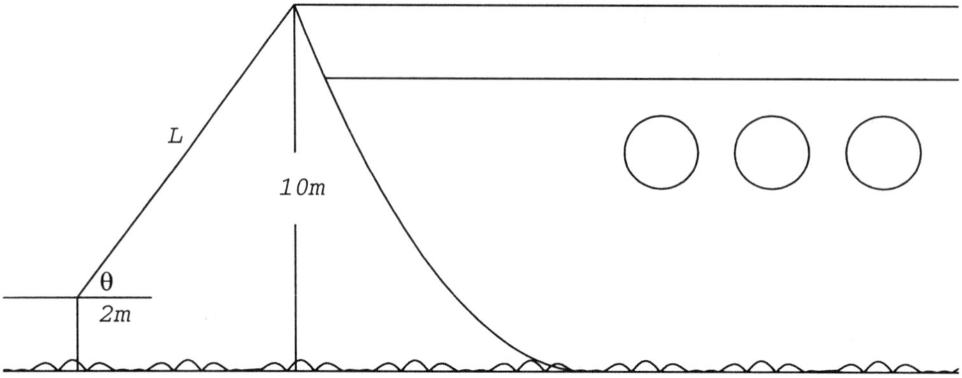

Figure 4.18 Diagram for ML4–13

ML4–13

A winch is pulling a large yacht toward a dock. The height of the dock is 2 meters above the waterline, and the rope is attached to a cleat on the bow of the yacht that is 10 meters above the waterline. Let the length of the rope be L. Assume that the winch is at the same height as the dock (Figure 4.18).

(a) What angle does the rope make with the horizontal when the length of rope between the winch and the cleat is 15 meters?

Answer: ______________________________

(b) The winch is drawing the rope in at a rate of 3 meters per minute. What is the rate of change of the angle that the rope makes with the horizontal when $L = 15$ meters?

Answer: ______________________________

(i) Is the angle θ getting bigger or smaller?

Answer: ______________________________

(ii) What is dL/dt in centimeters per second? Note that if L is decreasing with time, dL/dt will be negative.

Answer: ______________________________

(iii) Numerically, what is $d\theta/dt$ in radians per second when $L = 15$ meters? *Answer*:

(c) How fast is the yacht moving at this time? (Note that $\cos(\theta) = x/L$.) Apply the chain rule to find the speed dx/dt in terms of L and θ, and then substitute the known values for L and θ to obtain the final result.

Answer: ______________________________

(d) How fast is the yacht moving when $L = 10$ meters?

Answer: ______________________________

(e) Has the speed of the yacht increased or decreased as it approaches the dock?

Answer: ______________________________

(f) Draw the graph of the speed of the yacht for $L = 30$ meters to $L = 10$ meters.

ML4–14

Galileo (1564–1642) supposed that a falling body would obey the equation expressed in modern notation as $s = 1/2gt^2$, where g is the acceleration due to gravity and is approximately 10 meters per second squared.

(a) Graph the speed, $v = ds/dt$, as a function of time from $t = 0$ .. 10 seconds. Sketch the graph here. What is the shape of the velocity curve?

Answer: ______________________________

(b) Graph the acceleration, $a = dv/dt = d^2s/dt^2$, over the same time interval. What is the characteristic of this curve?

Answer: ______________________________

(c) State the formulas for the velocity and the acceleration.

Answer: ______________________________

ML4–15

Galileo's formula is accurate for an object falling in a vacuum over a short distance. However, air resistance causes an object to fall slower than this formula predicts. A slightly more realistic equation involving the case of a skydiver before she opens her parachute is $s = 38.1t + 148.e^{(-.257t)} - 148$, where s is in meters and t is in seconds. This formula contains a more accurate value for g, and it predicts a *terminal velocity*. As the skydiver's speed increases, the air pressure eventually produces enough force to balance gravitational force. At this speed, the skydiver's speed is a constant called the terminal velocity.

(a) Determine the expressions for the speed and the acceleration.

Answer:____________________ *Answer*:____________________

(b) Plot, on the same graph, the distance traveled in this case, where air resistance makes its effect apparent, and the distance traveled given by the expression in ML4–14 for free fall in a vacuum. Sketch the graph here and find how far the skydiver will fall after 10 seconds and without air resistance. ____________________

Answer: ____________________

(c) Plot the two velocities, one with air resistance and one without, on the same graph. Sketch the graph here and find how fast the skydiver will be going after 10 seconds with and without air resistance.

Answer: ____________________

(d) Plot the two velocities, one with air resistance and one without, on the same graph. Sketch the graph here and find the difference in the acceleration after 10 seconds with and without air resistance.

Answer: ____________________

(e) What is the terminal velocity, in meters per second and in miles per hour?

Answer:____________________ *Answer*:____________________

CHAPTER 5

Integration: Concepts and Techniques

Objectives for This Chapter

1. Learn to approximate the area of curved figures by rectangles
2. Define the definite integral
3. Distinguish between positive and negative areas
4. Understand the fundamental theorem of the calculus
5. Find functions that represent indefinite integrals
6. Identify the antiderivative as the indefinite integral
7. Review the various integral formulas for standard functions
8. Know that discontinuities can affect the validity of integral formulas
9. Learn standard techniques for integration
10. Apply techniques for checking the validity of Maple's integrals

Maple Commands Used in This Chapter

with(student):	Command to load the student package (library). Issue this command whenever you begin a new Maple worksheet when you are working through this chapter.
changevar(sqrt(1 - x^2) = u, Int(G, x), u);	Change the variable of integration from x to u.
combine(value(S));	*Combine* tries to join multiple terms into a single term.

expand(ex, trig);	Expand the expression *ex* using rules for trigonometric expansions.
expand(S);	Simpler form of the *expand* command.
factor(simplify(sum (ii, ii = 1 .. n)))	Factor a sum, after simplifying.
int(1/(x - 3)^2, x = 2 .. 4);	Maple's command for the definite integral.
int(1/(x - 3)^2, x);	Syntax for the indefinite integral.
Int(f(x), x = a .. b) = F(b) - F(a);	Maple's way of writing the fundamental theorem of the integral calculus.
Int(f(x), x = a .. b);	Inert form of the integration command.
Int(q, x) = int(q, x);	Both forms of the integration command in the same statement.
intparts(Int(x*ln(x), x), x);	The command to apply integration by parts to the integral, specified by the inert form of the *int* command.
isolate(p1, Int(x*ln(x), x));	Isolate the expression for the indefinite integral on the left-hand side of the equation $p1$.
lhs(p2) = rhs(p2);	The commands to specify the left hand side (*lhs*) and the right-hand side (*rhs*) of equation $p2$.
limit(S, n=infinity);	The limit, as n tends to infinity, of the expression S.
middlebox(x/2, x = 0 .. 5, 10, color = black);	Plots the line $y = x/2$ and draws ten rectangles under the line between $x = 0$ and $x = 5$.
middlesum(x/2, x = 0 .. 5, 10);	Sums the areas of the rectangles described in the *middlebox* command.
simplify((middlesum (f(x), x = 0 .. x+x/1000, 1001);	Simplify the summation expression.
simplify(c, symbolic);	The parameter *symbolic* is useful when a root sign (a radical) is present in expression c.
sqrt(1 - x^2);	The square root function.

subs(x = 2, F);	Substitutes 2 for x in expression F.
sum(ii, ii = 1 .. n);	Forms a sum over the index *ii*. The double letter is used to prevent potential conflicts, since i is a common name for a variable.
Sum(ii, ii = 1 .. n)	Inert form of the *sum* command.
value(n);	Used to evaluate inert expressions, like *Int*.
value(S), evalf(S)	These commands are *different*: *value* reduces an inert integral or sum (*Int* or *Sum*) to a formula, whereas *evalf* reduces the result to a decimal approximation.

Approximating the Area of Curved Figures by Rectangles

The essential idea behind the differential calculus is that of finding a formula for the slope of a tangent line to a continuous curve, given the point of tangency. The previous two chapters examined that idea and extended it to the solution of problems involving rates of change. An important result is that as long as a curve possesses a tangent, if sufficiently magnified, that portion of the curve will resemble a straight line. Thus, in a sense, all problems become linear provided these conditions hold and when only small changes are required.

Another type of linearization leads to a method of finding the area under a curve. The area under a curve is the central problem that the integral calculus addresses. In the early days of mathematics, various area formulas were known. Among them were the area of a square of side x, $A = x^2$, and the area of a rectangle with sides a and b, $A = ab$. The area of a triangle of height h and base b was also known: $A = 1/2bh$. These formulas remain true today, even though roughly 2,500 years have elapsed! The areas of polygons, many-sided figures whose sides are all straight lines, posed no problem for these early mathematicians. After all, they could be broken down into triangles, squares, and rectangles by drawing a few extra lines. The volumes of simple shapes soon came to be known. For example, the volume of a cube of side x is $V = x^3$, and the volume of a solid with rectangular faces is $V = abc$ (the sides being the lengths a, b, and c).

The areas of figures bounded by curved lines proved more of a problem, however, the most important example is that of the area of a circle. The ancients knew that the area of a circle is $A = \pi r^2$ (with r as the radius), but they had trouble calculating this number to more than a few decimal places. Of course, one of their problems was that they didn't have the benefit of decimal notation! The great Archimedes (ca. 287– 212 B.C.) determined that the area of a circle of radius 1 is less than 3 1/7 but greater than 3 10/71. He achieved this wonderful result by inventing the idea behind the integral calculus. Unfortunately, the lack of an adequate number system and the neglect of analytic geometry prevented him from going further.

We can go further because René Descartes (1596–1650) showed how to relate geometry to algebra. The result is called *analytic geometry*. As a result of his work, we have algebraic formulas for geometric figures like straight lines, circles, parabolas, and many others. With these formulas, the integral calculus can be developed. The basic goal is to find a way of calculating the area of a figure. We assume that the figure is bounded by sides that are expressible by an algebraic formula. We'll begin with the rectangle and the triangle, but we'll take the analytic geometry approach.

Example 5-1: The Area of a Rectangle

Find a general formula for the rectangle whose sides are bounded by the y axis, the line $y = b$, and the x axis. The remaining side is a vertical line (Figure 5.1).

Solution: The rectangle could be subdivided into smaller rectangles. There is no reason to perform this step when the objective is to find the area of a rectangle, but it illustrates the procedure that will be required later. Figure 5.2 shows the rectangle of Figure 5.1 divided into smaller rectangles.

The height of the rectangles remains at b, but the base of each rectangle is now $dx = a/10$. The area of each triangle is $a/10 \cdot b$, or $ab/10$. Since there are ten rectangles, the area of all

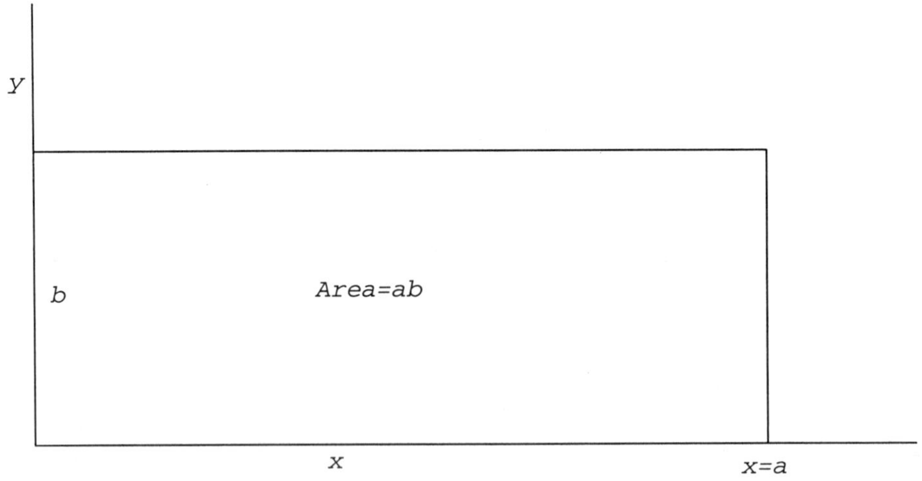

Figure 5.1 The Area of a Rectangle in the Style of Analytic Geometry

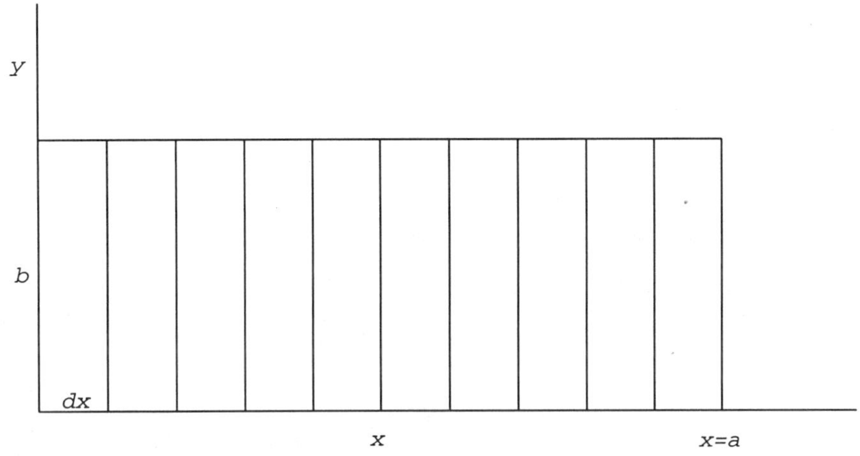

Figure 5.2 The Rectangle in Figure 5.1 Divided into Ten Smaller Rectangles

of them together is *ab*. The area of the rectangle in Figure 5.1 is *ab*, and the sum of the areas of the rectangles in Figure 5.2 is also *ab*. We would have a real problem with our mathematics if the two ways of obtaining the area were any different.

The technique of decomposing the figure into rectangles can be applied to other figures, like the triangle. Let us attempt to find the area of a triangle by this method of subdividing. The triangle is shown in Figure 5.3.

We know that the area of a triangle of base *a* and height *h* is *ah*/2. Let us try the technique of subdividing the figure into rectangles. We'll take a specific example so that Maple's plot routines will work. There is another reason why it helps to be able to have a numerical example: we want to get a sense of the correctness of our approximations. We'll start with a right-angled triangle. Let the base of our triangle be along the *x* axis and the hypotenuse be along the line $y = x/2$. In fact, that is how the triangle in Figure 5.3 was constructed. Maple contains three built-in routines for drawing rectangles: *leftbox, middlebox*, and *rightbox*. They are part of the student package. Here is how to reconstruct Figure 5.3 using the *middlebox* routine (Figure 5.4).

> middlebox(x/2, x = 0 .. 5, 10, color = black);

You will recognize the familiar plot options in the *middlebox* command. The parameter *10* tells Maple how many boxes to draw. The triangle has been approximated by a staircase of rectangles. How good is the approximation? We need to calculate the areas of the rectangles to find out. The hypotenuse of the triangle crosses the midpoint of the top of each rectangle.

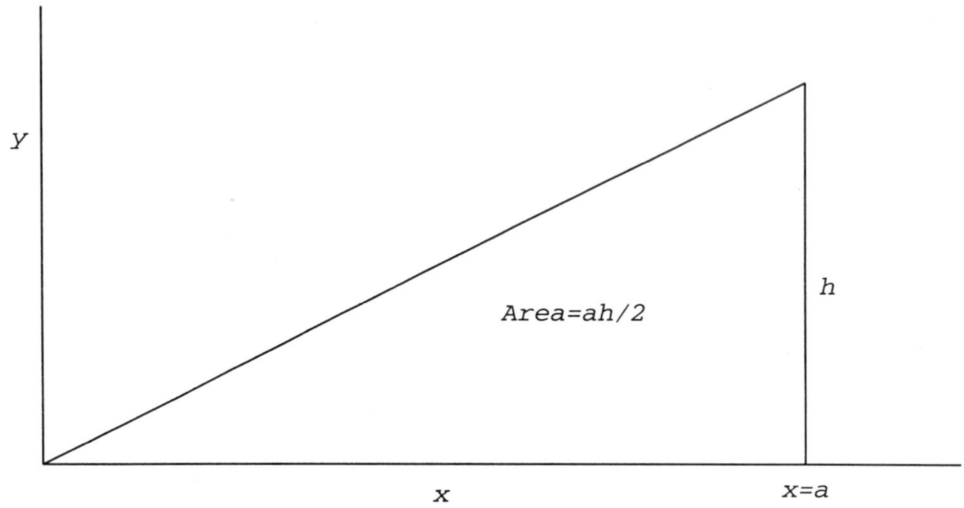

Figure 5.3 Area of a Triangle

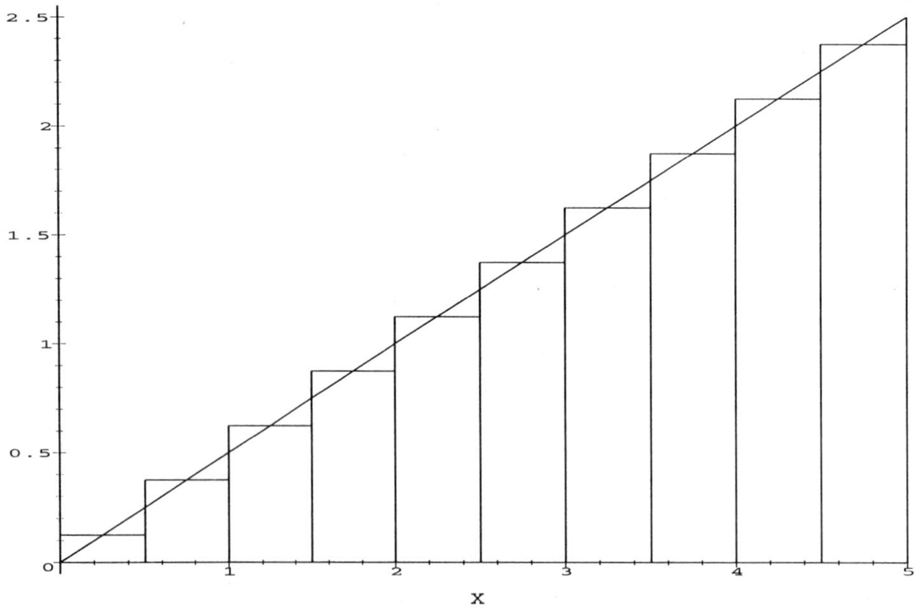

Figure 5.4 The Approximation to the Area of a Triangle Using Rectangles

The rectangles and the hypotenuse intersect at $x = 1/4, 3/4, 1\ 1/4, 1\ 3/4 \ldots 4\ 1/4$, and $4\ 3/4$. The heights at these points are $y = 1/8, 3/8, 5/8, 7/8 \ldots 2\ 1/8$, and $2\ 3/8$, since the equation $y = x/2$ must be satisfied. The base of each rectangle is 1/2, so the total area of all the rectangles is $1/2 \cdot (1/8 + 3/8 + \ldots + 17/8 + 19/8)$. It is nice to know that Maple can help with the sum. Corresponding to the graphing commands *leftbox, middlebox*, and *rightbox* are the summing commands *leftsum, middlesum*, and *rightsum*. Whichever one you use, the form of the command is basically the same as the corresponding *box* command. You don't put in any plot options, though, because these are not plot commands. In this case, since we used *middlebox*, we should now use *middlesum*.

> s := middlesum(x/2, x = 0 .. 5, 10);

$$s := \frac{1}{2}\left(\sum_{i=0}^{9}\left(\frac{1}{4}i + \frac{1}{8}\right)\right)$$

In case you are not familiar with it, the summation symbol ($\sum$) means that the terms inside the brackets are to be added. The index symbol, *i*, takes on all the integral values from $i = 0$ to $i = 9$. Thus, there are 10 terms in the sum, beginning with 1/8, for $i = 0$, and ending with $9/4 + 1/8 = 19/8$, for $i = 9$. You could find the numerical result for this sum by using the *evalf* function, but this would give you an approximate decimal as an answer. The exact result can be found using the *value* command.

> value(s);

$$\frac{25}{4}$$

If you have already worked out the area of the triangle in Figure 5.4 using the formula $A = ah/2$, you know that the exact answer is 25/4. The area of the ten rectangles is also 25/4! How is it that the sum of the rectangles gives the exact answer, even though the rectangles clearly make up a different figure? The answer is because the top of each rectangle forms two little triangles when it is cut by the hypotenuse of the large triangle. Moreover, the areas of the two little triangles are equal. Each base has length 1/4, and their altitudes are equal and opposite. The first half of the rectangle overestimates the area, and the second half underestimates the area by exactly the same amount. Altogether, the errors cancel each other exactly. That is why the area found by approximating a triangle by "middlebox" rectangles is exact.

We are now in a position to investigate curved lines, like parabolas, but before we do, we'll look at some properties of the summation symbol, which will appear when we attempt to use *middlesum* to add rectangles.

Formulas for Sums

As you probably expected, the Maple command to sum a number of related terms is *sum*. Actually the command comes in two forms, *sum* and *Sum*. Just like the *Limit* command, Maple writes the *Sum* command using the summation symbol. Then you can use *value* to get the answer. This allows you to express the summation symbol in regular mathematical notation. It comes in handy when you are writing reports to be able to use the same mathematical typesetting that appears in textbooks.

The formula for the sum of the first *n* whole numbers is found with the Maple command

```
> Sum(ii, ii = 1 .. n) = sum(ii, ii = 1 .. n);
```

$$\sum_{ii=1}^{n} ii = \frac{1}{2}(n+1)^2 - \frac{1}{2}n - \frac{1}{2}$$

There are two remarks that need to be made. The first is that the formula does not look like the formula found in most textbooks, where you will probably see the sum written as $\frac{1}{2}n(n+1)$. This is easy to fix. Simply work with the right-hand side of the equation, as in the following command:

```
> Sum(ii, ii = 1 .. n) = factor(simplify(sum(ii, ii = 1 .. n)));
```

$$\sum_{ii=1}^{n} ii = \frac{1}{2}n(n+1)$$

The second point concerns why we used *ii* as the index variable instead of the simpler choice, *i*. The reason here is that *i* is such a common symbol that you may have used *i* before in the same Maple session. In that case, the name *i* is no longer a free variable and you will get an error message. If you use double letters for index variables, there is less chance of making this type of mistake. The choice is yours.

Example 5-2: The Sum of the First *n* Squares

Find the sum of the squares of the first *n* whole numbers.

Solution:

> Sum(ii^2, ii = 1 .. n) = factor(simplify(sum(ii^2, ii = 1 .. n)));

$$\sum_{ii=1}^{n} ii^2 = \frac{1}{6}n(n+1)(2n+1)$$

Again we used *simplify* and then *factor* to express the answer in standard form, and we used both forms of the *sum* command. We will need this result shortly, when we investigate the area under the curve $y = x^2$.

Your Turn. Find the sum of the first *n* cubes and simplify the result.

Answer: ____

Combining Sums

When you work with the summation symbol, you need to realize that it is just a short form for writing the sum of many terms. For example, if you had 100 terms in your sum, you might decide to add the first 50 terms and remember the result. You might then add the last 50 terms together and remember that result, too. Finally, you might add the two remembered results to obtain your final answer for the complete sum. It doesn't matter what order you choose in which to add the terms, nor does it matter if you decide to find partial sums and then add these to get the final result. This can be summarized in the following Maple command. Note that the *combine* command is used to simplify the result. Try the command without using *combine* and you will see that the right-hand side remains the same as the left-hand side. Also note that the undefined function *ex*(*ii*) is used as a "dummy function" to represent any expression containing the index variable.

> S := Sum(ex(ii), ii = 0 .. n) + Sum(ex(ii), ii = n +1 .. N);

$$S := \left(\sum_{ii=0}^{n} \mathrm{ex}(ii)\right) + \left(\sum_{ii=n+1}^{N} \mathrm{ex}(ii)\right)$$

> S = combine(value(S));

$$\left(\sum_{ii=0}^{n} \mathrm{ex}(ii)\right) + \left(\sum_{ii=n+1}^{N} \mathrm{ex}(ii)\right) = \sum_{ii=0}^{N} \mathrm{ex}(ii)$$

Another property of the summation sign is its behavior when each term in the sum contains a common element.

> S := Sum(k*ex(ii), ii = 1 .. n);

$$S := \sum_{ii=1}^{n} k\ ex(ii)$$

> S = expand(S);

$$\sum_{ii=1}^{n} k\ ex(ii) = k\left(\sum_{ii=1}^{n} ex(ii)\right)$$

The common term *k* has been factored out, even though the *expand* command is used to accomplish the result. Obviously, Maple is treating the summation symbol differently than a sequence of additive terms, where the command *factor* would be appropriate.

We will apply the rules for summations to the problem of estimating the area under the parabola $y = x^2$. We'll start with a numerical example.

Example 5-3: The Area under the Curve y = x²

Estimate the area of the figure bounded by the *x* axis and the line $y = x^2$ from $x = 0$ to $x = 2$.

Solution: The area we want is shown in Figure 5.5.

We will attempt a solution using *middlebox* with ten rectangles. Figure 5.6 shows the rectangles.

> middlebox(x^2, x = 0 .. 2, 10);

Now we add the areas of the rectangles using *middlesum*.

> S := middlesum(x^2, x = 0 .. 2, 10);

$$S := \frac{1}{5}\left(\sum_{i=0}^{9}\left(\frac{1}{5}i + \frac{1}{10}\right)^2\right)$$

We'll calculate the value of this sum exactly, using *value*, and approximately, using *evalf*.

> value(S), evalf(S);

$$\frac{133}{50},\ 2.660000000$$

Actually, it doesn't do much good to calculate the sum exactly here, since as we can see from Figure 5.6 the upper and lower triangles do not have exactly the same area.[1] Therefore, the rectangles are truly an estimate of the area under the curve, not the exact result. We can

1. They are not true triangles—one of the sides is a curve, rather than a straight line.

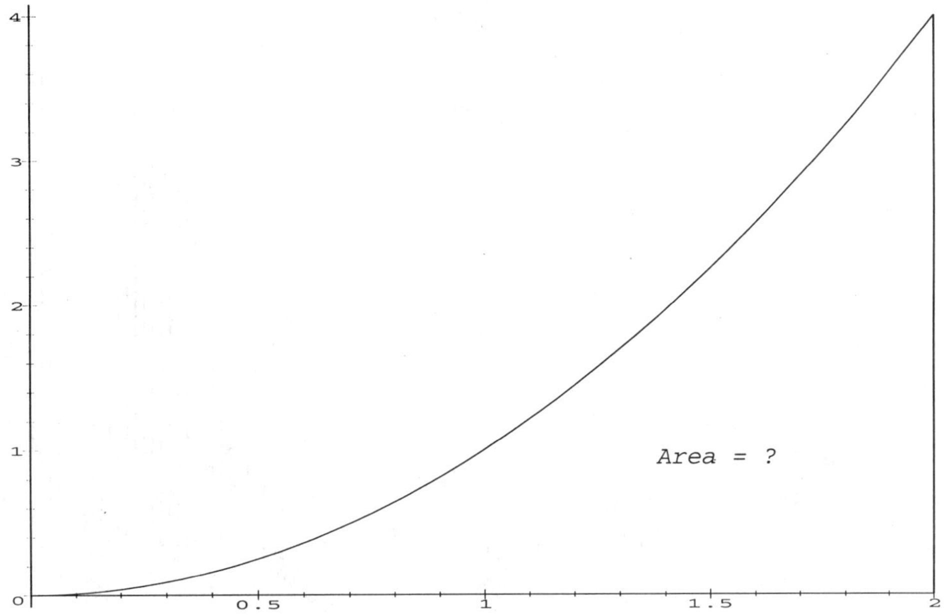

Figure 5.5 The Area under the Curve $y = x^2$ for the Interval $x = 0 .. 2$

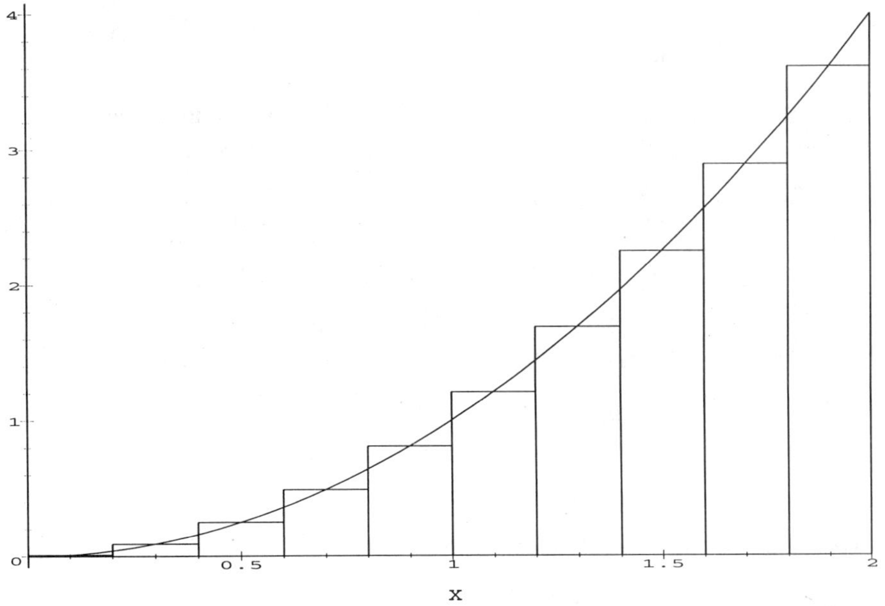

Figure 5.6 Approximation of the Area with Rectangles

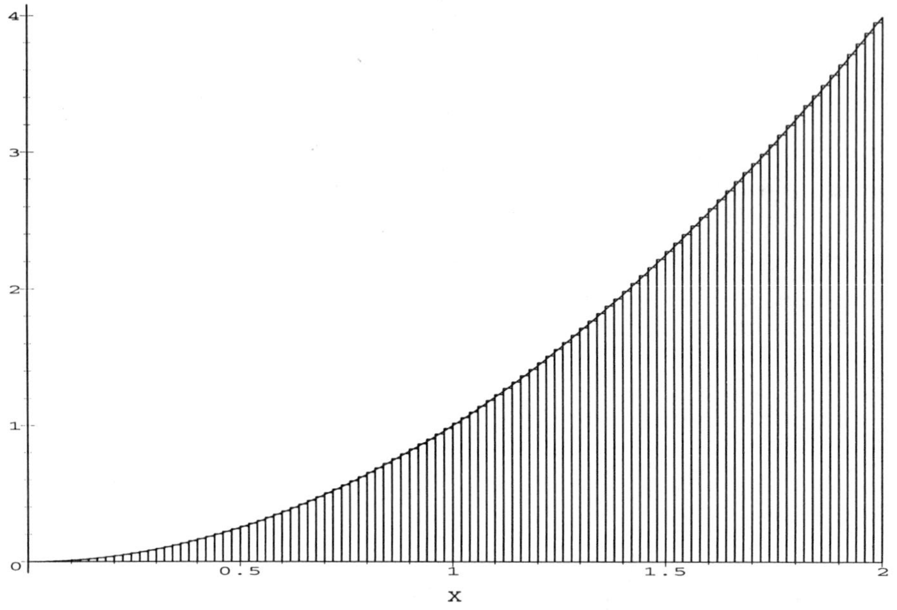

Figure 5.7 Approximating the Area with 100 Rectangles

make the estimate increasingly better by choosing smaller and smaller intervals for our rectangles.

> middlebox(x^2, x = 0 .. 2, 100, color=black);

The rectangles are so close together in Figure 5.7 that it is almost impossible to pick out the little triangles formed at the top of each rectangle. The total area of these triangles is

> S := middlesum(x^2, x = 0 .. 2, 100);

$$S := \frac{1}{50}\left(\sum_{i=0}^{99}\left(\frac{1}{50}i + \frac{1}{100}\right)^2\right)$$

The numerical value of the sum is

> value(S), evalf(S);

$$\frac{13333}{5000},\ 2.666600000$$

Finding Better Estimates

When you compare the decimal value of our area estimate in the two trials, one with 10 rectangles and one with 100 rectangles, it seems that the area is converging on 2.666666 . . ., or

8/3. It's harder to see that 13,333/5,000 is close to 8/3, yet it differs from 8/3 by only 1 part in 15,000. Check this statement by adding 1/1,5000 to 13,333/5,000. You get 8/3.

We can show that the area under the parabola is exactly 8/3 by applying the theory of limits. We make the number of rectangles a variable.

> S := middlesum(x^2, x = 0 .. 2, n);

$$S := 2\,\frac{\displaystyle\sum_{i=0}^{n-1}\left(4\,\frac{\left(i+\frac{1}{2}\right)^2}{n^2}\right)}{n}$$

Does *S* tend to a limit as the number of rectangles, *n*, tends to infinity? To find out, we issue the *limit* command.

> S1 := limit(S, n=infinity);

$$S1 := \lim_{n \to \infty} 2\,\frac{\displaystyle\sum_{i=0}^{n-1}\left(4\,\frac{\left(i+\frac{1}{2}\right)^2}{n^2}\right)}{n}$$

We evaluate the result just as we did in the previous two cases.

> value(S1), evalf(S1);

$$\frac{8}{3},\ 2.666666667$$

The result is exactly 8/3.

Definition of the Definite Integral

We have found the area under a specific curve, $y = x^2$, for a specific interval, $0 \le x \le 2$. Using the properties of the summation sign and the properties of limits, we can generalize these results to include many more functions and arbitrary intervals.

We will define the *definite integral* as the limit of the sum of the areas of rectangles in a situation where the base of these rectangles gets smaller and smaller. If *f* is a continuous function in the variable *x*, the area between *f*(*x*) and the *x* axis bounded by the points *x*1 and *x*2, is

> definite_integral = limit(middlesum(f(x), x = x1 .. x2, n), n = infinity);

$$definite\_integral = \lim_{n \to \infty} \frac{(x2 - x1)\left(\sum_{i=0}^{n-1} f\left(x1 + \frac{\left(i + \frac{1}{2}\right)(x2 - x1)}{n}\right)\right)}{n}$$

This is an imposing-looking formula! The key in integration, just as in differentiation, is to work out this formula for common functions. We can expect that the formulas will be no more difficult to master than the formulas for differentiation.

Positive and Negative Areas

In the examples we have discussed so far, the graph of the function has always been above the x axis. Consider the possible ways in which the integral could be defined when the graph of the function dips below the x axis. We will choose the function $5 - 4x - x^2$ over an interval of $x = 0 .. 2\sqrt{6} - 3$. The graph is shown in Figure 5.8, along with the rectangles produced by *middlebox*.

> ex := 5 - 4*x - x^2;

$$ex := -x^2 - 4x + 5$$

We want to define the integral as the area under the curve, but how should we define the area for the current example? Do we consider the area of each rectangle as a positive number (as we would in geometry), or do we take into account the fact that the y values are negative for $x > 1$? Perhaps, surprisingly, the choice is the latter. The area of any rectangle lying below the x axis is given a negative value. The reason for the choice is that it is the more natural definition for almost all practical cases. Because this is the accepted definition for the sum of the areas, *middlesum* yields the value

> evalf(value(middlesum(ex, x = 0 .. 2*sqrt(6) - 3, 10)));

$$.005706625152$$

By examining Figure 5.8, you can see that there are two quasi-triangles,[2] whose average height is about 5. The base of each quasi-triangle is 1. If we take the area of each triangle as positive, the combined area is about 10, nowhere near the *middlesum* value of 0.0057. We can calculate the exact area by taking the limit.

2. They are called *quasi-triangles* because the hypotenuse is not a straight line: instead, it has a slight curvature.

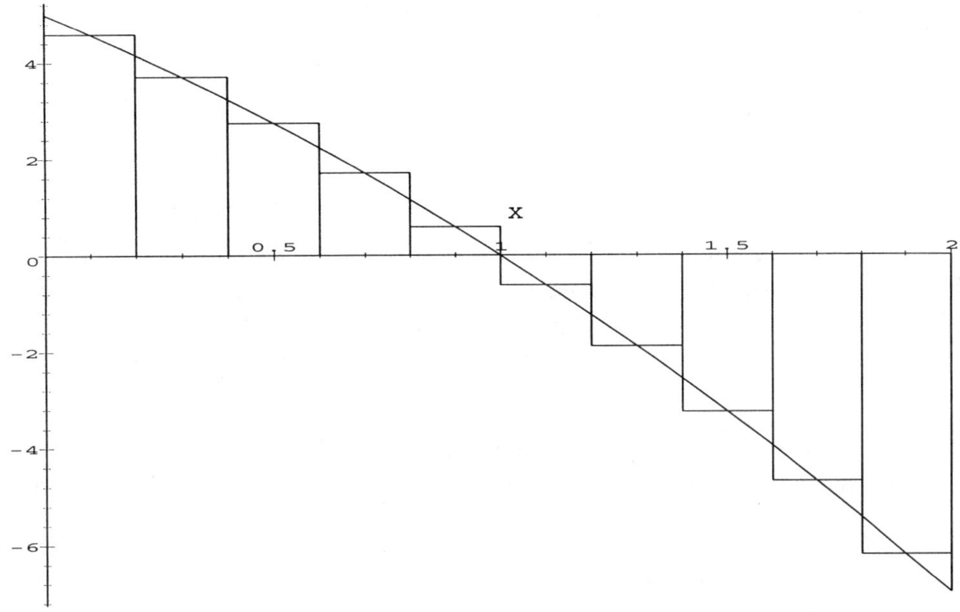

Figure 5.8 The Graph of $y = 5 - 4x - x^2$ and Ten Area-Approximating Rectangles

```
> value(limit(middlesum(ex, x = 0 .. 2*sqrt(6) - 3, n), n = infinity));
```

0

Obviously, Maple is following the standard rule of calculus that any area below the x axis is assigned a negative value. Once again, this is because the height of the rectangle, y, is negative when the rectangle lies below the x axis.

The Fundamental Theorem of the Calculus

We have considered the problem of finding the area under a curve. The area is approximated by drawing rectangles. As the number of rectangles increases, thereby making the base of each rectangle decrease, a better approximation is reached. The *definite integral* is defined as the limit of the sum of the areas of the rectangles. The symbol for the integral is an elongated S. In Maple, the definition of the definite integral of the function $f(x)$ between $x = a$ and $x = b$ is

> Int(f(x), x = a .. b) = Limit(middlesum(f(x), x = a .. b, n), n = infinity);

$$\int_a^b f(x)dx = \lim_{n \to \infty} \frac{(b-a)\left(\sum_{i=0}^{n-1} f\left(a + \frac{\left(i+\frac{1}{2}\right)(b-a)}{n}\right)\right)}{n}$$

We have used the capitalized forms of both the integral and the limit because we only want to show the formulas. There is nothing to evaluate at this stage. The left-hand side of the equation shows how the integral is written. The *lower limit* is a and the *upper limit* is b. The symbol dx is meant to refer to a small distance along the x axis. You can think of all the thin rectangles as having a base of length dx.

The right-hand side is another impressive-looking formula. Basically, it says that the interval $(b - a)$ is divided into n equal parts and that the function f is evaluated at the midpoint of each of these intervals. We identify dx with $(b - a)/n$, and $f(x)$ with $f\left(a + \frac{\left(i+\frac{1}{2}\right)(b-a)}{n}\right)$. It may take you some time to convince yourself that this is so, but just examine each part of the sum. The factor $(i + 1/2)$ selects the midpoint of the ith interval after it is multiplied by $(b - a)/n$ and added a. This positions us at the equivalent to point x. Thus, this whole term is equivalent to $f(x)$.[3]

Now we are ready to see a statement of the fundamental theorem of the integral calculus. Think of the upper limit of the integral as a variable and call this general point x. The integral is now a function of this upper limit x, and we ask for the derivative of this function. To get an indication of the answer, we choose a general function, $f(x)$, and define the integral as the sum of 1,000 rectangles. Call this sum $S(x)$. Then we add one more rectangle and extend the sum to include that rectangle. Call this sum $S(x + dx)$. Subtract the new sum from the old and divide by the length of the base of the rectangle, forming $(S(x + dx) - S(x.))/dx$. You recognize this as the difference quotient, which, in the limit as dx approaches zero, is the derivative. Thus, we are taking the derivative of the integral. Putting it all together in a Maple formula, we arrive at

> simplify((middlesum(f(x), x = 0 .. x+x/1000, 1001) - middlesum(f(x), x = 0 .. x, 1000))/(x/1000));

$$\left(\sum_{i=0}^{1,000} f\left(\frac{1}{2000}(2i+1)x\right)\right) - \left(\sum_{i=0}^{999} f\left(\frac{1}{2000}(2i+1)x\right)\right)$$

3. Really, the summation is equivalent to $f(x + dx/2)$, but we are approximating it to $f(x)$.

The first sum contains 1,001 intervals, whereas the second sum contains 1,000 intervals. The formula is exactly the same in both sums. Thus, every term but the last will drop out when they are subtracted. The remaining term is $f\left(\frac{1}{2{,}000}(2 \cdot 1{,}000 + 1)x\right) = f\left(\left(1 + \frac{1}{2{,}000}\right)x\right) \approx f(x)$. You can see that if 10,000 intervals were chosen, the result would be even closer to $f(x)$. In the limit, it would be at $f(x)$. We have made the following statement plausible. The *fundamental theorem of the integral calculus* states that the derivative of the integral of a function whose upper limit is x is $f(x)$.

Fundamental Theorem of the Integral Calculus (First Form)

$$\frac{d}{dx}\left(\int_a^x f(t)dt\right) = f(x)$$

The variable of integration, t, is called a dummy variable because the letter t could be changed to any other (unassigned) variable without changing the result.

The fundamental theorem states that differentiation is the inverse of integration. We will see soon that integration is "almost" the inverse of differentiation.

Indefinite Integrals (Antiderivatives)

The integral of the linear function $ax + b$ is

```
> expand(value(limit(middlesum(a*x + b, x = 0 .. x, n), n = infinity)));
```

$$xb + \frac{1}{2}ax^2$$

Thus, the form $ax + b$ transforms to $ax^2/2 + bx$ by integration. It is easily verified that the integral of b is bx and the integral of ax is $ax^2/2$.

We are close to our goal of finding formulas for the integrals of common functions. Since the integral is a function of its upper limit,

```
> Int(f(x), x= 0 .. a) = F(a), Int(f(x), x = 0 .. b) = F(b);
```

$$\int_0^a f(x)dx = F(a), \int_0^b f(x)dx = F(b)$$

These formulas state that the integrals are as-yet undetermined functions of their upper limits. However, both are the same function. That is what saying they are a function of their upper limits means. Since we know that integrals are additive because they are related to areas, we have, assuming $b > a$, the further relation:

```
> Int(f(x), x = a .. b) = F(b) - F(a);
```

$$\int_a^b f(x)dx = F(b) - F(a) \tag{5-1}$$

Equation 5-1 is also called the Fundamental Theorem of the Calculus. We can use this result to define the indefinite integral as

```
> Int(f(x), x ) = F(x);
```

$$\int f(x)dx = F(x) \tag{5-2}$$

Whenever a definite integral is needed, use Equation 5-1; when just the functional form for the integral is needed, use Equation 5-2. This presentation is slightly different from the usual calculus text, where the indefinite integral is given as

```
> Int(f(x), x ) = F(x) + constant;
```

$$\int f(x)dx = F(x) + \text{constant} \tag{5-3}$$

In Equation 5-1, the integral is the difference of two terms. The second term can be interpreted as the constant in Equation 5-3.

Equation 5-2 is presented without the so-called constant of integration because Maple writes its indefinite integrals in this way. You have to remember to add in the constant of integration if it is required by the problem you happen to be doing. You'll see many examples of this later on.

The foregoing discussion leads to the second way of stating the fundamental theorem.

Fundamental Theorem of the Integral Calculus (Second Form)

$$\int_a^b f(x)dx = F(b) - F(a),\ F' = f$$

where F' means the derivative of F.

Integral Formulas

We can use the definition of the integral as a limit to generate formulas for standard functions. Remember that the formulas below can contain an arbitrary constant of integration that is not shown (a feature of the Maple idiom).

The Integral of a Constant

```
> Int(C, x) = value(limit(middlesum(C, x = 0 .. x, n), n = infinity));
```

$$\int C\,dx = xC$$

The Integral of x

```
> Int(x, x) = value(limit(middlesum(x, x = 0 .. x, n), n = infinity));
```

$$\int x\,dx = \frac{1}{2}x^2$$

The Integral of x^n

```
> Int(x^n, x) = simplify(value(limit(middlesum(x^n, x = 0 .. x, r), r = infinity)));
```

$$\int x^n\,dx = \frac{x^{(n+1)}}{n+1}$$

Note that we had to change the variable for the number of intervals to r so that it would not conflict with the power to which x is raised, which is traditionally called n. This formula is good for $n \neq -1$. If n equals -1, the following formula, for the integral of $1/x$, must be used.

From now on, we will use Maple's integration capabilities to list the integration formulas rather than evaluating them as limits of sums.

The Integral of 1/x

```
> Int(1/x, x) = int(1/x, x);
```

$$\int \frac{1}{x}\,dx = \ln(x)$$

The Integral of e^x

```
> Int(exp(x), x) = int(exp(x), x);
```

$$\int e^x\,dx = e^x$$

The Integral of the Trigonometric Functions

```
> Int(sin(x), x) = int(sin(x), x);
```

$$\int \sin(x)\,dx = -\cos(x)$$

> Int(cos(x), x) = int(cos(x), x);

$$\int \cos(x)\,dx = \sin(x)$$

> Int(tan(x), x) = int(tan(x), x);

$$\int \tan(x)\,dx = -\ln(\cos(x))$$

> Int(cot(x), x) = int(cot(x), x);

$$\int \cot(x)\,dx = \ln(\sin(x))$$

> Int(sec(x), x) = int(sec(x), x);

$$\int \sec(x)\,dx = \ln(\sec(x) + \tan(x))$$

> Int(csc(x), x) = int(csc(x), x);

$$\int \csc(x)\,dx = \ln(\csc(x) - \cot(x))$$

All of the formulas for the integrals of common functions except the last three should be memorized. They are the basic building blocks for many of the frequently encountered problems in integration.

Compound Integrals

The rules for simplifying integrals are the same as those for sums. The integral is the area under a curve; it has a couple of special properties. You have seen one of these properties already, that is, any areas below the x axis are negative. The other special property of the integral concerns the interval of integration. Up to now, we have assumed that the lower limit, a, lies to the left of the upper limit, b. If the limits are interchanged, we have the rule that reversing the limits of integration reverses the sign of the integral.

> Int(f(x), x = a .. b) = -Int(f(x), x = b .. a);

$$\int_a^b f(x)\,dx = -\int_b^a f(x)\,dx$$

Example 5-4: Reversing the Limits of Integration

Find out if Maple knows the rule about reversing the limits. Find the integral of sin(x) for x = 0 .. Pi and for x = Pi .. 0.

Solution: Write out the integrals using both the uppercase and lowercase forms. Use the *copy* and *paste* commands to minimize your typing.

> Int(sin(x), x = 0 .. Pi)= int(sin(x), x = 0 .. Pi), Int(sin(x), x = Pi .. 0) = int(sin(x), x = Pi .. 0);

$$\int_0^{\pi} \sin(x)dx = 2, \int_{\pi}^{0} \sin(x)dx = -2$$

The integrals are negatives of each other, as expected.

Your Turn. Evaluate the integral of $1/x$ from $x = 1$ .. exp(1). Show that the integral of $1/x$ from $x = \exp(1)$ .. 1 is the negative of the first integral.

Answer: ____________________

Other Useful Rules for Simplifying Integrals by Hand

The integral of the product of a constant and a function. The following result shows that the constant can be factored out of the integral sign.

> Int(c*f(x), x) = expand(Int(c*f(x), x));

$$\int c\ f(x)dx = c \int f(x)dx$$

Thus, the integral of the sum of functions is the sum of the integrals.

> Int(f(x) + g(x), x) = expand(Int(f(x) + g(x), x));

$$\int f(x) + g(x)dx = \int f(x)dx + \int g(x)dx + C$$

Note the appearance of a constant term in the result. The indefinite integral is any function whose derivative equals the function that is being integrated. Since the derivative of a constant is zero, the functions $F(x)$ and $F(x) + C$ have the same derivative, $f(x)$. When you use Maple to find the integral of a sum and then use it to find the sum of the integrals, you may think that you are getting two different results because the expressions don't look the same. However, if you plot both answers, you will find that the shapes of the curves are the same, but one lies above the other, indicating that they differ by a constant amount. This is permissible, because two indefinite integrals that differ by a constant are considered equivalent.

Discontinuities

Be careful to examine the domain of the function for discontinuities when doing a definite integration. If the indefinite integral of $f(x)$ is $F(x)$, you know that

$$\int_a^b f(x)dx = F(b) - F(a)$$

However, the formula is true only if the function f is continuous over the domain of x. Watch what terrible things happen if this is forgotten. Imagine that the problem is to find the area under the curve $(x-3)^{-2}$ for $x = 2 .. 4$. If the decision is to apply the previous formula, the indefinite integral is found and named F. Then, the values for the end-points are substituted in F and subtracted. The result for the area under the curve is –2.

> F := int(1/(x - 3)^2, x);

$$F := \frac{-1}{(x-3)^2}$$

> subs(x = 4, F) - subs(x = 2, F); (Wrong answer!)

$$-2$$

The correct approach is to ask Maple for the definite integral. The answer is infinity, not –2.

> int(1/(x - 3)^2, x = 2 .. 4);

$$\infty$$

A plot of the function shows the discontinuity at $x = 1$ (Figure 5.9). It is obvious that the rectangles built around $x = 3$ will be very high, resulting in a large contribution to the area.

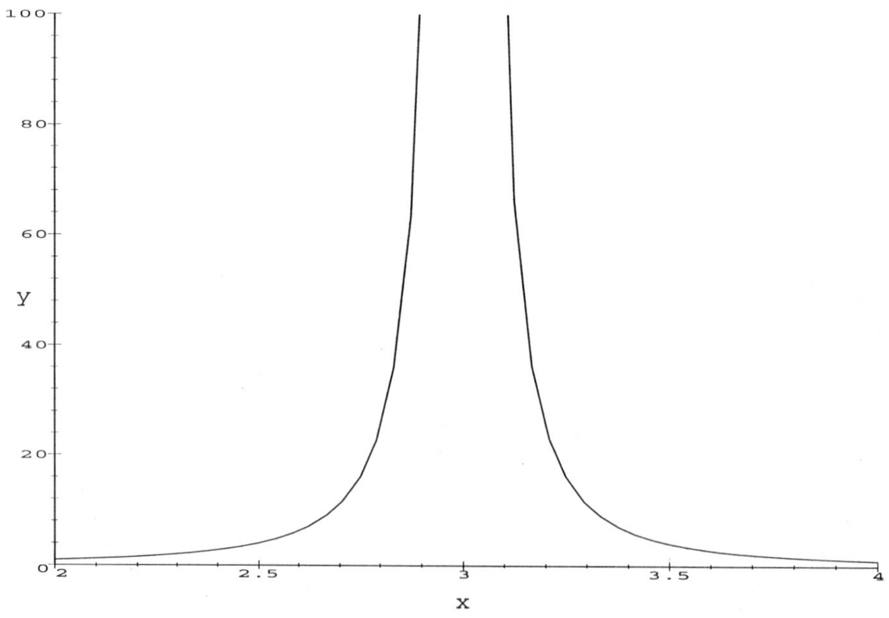

Figure 5.9 The Area of This Discontinuous Function Is ∞

Techniques of Integration

The chain rule guarantees that the derivative of any formula can always be found as long as the derivatives of all the parts of the function are known. Give Maple any formula made up of the common functions, no matter how complicated, and Maple will find an answer. The situation is different when it comes to integration. Here, the problem is to find a function whose derivative matches the *integrand* (the name given to the function to be integrated). Not every combination of common functions has an integral that is also a common function. For this reason, integration is a more difficult problem than differentiation. You have no guarantee that there is a way of expressing the integral of a given function in terms of known functions. Even worse, Maple may know some functions that you do not, and it may write its answers in terms of those functions.

This means that there are two difficulties. Maple may report an answer that you do not recognize, or Maple may not be able to find the answer at all. The cure for the first difficulty is to look up the unrecognized function in some reference texts, starting with Maple's own Help pages. The second difficulty has occupied mathematicians over the centuries. In general, integration is difficult. Various techniques have been invented to cover certain broad classes of problems. We will introduce a few of the techniques that you can employ to assist Maple in finding the integral, but you may never need them. Maple is quite good at finding the integrals that present themselves in introductory calculus courses.

Change of Variable

You can use Maple to help you check your work as you learn the standard techniques for integration. Often you will want to check the solution to a problem in a step-by-step fashion, besides checking just the answer. Let us say that you want to find the integral of

```
> q := x*sqrt(1 - x^2);
```

$$q := x\sqrt{1-x^2}$$

If you ask Maple for the integral of this function, you get a direct answer:

```
> Int(q, x) = int(q, x);
```

$$\int x\sqrt{1-x^2}\,dx = -\frac{1}{3}(1-x^2)^{3/2}$$

This is wonderful as long as all you want is the final result. But what if you want to apply a method of integration, such as the *change of variable* method? The student package contains a command called *changevar* to assist you. The command has three parameters. The first is an equation specifying the substitution to be made for the old variable in terms of the new variable. The middle parameter is the integral in which the substitution is to take place. The last parameter is the dummy variable for the integration, in other words, it is the new variable. Here is an example of its use:

> n := changevar(sqrt(1 - x^2) = u, Int(q, x), u);

$$n := \int -u^2 \, du$$

The integral has been transformed into something simple enough to become a mental calculation. The integral, in terms of the new variable, is

> n1 := value(n);

$$n1 := -\frac{1}{3}u^3$$

This is not the final answer, because we want the integral in terms of the old variable. The *subs* command gets us the final result.

> subs(u = sqrt(1 - x^2), n1);

$$-\frac{1}{3}(1 - x^2)^{3/2}$$

This is the same result as the one obtained by direct integration.

Your Turn. You may be wondering how to make the choice for the substitution formula. Again, the answer is that there is a certain amount of creativity involved. Pick a group of terms (a subexpression) that looks promising or a transformation that you recognize. If you end up with a more complicated integral, you will know you made a bad choice—try again. Here are two other possibilities. Write the transformed integral in each case.

(a) $u = 1 - x^2$. *Answer*: The integral, in terms of u, is ________________

(b) $u = \cos(x)$. *Answer*: The integral, in terms of u, is ________________

(c) Which is the useful substitution? *Answer:* ____________________

Integration by Parts

The formula for integration by parts is derived from the product rule for differentiation. We will not reproduce the derivation here, but merely state the result.

> Int(u, v) = u*v - Int(v, u);

$$\int u \, dv = uv - \int v \, du$$

The technique is best explained with an example. Suppose the problem is to find

> r := Int(x*ln(x), x);

1. The quantities of interest are *u*, *du*, *v*, and *dv*. Make up a table containing these quantities, like this one.

u =	dv =
du =	v =

2. Choose the expression you want to use for *dv*. Whatever you choose for *dv* must be easy to integrate. The quantity *u* is whatever factor or factors remain once *dv* is chosen. Enter the expressions for *u* and *dv* in the table below. In our current example, the choices for *u* and *dv* might be $u = \ln(x)$ and $dv = x$.

u = ln(x)	dv = x dx
du =	v =

3. The differential *du* is found by differentiating ln(*x*). The table entry for *v* is found by integrating the differential *dv*. The completed table is shown in Table 5.3.

u =ln(x)	dv = x dx
du = (1/x) dx	v = $x^2/2$

4. All the terms in $\int u\,dv = uv - \int v\,du$ are now specified, so the original integral is transformed into $\int x \ln(x)dx = \frac{1}{2}x^2\ln(x) - \int \frac{1}{2}x\,dx$.

5. The remaining integral on the right-hand side is a simple one, so the final result is $\int x \ln(x)dx = \frac{1}{2}x^2\ln(x) - \frac{1}{4}x^2$

6. The result in step 5 is the indefinite integral. An arbitrary constant can be added to the right-hand side, which will result in the indefinite integral as well. Remember: the indefinite integral, or antiderivative, is the function whose derivative yields the integrand.

The wrong choice for dv can make the problem harder rather than easier. In other cases, it may *seem* to make the problem harder but prove to be the way to a solution. You must gain some experience with this technique before you can become comfortable with it. Our next example shows what happens when the poorer choice is made.

Example 5-5: Integration by Parts of $x \cdot \ln(x)$, Choosing $dv = \ln(x)$

Use the method of integration by parts on $\int x \ln(x)dx$, choosing $dv = \ln(x)$

Solution: Maple's student package contains commands that help break down calculus in a nearly step-by-step fashion. This problem requires the commands *intparts* and *isolate*. We will follow the same steps as before, but this time we will choose $u = x$. The *intparts* command requires the integral as its first parameter and the function u as the second parameter. The next command constructs an equation whose left side is the unevaluated integral and whose right side is the *intparts* command. The equation is given the name p so that we can refer to it later.

> p := Int(x*ln(x), x) = intparts(Int(x*ln(x), x), x);

$$p := \int x \ln(x)dx = x(x \ln(x) - x) - \int x \ln(x) - x\, dx$$

The integral on the right can be expanded.

> p1 := expand(p);

$$p1 := \int x \ln(x)dx = x^2 \ln(x) - x^2 - \int x \ln(x)dx + \int x\, dx$$

Our choice for u as x rather than $\ln(x)$ results in the integral that we are looking for appearing on the right-hand side of the equation as well as on the left-hand side. A situation like this is not hard to deal with. It is similar to the equation $x = a - b - x + c$, and you can solve it for x. In the same fashion, we can isolate the integral we are trying to evaluate on the left-hand side using *isolate.*

> p2 := isolate(p1, Int(x*ln(x), x));

$$p2 := \int x \ln(x)dx = \frac{1}{2}x^2 \ln(x) - \frac{1}{2}x^2 + \frac{1}{2}\int x\, dx$$

The left-hand side contains an integral, but it is a very simple one and can be evaluated easily. Even with the poorer choice for the function u, integration by parts has produced a solution. The value command is used to reduce the right-hand side to an expression that does not contain an integral sign.

> lhs(p2) = value(rhs(p2));

$$\int x \ln(x)dx = \frac{1}{2}x^2\ln(x) - \frac{1}{4}x^2$$

The answer is the same as was obtained by choosing $u = \ln(x)$.

Your Turn. Find the integral of $x^2 \ln(x)$ using integration by parts.

Answer: ______________________________

Trigonometric Substitutions

The method of trigonometric substitution is a special case of a change of variable. The *changevar* command is used for trigonometric substitutions, just as it is for other substitutions.

Example 5-6: The Integral of $\frac{x^2dx}{\sqrt{25 - x^2}}$

Find the integral by means of a trigonometric substitution.

Solution: The trick here is to recognize the form $25 - x^2$ as similar to $1 - \sin^2(\theta)$. The required substitution is $x = 5\sin(\theta)$. We start by naming the integral c.

> c := Int(x^2/sqrt(25 - x^2), x);

$$c := \int \frac{x^2}{\sqrt{25 - x^2}}dx$$

Then we make a change of variable based on the equation $x = 5\sin(\theta)$:

> c := changevar(x = 5*sin(theta), c, theta);

$$c := \int 125\frac{\sin(\theta)^2\cos(\theta)}{\sqrt{25 - 25\sin(\theta)^2}}d\theta$$

Note first that we have changed the meaning of the name c by using it twice in the assignment. This is permissible. The right-hand side of the assignment statement is evaluated using the old value for c, and the result is given the same name. Thus, the name c has been redefined. The main thing to ask is whether the integral has been simplified by the substitution.

The quantity in the radical sign can be simplified. There is a common factor of 25, and $1 - \sin^2(\theta) = \cos^2(\theta)$. Issue Maple's simplification command for potentially double-valued quantities, like square roots, using the *symbolic* parameter.

```
> c := simplify(c, symbolic);
```

$$c: := 25\int -\cos(\theta)^2 + 1 d\theta$$

Has the substitution helped? We now need the integral of $\cos^2(\theta)$. Using the double angle formula for $\cos^2(\theta)$, we perform the next substitution:

```
> c := subs( cos(theta)^2 =1/2*1 + cos(2*theta), c );
```

$$c := 25\int \frac{1}{2} - \cos(2\theta) d\theta$$

Finally, we have an integral that is easy to evaluate. The result is

```
> c1 := value(c);
```

$$c1 := \frac{25}{2}\theta - \frac{25}{4}\sin(2\theta)$$

Lastly, back-substitute for θ.

```
> c2 := Int(x^2/sqrt(25 - x^2), x) = subs(theta = arcsin(x/5), c1);
```

$$c2 := \int \frac{x^2}{\sqrt{25 - x^2}} dx = \frac{25}{2} \text{ arc } \sin\left(\frac{1}{5}x\right) - \frac{25}{4} \sin\left(2 \arcsin\left(\frac{1}{5}x\right)\right)$$

Compare this answer to Maple's by direct integration.

```
> c4 := int(x^2/sqrt(25 - x^2), x);
```

$$c4 := -\frac{1}{2}x\sqrt{25 - x^2} + \frac{25}{2} \arcsin\left(\frac{1}{5}x\right)$$

As we mentioned before, these two results look different but are really the same.

Your Turn. Check that the results are the same by plotting the two indefinite integrals over the interval $x = 0 .. 5$. Do you see one plot or two? Add 0.2 to one of the expressions and plot both of them again. Do you see two lines now?

We'll come back to the question of verifying results in the section after next, "Checking Maple's Integrals."

Numerical Approximations

There are two cases to consider when discussing numerical approximations. In the first case, Maple has (or you have) worked out the integral as a formula. All that is required in this case is to put in the limits of integration. Using Example 5-6, if we want to find the area under the curve $y = x^2/\sqrt{(25 - x^{2)}}$ between the values $x = 1$ and $x = 5$, the command is simply

```
> evalf(subs(x=5, rhs(c2)) - subs(x=1, rhs(c2)));
```

19.56746982

As a check on our work in the last section, we will evaluate the area using the directly integrated expression:

```
> evalf( int(x^2/(sqrt(25 - x^2)), x = 1 .. 5));
```

19.56746982

The results are the same. We have found the area under the curve to ten decimal places, and we have checked our work by doing the problem in two different ways and getting the same answer.

The next case involves cases where Maple cannot find a formula for the integral. If you try to integrate the function sin(sin(x)) for $x = 0 .. 1$, Maple merely repeats the question:

```
> c := int(sin(sin(x)), x=0..1);
```

$$c := \int_0^1 \sin(\sin(x))dx$$

All you have to do in this case is to evaluate the result using *evalf.*

```
> evalf(c);
```

.4306061031

Alternately, you can do a problem in one step provided you don't need to know whether Maple can find an antiderivative.

```
> evalf(Int(cos(cos(x)), x = 0 .. 1));
```

.6597810536

That's all there is to numerical integration of simple functions in Maple! You are using Maple's powerful numerical integration routines to assist you without your having to know the details of the numerical work.

Checking Maple's Integrals

Often, two different ways of doing an integration problem will give two apparently different results. You could decide to check the answer by taking its derivative, but it, too, may not resemble the integrand. The best way to make a quick check is to plot the two functions. These may be either the answers that you obtained by two different methods or the integrand and the derivative of Maple's answer to the integral. If you choose a representative range and plot both functions on the same graph, you should see just one line, because they overlap.

Next, shift one of the functions upward by adding a small amount to the function. The amount you add depends on the scale of the *y* axis. It should be just enough to make the two plots distinct. If you do see two closely separated curves on the new plot, you are assured that your answers are at least very close, and probably exactly the same.

If you really want to show that two forms for an integral are the same, persistence will usually pay off. You are probably convinced that the two formulas for the integrals in Example 5-6 in the section on trigonometric substitution are identical, but how do you get Maple to confirm it? One good method is to subtract the two expressions. If they are truly equal, the result after simplification will be zero.

Returning to Example 5-6 and assuming that the definitions of *c*2 and *c*4 are still valid,

```
> ex := simplify(rhs(c2) - c4, symbolic);
```

$$ex := -\frac{25}{4}\sin\left(2\arcsin\left(\frac{1}{5}x\right)\right) + \frac{1}{2}x\sqrt{25 - x^2}$$

The output is disappointing. The result should be zero, yet we get a complicated trigonometric term containing the sine of an arcsine. The path ahead lies in recognizing that arcsin(x/5) is an angle. The formula says to take twice the sine of that angle, in other words, we need sin(2θ). Perhaps we should tell Maple to *expand* the expression.

```
> expand(ex, trig);
```

$$0$$

There it is. The two forms of the integral are shown to be equal, so their difference is zero. In general, the problem of finding the closed form of an integral, if it exists, is far more difficult than most elementary texts let on. Understand that the integral is the area under the curve, and let Maple attempt to find an antiderivative. If it can't, you're in for some work. Cunning may be necessary!

Paper and Pencil Exercises

PP5–1

You need to become familiar with the standard formulas for antiderivatives. As well, you need practice in applying these formulas to simple concrete examples. To help you in the process of mastering basic integration, do the following problems mentally. You may need to jot down some intermediate steps, but you should be able to do most of these problems with minimum paperwork. Always match the variable of integration to the differential. If a symbol is different than the differential, treat it as a constant that can be factored outside the integral sign.

Usually, you can apply the standard formulas for the antiderivative only after doing some type of rearrangement on the terms, so that the expression is recast in standard form. Even though Maple doesn't do so, you should include the arbitrary constant of integration in your answers.

(a) $\int 3p\,dp$ *Answer*: ____________________

(b) $\int 3p\,dx$ *Answer*: ____________________

(c) $\int 3\,x^2dx$ *Answer*: ____________________

(d) $\int \frac{1}{4}x^3dx$ *Answer*: ____________________

(e) $\int 2\sin(\theta)d\theta$ *Answer*: ____________________

(f) $\int 2\sin(\theta)dx$ *Answer*: ____________________

(g) $\int 6\cos(x)dx$ *Answer*: ____________________

(h) $\int \frac{10}{x}dx$ *Answer*: ____________________

(i) $\int \frac{10}{x} + 10dx$ *Answer*: ____________________

(j) $\int 1 + x + x^2dx$ *Answer*: ____________________

(k) $\int \frac{x + 2x + 3x}{3}dx$ *Answer*: ____________________

PP5–2

Continue practicing, this time by doing some definite integrals.

(a) $\int_{-5}^{5} 5dx$ *Answer*: ____________________

(b) $\int_{-5}^{5} -5dx$ *Answer*: ____________________

(c) $\int_{0}^{5} 5 - x\,dx$ *Answer*: ____________________

(d) $\int_0^{10} 5 - x dx$ *Answer:* ______________

(e) $\int_0^{\pi} \sin(\theta) d\theta$ *Answer:* ______________

(f) $\int_{-\pi}^{\pi} \sin(\theta) d\theta$ *Answer:* ______________

(g) $\int_0^{\frac{1}{2}\pi} \cos(\theta) d\theta$ *Answer:* ______________

(h) $\int_{-\frac{1}{2}\pi}^{\frac{1}{2}\pi} \cos(\theta) d\theta$ *Answer:* ______________

(i) $\int_1^{e} \frac{1}{x} dx$ *Answer:* ______________

(j) $\int_1^{10e} \frac{1}{x} dx$ *Answer:* ______________

(k) $\int_1^{2} \cos(\theta) d\theta$ *Answer:* ______________

(l) $\int_{\frac{1}{2}\pi - x}^{\frac{1}{2}\pi + x} \cos(\theta) d\theta$ *Answer:* ______________

Maple Lab

Note: Before doing any of the problems in this lab, load the student package by executing the following command. Some of the lab commands are part of the student package, which is thus required in the Maple session.

> with(student):

ML5-1: Comparing Estimates of the Area under a Curve

You saw at the beginning of this chapter that the area under a curve can be approximated by rectangles. The accuracy of the estimate depends on how many rectangles are chosen and how the height of the rectangle is chosen. In Maple, you have three choices for the height of the rectangles. The commands that show the placement of the rectangles are *leftbox*, *middlebox*, and *rightbox*. The corresponding three commands to calculate the total area of the rectangles are *leftsum*, *middlesum*, and *rightsum*.

Given a function, $f(x)$, the area under the curve may or may not exist. If it does exist, then all three methods of finding the area will converge to the same result as the number of rectangles approaches infinity.

You will investigate which method gives the most accurate result for a given number of rectangles.

(a) Let the curve be defined by the expression $y := \dfrac{19x(1-x)}{50}$.

(i) Plot the curve from $x = 0$ to $x = 1$ and observe its shape. You will investigate the area of the part above the x axis.

(ii) Choose 20 rectangles and plot the approximate area under the curve using the three "box" commands. For example, the *middlebox* command is

> y := 19/50*x*(1 - x);

$$y := \frac{19}{50}x(1 - x)$$

> middlebox(y, x = 0 .. 1, 20, thickness=0, color=black):

> (middlesum(y, x = 0 .. 1, 20));

$$\frac{1}{20}\left(\sum_{i=0}^{19}\left(\frac{19}{50}\left(\frac{1}{20}i + \frac{1}{40}\right)\left(\frac{39}{40} - \frac{1}{20}i\right)\right)\right)$$

Examine all three results. Which one looks like it will give the most accurate estimate for the area?

Answer: ____

(iii) Compute the corresponding sum commands. For instance, the *leftsum* command is

> leftsum(y, x = 0 ..1, 20);

Record all three answers.

Answer: ____

(iv) Change the number of rectangles to 200 and repeat the sum calculations. Examine the results and compare them to (iii). Convert the numbers to decimal form and guess the exact area.

Answer: ______________________________

(v) Choose any of the *sum* commands and form the limit as the number of rectangles increases to infinity. Record your result as a fraction.

Answer: ______________________________

(vi) Convert your answer to a decimal number. Was your guess in (iv) correct?

Answer: ______________________________

(b) Repeat the steps in ML5–1 to estimate the area under the sine function from $x = 0$ to $x = \pi$. Choose 20 rectangles, and then 21 rectangles. You might expect that by adding 1 more rectangle, the estimates for the area under the curve would get better. Is this what happens?

Answer: ______________________________

(c) What changes would you make to deal with a function? Since *middlesum* and the related commands require expressions, you would change any functions into expressions. In this problem, we'll show you how to deal with a function that is described by different expressions for different domains of x. Maple's *piecewise* function is built to assist with this task. It allows you to define a function described by different expressions for different domains of x. Let the function be the "tent" function, defined by

> f := x->piecewise(x>=0 and x<1/2,x,x>=1/2 and x<1,1-x);

$$f := x \rightarrow \text{piecewise}\left(0 \leq x \textbf{ and } x < \frac{1}{2},\ x,\ \frac{1}{2} \leq x \textbf{ and } x < 1,\ 1 - x\right)$$

The above command is equivalent to the statement $f(x) = x$ for $0 \leq x < 1/2$, and $f(x) = 1 - x$ for $1/2 \leq x \leq 1$.

(i) Plot the tent function from $x = -1$ to $x = 2$. Do you get a plot even though the function is not defined outside the interval $0 \leq x \leq 1$?

Answer: ______________________________

(ii) Use one of the *box* commands to construct ten rectangles under the curve from $x = 0$ to $x = 1$.

(iii) Use the corresponding *sum* command to estimate the area of the tent function.

Answer: ______________________________

(iv) The figure contains only straight lines. Calculate the area using geometry (or trigonometry). Is it the same as you found in (iii)?

Explain: ______________________________

ML5–2: The Definite Integral

Recall that if $f(x)$ is negative for $x = a .. b$, then the integral $\int_a^b f(x)dx$ is also negative. This explains why the integral of x - 1 from $x = 0$ to $x = 2$ is 0: $\int_0^2 x - 1dx = 0$. Check that Maple gives this answer by issuing the command > *int(x – 1, x = 0 .. 2);*. The result tells you that the area between the line and the x axis from $x = 0$ to $x = 1$ is equal to the area between the x axis and the line for $x = 1$ to $x = 2$. You will extend this idea of "balancing areas" to more complicated shapes in the following examples.

(a) Find the smallest positive, nonzero value of x for which $\int_0^x \cos(\theta)d\theta = 0$.

Answer: ______________________________

Could you verify the result using an argument based on trigonometry, without the need to do any calculation?

(b) Find the smallest positive, nonzero value of x such that $\int_1^x 3 + t - t^2 dt = 0$.

Answer: ______________________________

(c) Find the value of x such that $\int_{-x}^x \sin(\theta)d\theta = 0$.

Answer: ______________________________

ML5–3

How would you demonstrate that Maple knows about the fundamental theorem of the calculus?

(a) What command would you use to differentiate the integral of an arbitrary function?

Answer: ______________________________

(b) What happens if you integrate the derivative of an arbitrary function? Is this the same as (a)? If not, explain the difference.

Answer: ______________________________

(c) Note that the result in (b) happens every time you ask Maple to integrate an expression for which it knows the answer. In general, there is a *constant of integration* that Maple expects *you* to put in. This constant can be determined if you know some fact about the integral. Use the fact that the velocity $v(0) = 20$ meters per second (m/s) at $t = 0$ to find the formula for the velocity of an object, knowing that its acceleration is given by $a = 100t - t^2$, from $t = 0$ to $t = 10$ s.

Answer: ______________________________

ML5–4

Find the area of a circle by integration. The formula for a circle of radius r centered at the origin is $x^2 + y^2 = r^2$.

(a) Solve this equation for y to yield the expression for the quarter circle in the first quadrant. Name this expression y. The circle exhibits a high degree of symmetry. Certainly, the x and y axes divide the circle in four equal parts, so the area between the upper circumference and the positive x axis is one-quarter the area of the whole circle. State the command to find the area of this quarter circle.

Answer: ______________________________

(b) Have Maple perform the integration and state the result.

Answer: ______________________________

(c) What is the area of the whole circle? Use A for the symbol of the whole area.

Answer: ______________________________

(d) At this point, Maple has given you an answer to an integration problem that you might not be able to do by paper and pencil methods. List some ways in which you could check the validity of the answer—besides stating that you have been taught this formula in the past!

Answer: ______________________________

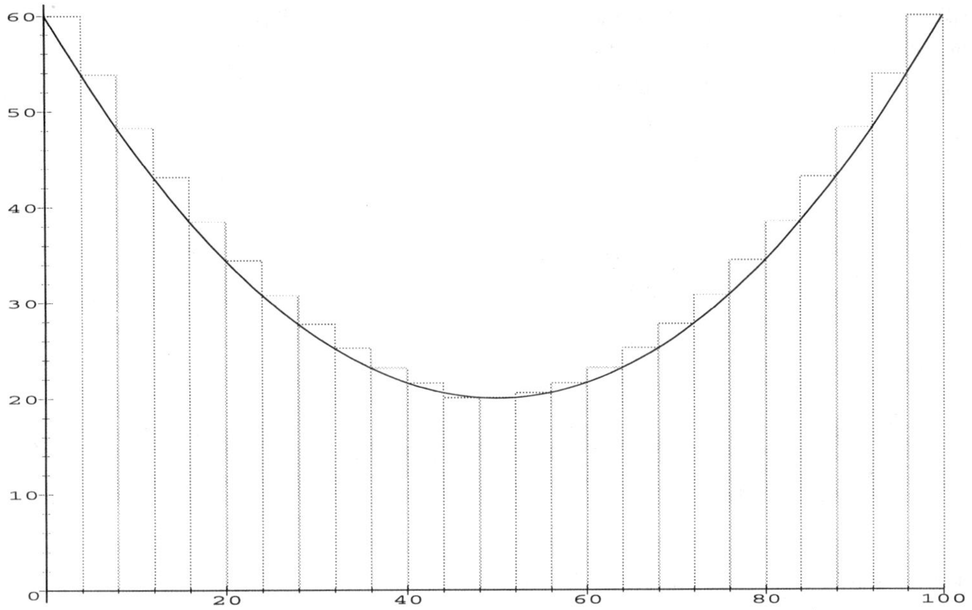

Figure 5.10 Banners Running from a Bridge's Suspension Cable to the Roadway

ML5–5

Civic planners decide to hang banners from the support cables of the bridge spanning the river that runs through the center of town. They want to cover one side of the bridge completely. The diagram in Figure 5.10 gives you an idea of their plan.

The material for the banners comes in a roll 4 feet wide. Each banner will be run from the roadway up to the suspension cable, which is a parabolic arch. One corner of the banner just touches the arch, and the excess triangular bit will be wrapped around the cable to help secure it.

(a) If one corner of the banner just touches the parabolic cable, as shown in Figure 5.10, how many feet of banner cloth are required? (The roll of cloth is 4 feet wide. You are asked to find the total length of cloth needed.)

Answer: ________________________________

(b) A cost-conscious critic raises an objection: some cloth is being wasted because the triangular bits at the top are to be wrapped around the cable instead of being cut off and saved. How many square feet of cloth are being wrapped around the cable?

Answer: ______________________________

(c) Compare the area of the triangular bits to the total area of the cloth. What percentage of the total area do the triangular pieces represent?

Answer: ______________________________

ML5–6: The Ellipse

Derive the formula for the area of an ellipse, given that the equation for an ellipse is $\frac{x^2}{a^2} + \frac{y^2}{b^2} = 1$.

(a) Plot the relation, given $a = 5$ and $b = 4$. From the plot, determine the interpretation of the constants a and b. *Hint:* use *plots[implicitplot]* to plot a relation.

(b) Solve the equation for y. There are two solutions, one for the part of the ellipse above the x axis and one for the part below.

Find the area in the first quadrant by integration. What are the limits for the x? Be careful! If you make a mistake in the limits of integration, your answer will be meaningless. Can you use a symmetry argument to show that the area of the ellipse is four times this area? Alternatively, you could find the area of each of the four pieces by integration and add them together, remembering to change the sign of the areas of the pieces below the x axis to make them positive.

(c) Calculate the area of the ellipse given by $\frac{x^2}{2^2} + \frac{y^2}{\left(\frac{1}{2}\right)^2} = 1$ using the *middlesum* method.

Compare your answer to the digits for π.

CHAPTER 6

Applications of Integration

Objectives for This Chapter

1. Survey some types of problems that require integration as part of their solution
2. Learn how to formulate the problems in terms of integrals
3. Use Maple to perform the integrations

Maple Commands Used in This Chapter

Int(f(x), x)	The inert form of the integral command.
int(f(x), x)	The command to perform symbolic integration.
plot(f(x), x = a .. b)	Plot the expression $f(x)$, from $x = a$ to $x = b$.
sin(x)	The sine of angle x, (x in radian measure).
solve(f(x) = c, x)	Solve the equation for the unknown, x.
subs(x = c, f(x))	Substitute the value c for x in expression $f(x)$.
diff(f(x), x)	Take the derivative of $f(x)$ with respect to x.
sqrt(x)	Compute the symbolic square root of quantity (or expression) x.
evalf(expr)	Reduce, as much as possible, expression *expr* to numeric form.
simplify(s, symbolic);	Simplify an expression containing a square root.

Applications of Integration

The need for the integral arises whenever anyone needs to find the area under a curve. This chapter illustrates some common situations in which an area under a curve needs to be found in order to solve a problem. The need for integration first arose in physics, and we will repeat history by giving as the first example a derivation of some of the basic formulas of motion. Other examples follow, taken from the disciplines of electronics, geometry, mechanics, hydraulics, economics, finance, ecology, and probability. Even if you are studying only one (or none!) of these subjects, please work through each example. The methods illustrated in the examples are almost certainly applicable to your field of study.

You may find it curious that this chapter comes before the one on techniques of integration, since some of the examples in this chapter contain integrals that are difficult to evaluate. The point is that Maple is available, and it knows how to apply all of the techniques discussed in the next chapter. It's better to see how integrals arise in various problem situations before learning specialized techniques for evaluating complicated integrals. In more complicated cases, Maple may need you to give it some help before it can find a solution, which is what the next chapter is about.

It's worth repeating that although the integral is the area under a curve, the area in question will not necessarily have units of area. For instance, if the topic under consideration specifies a force being exerted on an object as a function of distance, the y axis measures the force in newtons, and the x axis measures the distance in meters. The area under the force-distance curve has newton-meter (Nm) units. This is an energy unit, because 1 newton-meter = 1 joule (1 Nm = 1 J).

Motion under an Applied Force

There are a number of basic formulas for motion that can be derived by integration. We will begin by considering the motion of an object acted on by a constant force. Imagine Galileo[1] dropping a cannonball from the Leaning Tower of Pisa. Since the cannonball is both heavy and aerodynamic, we can ignore air resistance during its rather short trip from the top of the tower to the ground below. The cannonball is acted on by a constant gravitational force. Galileo anticipated Newton's[2] laws of motion, which, for the purposes of this example, can be stated as $F = ma$, and $F = mg$, where F is the force acting on the cannonball, m is the mass of the cannonball, g is a constant representing the acceleration due to gravity, and a is the acceleration of the cannonball. Equating ma and mg and canceling m, we find the extremely simple relation $a = g$. Moreover, the definition of acceleration is the rate of change of velocity,

1. Galileo Galilei (1564–1642), Italian astronomer and mathematician.
2. Sir Isaac Newton (1642–1727), English mathematician and physicist.

so $a = dv/dt$. Since a is a derivative, the velocity must be the antiderivative. In other words, $dv = g\,dt$, so $v = \int g\,dt$.

> v = int(g, t);

$$v = gt$$

We have stated that g is a constant. In the metric system, $g \cong 10$ m/s^2 (meters per second squared). Thus, the velocity is a linear function of time. Our analysis is correct if we take our time measurements from the instant Galileo dropped the cannonball and assume that its initial velocity was 0 m/s. Remember that the antiderivative contains an arbitrary constant that you must put in because Maple doesn't. In this case, the constant is 0 because $v = 0$ when $t = 0$, just as our solution says. Note also that the velocity is a positive number. Effectively, what we have done is to assign positive values to the downward direction. Not only will downward velocities be positive, but distances measured downward will be positive as well. Consequently, velocities and distances in the upward direction will be negative because of our convention.

If Galileo had somehow managed to throw the cannonball upward, it would have had a velocity at the moment of release. The equation of motion remains the same, but this time we must put in the arbitrary constant:

> v = int(g, t) + v[0];

$$v = gt + v_0$$

Let us say that Galileo was able to give an initial velocity of –2 m/s to the cannonball. The velocity is negative because it is in the upward direction and our convention specifies that positive velocities are in the downward direction. Knowing this *initial condition*, that $v_0 = -2$ m/s when $t = 0$, we find that the formula for the velocity as a function of time is $v = gt - 2$. This equation tells us that the velocity is zero at $t = 0.2$ s. The interpretation is that the cannonball moves upward for a short time until gravity reduces the upward velocity to 0. After $t = 0.2$ s, the cannonball has a positive velocity, which means that it is moving in the downward direction.

Next, we ask for the distance the cannonball travels as a function of time. Since the distance, s, in meters is related to the velocity, v, by $v = ds/dt$, the distance is given by the antiderivative $s = \int v\,dt$. In the case of a constant velocity, the integral is:

> s = int(v, t);

$$s = vt$$

This is the well-known formula that states that distance traveled is the product of the velocity multiplied by the time. It is valid so long as the velocity is a constant.

In the case of Galileo dropping the cannonball, the velocity is a function of time; the integral is:

> s = int(v(t), t);

$$s = \int v(t)dt$$

Note the difference in the last two commands. In the former, Maple assumes that v is a constant, independent of time. In the latter, the velocity is given as an undetermined function of time. Since Maple does not know a specific formula for $v(t)$, it cannot perform the integration and simply rewrites the input in standard mathematical notation. We know that in the case of a cannonball dropped from a tower, the velocity is given by $v = gt$, so the integral formula becomes:

> s = int(g*t, t);

$$s = \frac{1}{2}gt^2$$

This is an equally famous formula, which states that the distance traveled by a falling object that starts from a resting position and moves with a constant acceleration is one-half its acceleration times the square of the elapsed time.

Your Turn

How far has the cannonball moved when $t = 0$? *Answer*: ____________________

How far has the cannonball moved when $t = \sqrt{2}$ s? *Answer*: ____________________

You can do a surprising amount of analysis using these formulas as long as you take the time to understand how they are applied in various situations dealing with constant acceleration. PP6–1 asks you to develop other useful formulas. This problem, along with PP6–2, helps you to analyze problems using a blend of graphical analysis and algebra. The integral calculus helps you to understand how the formulas of motion follow rather naturally from the basic assumption of constant acceleration.

Example 6-1

Examine the case of Galileo dropping his cannonball once more. Plot the curves for acceleration, velocity, and distance traveled as a function of time on the same graph (Figure 6.1). Where do the curves intersect, and is there any significance to these points of intersection?

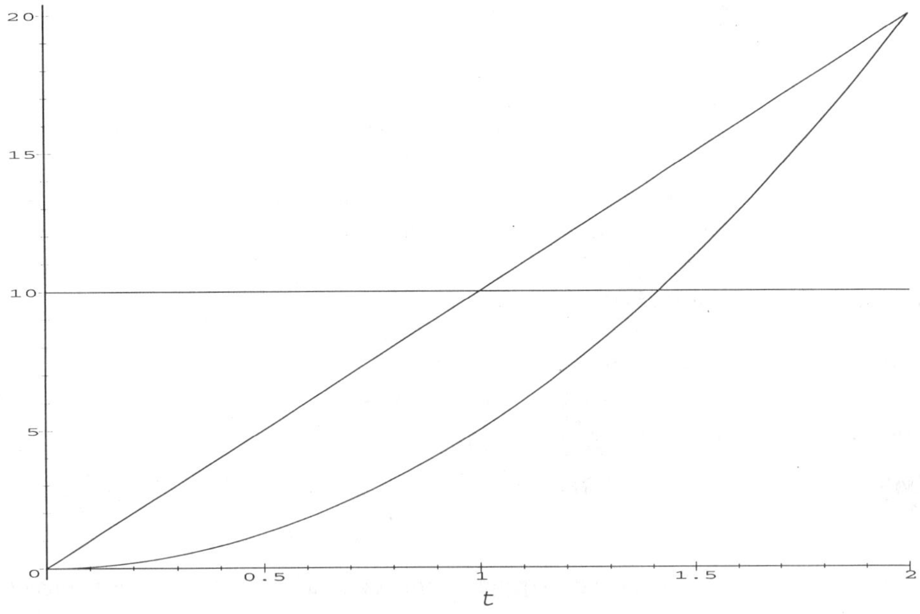

Figure 6.1 Acceleration, Velocity, and Distance Traveled

Solution:

```
> plot({10, 10*t, 5*t^2}, t = 0 .. 2);
```

Can you pick out which curve is which? The acceleration is 10, a constant. Its graph is the horizontal straight line. The velocity is a linear function, which makes it easy to pick out once the plot for the acceleration is eliminated. The curved line must be the distance traveled, which is a quadratic function of time.

The distance curve intersects the velocity line at $t = 2$ s, and the acceleration line at $t = \sqrt{2}$ s. There is no great significance to these intersection points. The velocity line intersects the acceleration line at $t = 1$ s. The significance of this fact is that at this point, the velocity is numerically equal to the acceleration. Basically, this is the definition of acceleration: it is the amount by which the velocity changes in 1 second.

Your Turn. How far does an object travel if it is subjected to an acceleration of 33 m/s^2 for 4 seconds? Assume that the initial speed of the object is 0.

Answer: ______________________________

Example 6-2

Normally, a rocket motor produces a constant thrust. A spacecraft on a mission to the planetoid system TH develops an oscillating thrust component due to a malfunction. The spacecraft was initially coasting at a speed of 15,000 kilometers per hour before the rocket engines turn on. The acceleration of the spacecraft with the motor on is given by $a = 8 + 0.5\sin(t/10)$ m/s^2, where t is in seconds. The engine produces this thrust for one hour. Find the distance the spacecraft travels during this time, and find the distance it would have traveled if the rocket motor had never turned on.

Solution: First, convert to a consistent system of units (15,000 km/hr = 4,167 m/s, and 1 hr = 3,600 s). The velocity as a function of time is found by integrating the acceleration and applying the initial condition:

> v:= int(8 + 0.5*sin(t/10), t) + 4167;

$$v := 8.t - 5.\cos(.1000000000t) + 4167$$

The velocity contains the same type of ripple as the acceleration, as seen by the inclusion of the cosine term in the expression for v. The distance traveled is found by integrating this expression:

> s = int(v, t = 0 .. 3600);

$$s = .6684115205\ 10^8$$

The spacecraft has traveled 66.84 million meters, or 66,840 km. If the spacecraft had merely coasted for 1 hour, the distance traveled would have been 15,000 km, since its speed does not change unless the motor turns on.

As a matter of interest, we can calculate the final velocity of the spacecraft.

> Vf := int(8 + 0.5*sin(t/10), t = 0 .. 3600) + 4167;

$$Vf := 32973.41846$$

Here, *Vf* represents the final velocity after the rocket engine has been on for an hour. This speed, in meters per second, converts to nearly 119,000 km/hr.

Your Turn. A spacecraft with an experimental engine is being tested just outside the Deep Space 99 test site, far from gravitationally massive objects. The spacecraft is initially at rest with respect to the Doppler radar speed meter in the lab on Deep Space 99. The engine turns on and the acceleration is measured as $a = 10e^{-t/100}$ m/s^2 during the test, which lasts 10 seconds. Then the engine is shut down, and immediately the reverse thrusters are turned on to slow the spacecraft. The acceleration provided by the thrusters is $a = -30/(t + 1)$, where, in this formula, $t = 0$ is the time when the thruster is first turned on. How long do the thrusters have to remain on to bring the spacecraft to a halt, and how far has it traveled?

Answer: $t =$ ____

Answer: $s =$ ____

Work and Energy

Physics tells us that mechanical work is one of the many forms of energy. Mechanical work must be done to stretch a coil spring. The definition of mechanical work is $W = \int F\,dx$, where W is the energy in joules, F is the applied force in newtons, and dx is the small distance through which the force moves, in meters. The reason that the integral is part of the definition is that the force may change with distance instead of remaining constant. In the case of a stretched spring, this energy is not lost; instead, it is stored in the potential energy of the stretched spring. Many springs obey *Hooke's law*, which states that the return force in a stretched spring is proportional to its elongation. The formula is $F = kx$, where k is the force constant, in Nm, and x is the amount by which the spring is stretched when the force, F, in newtons (N), is measured.

Example 6-3

It takes a force of 2,000 N to stretch an automobile spring by 3.65 cm. How much work is done in stretching this spring to a distance of 10 cm?

Solution: The force constant, k, is determined from the data. A force of 2,000 N stretches the spring 3.65 cm, so we substitute these values in $F = kx$ and, after converting centimeters to meters, we solve for k.

```
> solve(2000 = k*3.65e-3, {k});
```

```
{ k = 547945.2055}
```

The work done is $W = \int F\,dx$. Therefore, form the command

```
> int(547945.2055*x, x = 0 .. 0.1);
```

```
2739.726028
```

Thus, 2,740 joules of work is done in stretching the spring a distance of 10 cm. Notice that the limits for x in the integration must be in meters, not centimeters.

Your Turn. Assume that the same automobile spring has been stretched 2 cm. How much more work is done if the spring is further stretched from $x = 2$ cm to $x = 7$ cm?

Answer: ______________________________

Electronics: The Voltage across a Capacitor

A capacitor is a device for storing electrical charge on its plates. The voltage V_C across the plates of a capacitor is given by $V_C = 1/C\int i\,dt$, where C is the capacitance in farads (F), i is the current in amperes (A), and t is the time in seconds during which the electrical current flows.

Example 6-4

An alternating current, given by the formula $i(t) = 1.55sin(377t)$ amperes, charges a capacitor from $t = 0$ to $t = 3$ milliseconds (ms). Find the voltage across the capacitor if its capacitance value is 10 μF (1 μF = 10^{-6} F).

Solution: Substitute the known values into the integral formula:

```
> V[C] = 1/10e-6*int(1.55*sin(377*t), t = 0 .. 3e-3);
```

$$V_C = 236.0953494$$

The voltage across the capacitor is 236 V. Note that the only difficult thing about this problem is the arithmetic, which Maple handles very nicely.

Your Turn. Find the voltage across the capacitor when the current has been flowing for 6 ms, and then for 15 ms. Compare your answers to V_C at 3 ms and explain.

Answer: V_C at 6 ms = __________ *Answer*: V_C at 15 ms = __________

Answer: At $t = 15$ ms, V_C is (greater/less) __________ than at $t = 3$ ms because ________

____

Volumes by Integration

It is possible to find the volume of an object with a single integration so long as the object has an axis of symmetry. There are two basic methods: the method of disks and the method of cylindrical shells. Usually, one method results in a simpler integral than the other.

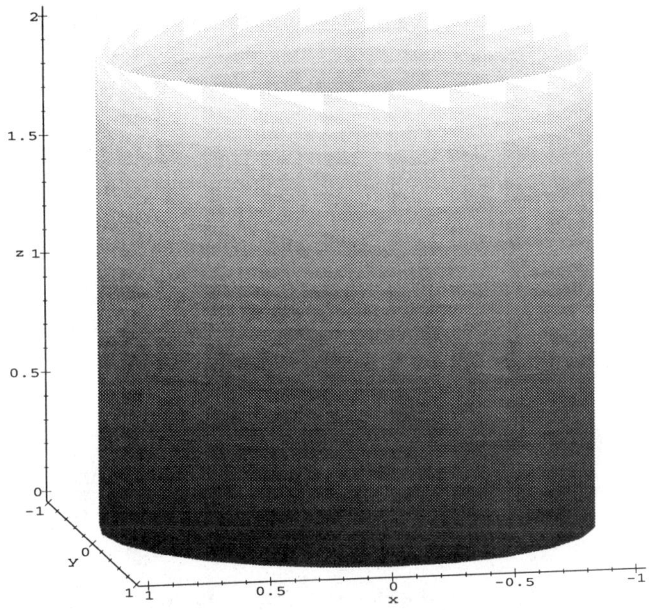

Figure 6.2 A Cylinder of Radius 1 and Height 2

Example 6-5

Find the volume of a cylinder by the method of disks and check the result by deriving the formula from elementary considerations.

Solution: Form a cylinder of radius r by choosing the circle $x^2 + y^2 = r^2$, and move this circle along the z axis, from $z = 0$, the base of the cylinder, to $z = h$, the top of the cylinder. Figure 6-2 shows a cylinder of radius 1 and height 2.

Take a slice of thickness dz parallel to the x-y plane to form a disk. The area of one face of the disk is πr^2, and the height of the disk is dz. Thus, the volume, V, of the disk is $dV = \pi r^2 dz$. The total volume of the cylinder is found by adding the volumes of all the disks together. Since the thickness, dz, is a differential, integration is required. The integral is $V = \int_0^h \pi r^2 dz$.

> V = int(Pi*r^2, z = 0 .. h);

$$V = h\pi r^2$$

Because the cylinder is formed from a circle whose radius is constant, the volume can be deduced in one step by observing that the volume of an object of constant cross-sectional area is found by multiplying that cross-sectional area by the height.

Your Turn. Calculate the volume, in cubic centimeters, of a mile of copper wire whose diameter, including its insulating covering, is 0.15 cm.

Answer: ______________________________

The more challenging problems, of course, arise when the cross-sectional area of the solid or figure is not a constant. The next example shows you how to use the integral calculus to derive the formula for the volume of a cone.

Example 6-6

A cone has a circular base of radius r. The height of the cone is h. Derive the formula for the volume, V.

Solution: We will use the method of disks. From the cone in Figure 6.3, think of taking a small slice whose thickness is dz. The slice is parallel to the y-z plane.

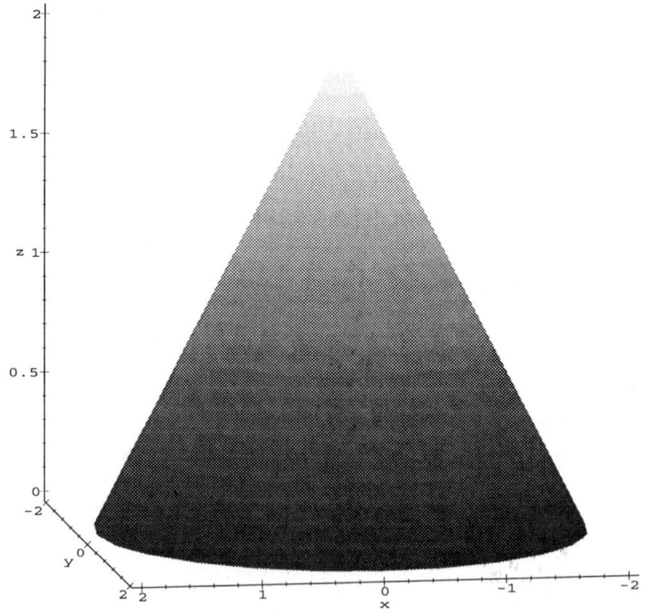

Figure 6.3 A Right Circular Cone of Base Diameter 4 and Height 2

The slice forms a thin disk whose radius depends on the height, z, where the slice is taken. Our challenge is to build a formula for this radius. From the diagram, you can see that the cone looks like a triangle when it is viewed directly from the side. The base of the triangle is $2r$, and this dimension decreases linearly to 0 at height h. Thus, at any height, z, the radius of the disk is $r(1 - z/h)$, which you verify by observing that it is a linear function of z, having the correct values at $z = 0$ and at $z = h$. The representative disk has volume $dV = \pi(\ r(1 - z/h))^2 dz$.

We must add the disks from $z = 0$ to $z = h$, which means that the integral we want is:

$$V + \int_0^h \pi\left(r\left(1 - \frac{z}{h}\right)\right)^2 dx$$

> V = int(Pi*(r*(1 - z/h))^2, z = 0 .. h);

$$V = \frac{1}{3} h \pi r^2$$

The volume of a cone is one-third the volume of the cylinder on the same base and height.

Your Turn. Some ancient mounds are roughly the shape of a truncated cone, one that has had its pointed end cut off. Such a cone can be characterized by its height, h; its bottom radius, r_1; and its top radius r_2. Derive the formula for the volume of the truncated cone (Figure 6.4).

Hint: It is easy to find a formula for the volume of the truncated cone by subtracting the volume of the missing top portion from the volume of the complete cone, including the top portion. The difficulty comes in labeling the diagram. Let the height of the truncated cone be h and the height of the complete cone be $h + x$. Relate x to h, r_1, and r_2 by showing that $x = \dfrac{r_2 h}{r_1 - r_2}$. Next, perform the following substitution and simplify the result:

> subs(x = r2*h/(r1 - r2), 1/3*Pi*r1^2*(h + x) - 1/3*Pi*r2^2*(x));

Answer: ______________________________

Example 6-7

Verify that the volume of a sphere is $V = \dfrac{4\pi r^3}{3}$.

Solution: Center a sphere of radius r at the origin. The equation for the points on the surface of the sphere is $x^2 + y^2 + z^2 = r^2$. We will use the method of disks, again parallel to the x-y plane. Since the disks cut the sphere, they are circles of radius $\sqrt{r^2 - z^2}$. Sketch the sphere intersecting the z-x plane. The equation of this circle is $x^2 + z^2 = r^2$. Therefore, the line parallel to the x axis and joining the z axis to the circle at height z has length $\sqrt{r^2 - z^2}$. The volume

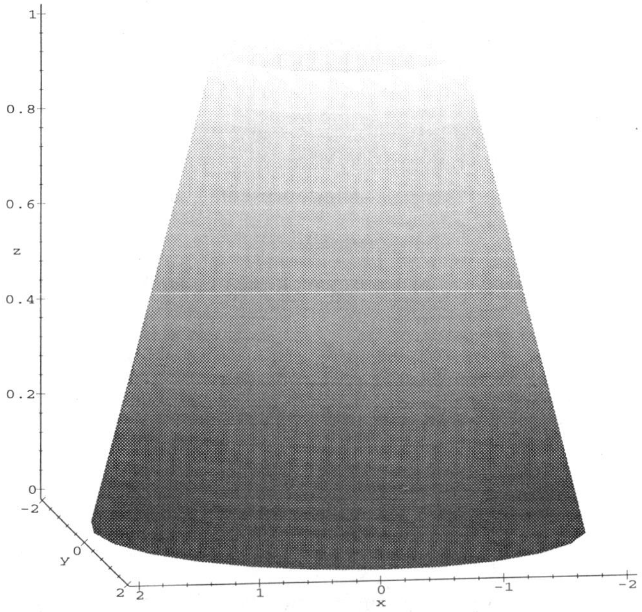

Figure 6.4 The Truncated Cone

of the disk is $dA = \pi(r^2 - z^2)dz$. These slices are taken from $z = -r$ to $z = +r$. Thus, we have the integral

> V = int(Pi*(r^2 - z^2), z = -r .. r);

$$V = \frac{4}{3}r^3\pi$$

The volume of a sphere by integration is $V = \frac{4}{3}r^3\pi$, the same result that you have seen many times before.

Your Turn. Find the volume of the "flowerpot" generated by rotating the curve $x = e^Z$ around the z axis (Figure 6.5).

Hint: The first step is to draw the curve to be rotated. The curve and a side view of the slice (a disk) are shown in Figure 6.6.

Answer: ____

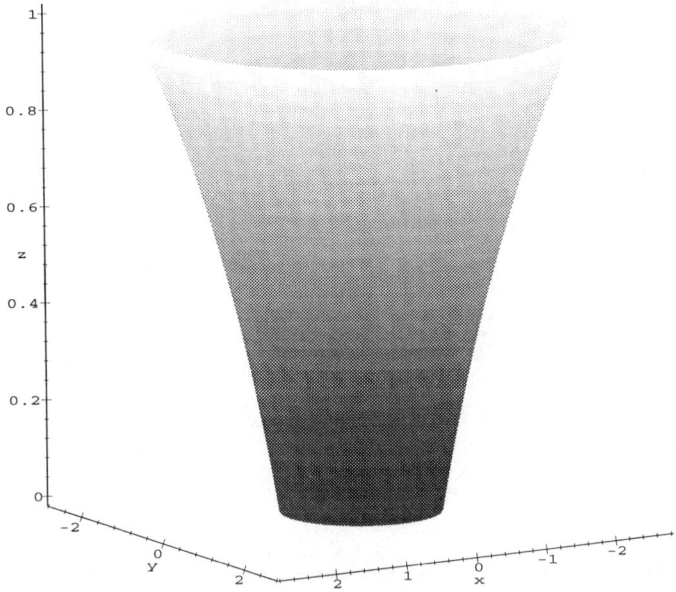

Figure 6.5 The Flowerpot Generated by e^{Z}

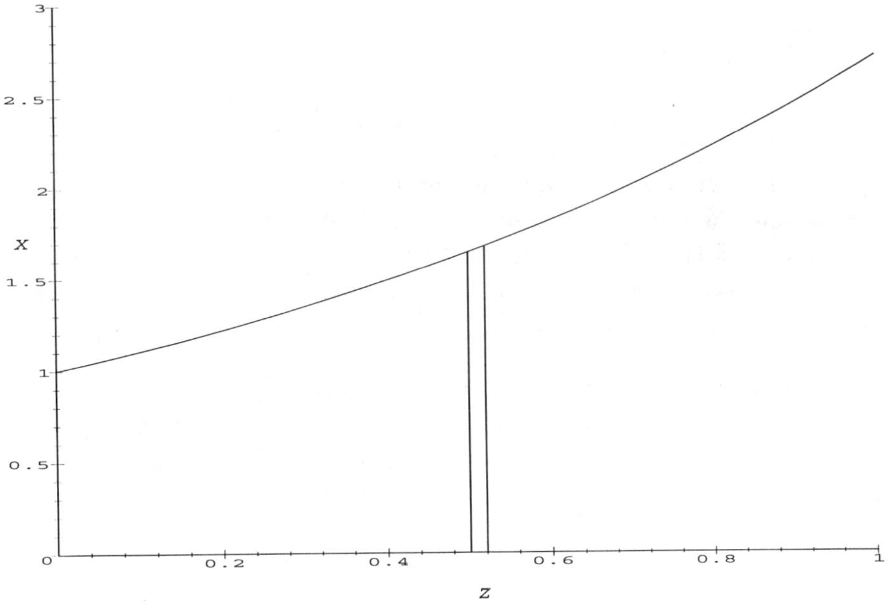

Figure 6.6 The Curve That Is to Be Rotated around the *z* Axis

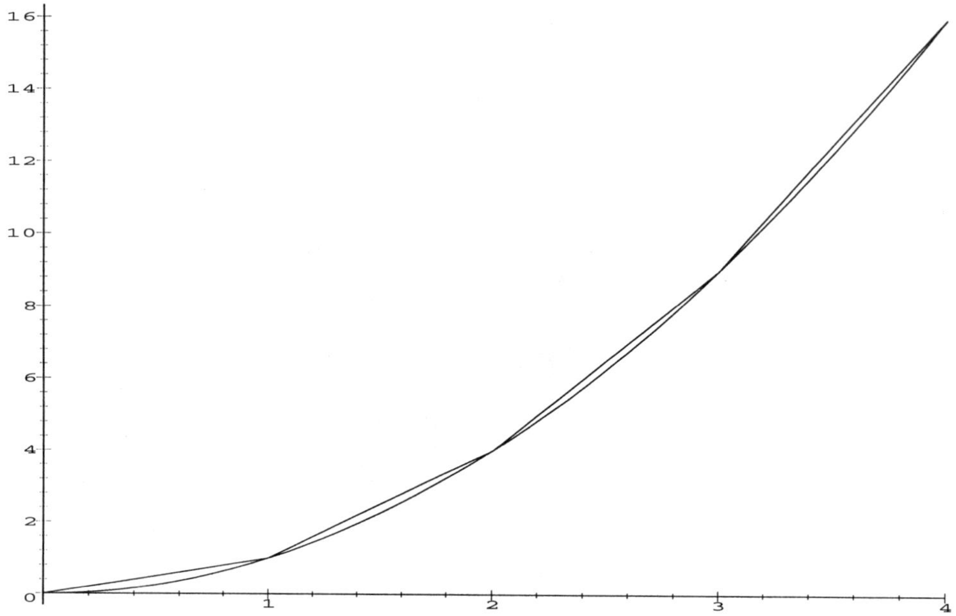

Figure 6.7 A Parabola Approximated by Four Straight Lines

Arc Length

The distance between any two points (x_1, y_1) and (x_2, y_2) on the plane is calculated using the Pythagorean formula $d = \sqrt{(x_2 - x_1)^2 + (y_2 - y_1)^2}$. This is the straight-line distance between the given points, but what if we require the length of a curved line joining the two points? One solution would be to fit a string so that it follows the curve, then straighten the string, and measure the distance. The mathematical equivalent of fitting a string to the curve is to approximate the curve with a sufficient number of straight lines. Figure 6.7 shows a curve and its approximation by straight lines.

It is difficult enough to see the difference between the curve and the straight-line approximation when only four straight lines are used. However, if ten straight lines were used, it would be even more difficult to detect any difference. This observation gives us the hint we need to develop a formula for arc length.

From the diagram in Figure 6-8, we derive the following relations

$$ds = \sqrt{dx^2 + dy^2}$$

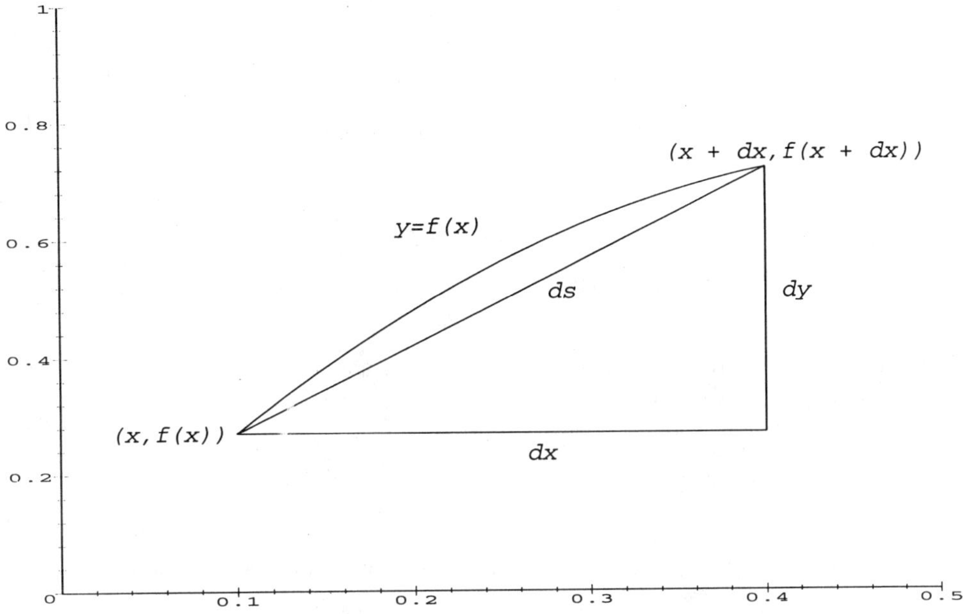

Figure 6.8 Diagram for the Derivation of the Formula for Arc Length

$$ds = \sqrt{dx^2 + dy^2} = dx\sqrt{1 + \frac{dy^2}{dx^2}}$$

$$\frac{ds}{dx} = \sqrt{1 + \left(\frac{dy}{dx}\right)^2} \tag{6-1}$$

The arc length must be the integral on the right-hand side of the last equation. Therefore, the formula for arc length is

$$s = \int\sqrt{1 + \left(\frac{dy}{dx}\right)^2}\,dx \tag{6-2}$$

The integrand is complicated by the fact that it contains a square root of 1 plus the square of the derivative of $f(x)$. Integrals as complicated as this may not have a solution in terms of elementary functions. Nevertheless, we will press on and see if we can find the lengths of some common figures. The first example in this section will test the formula we have derived by applying it to a straight line, whose length can be calculated easily without integration.

Example 6-8

Find the length of the "curve" $y = 3x + 4$, from $x = 1$ to $x = 3$, using the arc length formula.

Solution: The derivative of $y = 3x + 4$ is $dy/dx = 3$. Substituting this into the arc length formula yields

$$s = \int_1^3 \sqrt{1 + 3^2}\,dx$$

$$s = \sqrt{40}$$

The end-points of the given segment of the straight line are (1, 7) and (3, 13). Applying the Pythagorean formula for the length yields

$$s = \sqrt{(3-1)^2 + (13-7)^2}$$

$$s = \sqrt{40}$$

Since both methods give the same result, we can be confident that the formula we derived for arc length in terms of an integral is correct.

Your Turn. Find the length of the "curve" $x = 3y + 4$, from $y = 1$ to $y = 3$, using the arc length formula.

Hint: Using Figure 6.8, you can derive a formula involving dx/dy and integrate with respect to y. Alternatively, you can solve the equation for y and determine the x values of the end-points.

Answer: ____

Example 6-9

Show that the arc length formula correctly finds the length of an arbitrary straight line.

Solution: The equation for the general straight line can be given using the point-slope form

$$y = mx + b$$

The slope is $dy/dx = m$. Substituting this into the formula for arc length yields

$$s = \int_{x_1}^{x_2} \sqrt{1 + m^2}\,dx \tag{6-3}$$

You don't need Maple to do this integration! The square root term is a constant and can be factored outside the integral sign. The integral of dx is x. Why is this so? This integral is the same as $\int 1dx$, and the antiderivative of 1 is x. Therefore, the length of the line between end-points given by x_1 and x_2 is

$$s = \sqrt{1 + m^2}(x_2 - x_1) \tag{6-4}$$

Applying the Pythagorean formula, we have

$$s = \sqrt{(x_2 - x_1)^2 + m^2(x_2 - x_1)^2} \tag{6-5}$$

Factoring the common term out of the square root sign gives us

$$s = (x_2 - x_1)\sqrt{1 + m^2} \tag{6-6}$$

Thus, the calculation using the integral formula produces the same result as the Pythagorean formula.

Your Turn. A straight line is defined by the equation $y = 3x - 3$. Find the length of the line segment between $x = -2$ and $x = 3$.

Answer: ______________________________

Example 6-10

Find the length of $y = x + \sin(x)/100$ from $x = 0$ to $x = 2\pi$. Compare this length to the straight-line distance between the end-points of the curve (Figure 6.9).

Solution: Sketch the curve to check that it is close to a straight line.

```
> plot(x + sin(x)/100, x = 0 .. 2*Pi);
```

The curve looks almost straight. We can easily calculate its approximate length by assuming that it is the straight line given by $y = x$. The length of the line segment is just $2\sqrt{2}\pi$, which is approximately 8.8858.

We will show a step-by-step approach to the solution using the arc length formula.

1. Find dy/dx, which is needed for the arc length formula.
2. Form the integral for arc length and attempt to evaluate it using Maple's *int* command.
3. If Maple can't do the integral, use *evalf(Int...;* instead. This command computes the numerical approximation to the integral.
4. Compare the answer to an estimate of the length that you obtain by other means.

Applying these steps to the current problem, we obtain

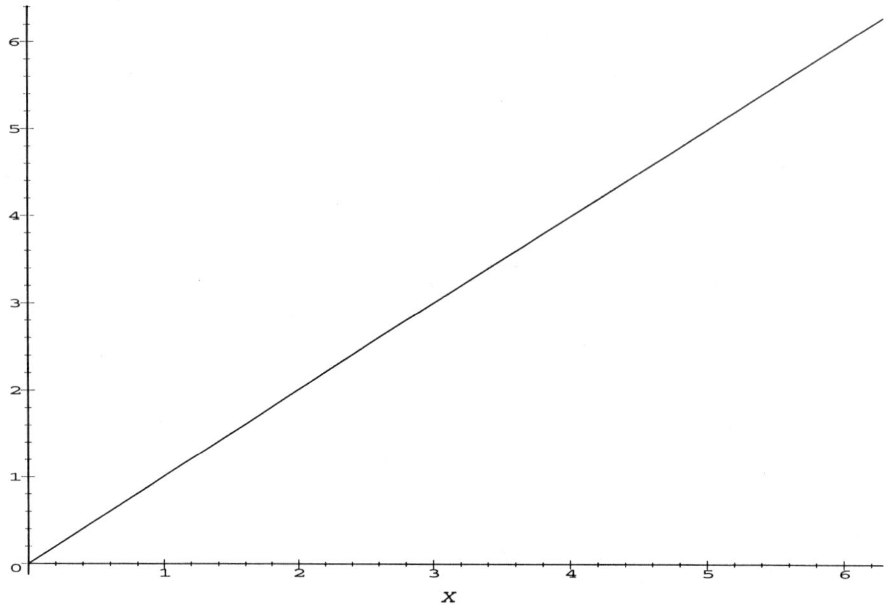

Figure 6.9 The Graph Showing That $x + \sin(x)/100$ Looks Like a Straight Line

> **yp := diff(x + sin(x)/100, x);**

$$yp := 1 + \frac{1}{100}\cos(x)$$

> **s = int(sqrt(1 + yp^2), x = 0 .. 2*Pi);** (Be prepared to wait!)

Maple takes a long time to find an answer to this integral. The answer contains large integers, strange functions (*EllipticPi, EllipticF, EllipticE*) and the imaginary quantity *I*. For practical purposes, the answer is useless. It is equivalent to getting no answer at all. Since our objective was to get a numerical answer anyway, we issue the command for numerical integration:

> **evalf(Int(sqrt(1 + yp^2), x = 0 .. 2*Pi));**

$$8.885821413$$

This takes Maple only a fraction of a second to compute, and it checks very well with our initial estimate.

Your Turn. Compute the length of the curve given by the equation $y = x + sin(100x)/100$. First make an estimate of the length. What difference is there in the curve if it oscillates 100 times faster than in Example 6-10? What is your estimate for the length of the curve $y = x + \sin(kx)/100$ as k becomes larger?

Answer: ____

The ratio of the circumference of a circle to its diameter is a constant, no matter what size circle is chosen. Let us see how this constant can be computed using the formula for arc length. We would like to verify that $C = \pi D$, where C is the circumference of a circle of diameter D. Since the radius of a circle is half its diameter, the formula is equivalent to $C = 2\pi r$. Any point on the circumference of a circle of radius r, centered at the origin, obeys the relation $x^2 + y^2 = r^2$. Therefore, the equation for the part of the circumference in the first quadrant is $y = \sqrt{r^2 - x^2}$. Since the x-y axes divide the circle in four equal portions, the length of this curve is one-quarter the whole circumference. We calculate the length of this portion of the circle in exactly the same way as in the previous example.

> y := sqrt(r^2 - x^2);

$$y := \sqrt{r^2 - x^2}$$

> yp := diff(y, x);

$$yp := -\frac{x}{\sqrt{r^2 - x^2}}$$

> s := int(sqrt(1 + yp^2), x = 0 .. r);

$$s := \frac{1}{2} \frac{\sqrt{r}\,\sqrt{\frac{1}{r}}\left(\ln\left(-\frac{1}{r^2}\right) + 2\ln(r)\right)}{\sqrt{-\frac{1}{r^2}}}$$

When Maple gives us a complicated-looking result, we should suspect that some simplification is possible. The expression for s contains square roots, so we should realize that Maple is reluctant to take the square root until it knows that r is a positive real number. We could inform Maple of this fact, or we could try this version of the *simplify* command:

> q := simplify(s, symbolic);

$$q := \frac{1}{2} r\,\pi$$

The *simplify* command has done its job quite nicely. Since q is one-quarter the circumference, the full circumference is

> C = 4*q;

$$C = 2 r \pi$$

We have derived the length of the circumference of a circle using the formula for arc length. Buoyed by this success, let us try to perform the same feat for the circumference of an ellipse. The ellipse is described by the relation $\frac{x^2}{a^2} + \frac{y^2}{b^2} = 1$. The ellipse cuts the x axis at a, and it cuts the y axis at b. The ellipse is symmetric about both the x and the y axes, so once again, the part of the curve in the first quadrant is one-quarter the whole curve. Solving for y, the expression for the curve in the first quadrant is

> y := b*sqrt(1 - x^2/a^2);

$$y := b\sqrt{1 - \frac{x^2}{a^2}}$$

Following the steps for the calculation of the arc length once more, we find

> yp := diff(y, x);

$$yp := -\frac{b x}{\sqrt{1 - \frac{x^2}{a^2}}\, a^2}$$

> s := int(sqrt(1 + yp^2), x = 0 .. a);

$$s := \int_0^a \sqrt{1 + \frac{b^2 x^2}{\left(1 - \frac{x^2}{a^2}\right) a^4}}\, dx$$

This time, Maple copies the integral in typeset format without further evaluation. Maple is telling us that it needs help to calculate this integral. In fact, there is no way of writing this integral in terms of the functions we have studied so far. A new function must be invented (called, not surprisingly, the *elliptic integral*) to represent the answer to this integration problem.

Example 6-11

Even though Maple is unable to find a general formula for the circumference of an ellipse, it can approximate the circumference if a and b are definite numbers. Let $a = 3$ and $b = 2$. Find the circumference accurate to ten decimal places.

Solution: The circumference of this ellipse is given by the command

> **evalf(subs(a = 3, b = 2, s));**

```
3.966359897
```

Your Turn. Find the circumference of the ellipse defined by $a = 99$, $b = 101$, and compare it to that of a circle of radius 100.

Answer: Area of ellipse = _________ Area of circle = _________

Paper and Pencil Exercises

PP6–1

In the section, "Motion under an Applied Force," the basic formulas for motion in the case of constant acceleration were given. These are $s = \frac{1}{2} at^2$ and $v = at$. This problem asks you to develop more general formulas for the case of constant acceleration.

(a) If an object is subject to constant acceleration, its velocity-time graph is a straight line. Use this fact to prove that the distance traveled is the product of the average velocity multiplied by the time. Let the initial velocity be v_0 and the final velocity be v_f. Sketch the velocity-time graph.

Answer: ______________________________

(b) Calculate the area under the velocity-time graph by using the formula for the area of a trapezoid or by subdividing the graph into a triangle and a rectangle and summing the two areas.

Answer: ______________________________

(c) Does the expression for the average velocity appear in the answer to (b)?

Answer: ______________________________

(d) State the general formula for the velocity as a function of time. Use v_0 for the initial velocity and v_f for the final velocity.

Answer: ______________________________

(e) Find the general formula for the distance as a function of time. To get the answer, you must integrate your answer to (d) and add a constant of integration. Use s_0 as the symbol for the initial displacement.

Answer: ______________________________

(f) Derive a formula that relates the distance s to the initial velocity, v_0, the final velocity v_f, and the acceleration, a. Start with the formula for s in terms of v_0, v_f, and t; that is, with the formula you derived in (b). Eliminate t using (d). Express your result in a form that contains no fractions.

Answer: ______________________________

PP6–2

An automobile's acceleration is not constant. Let the acceleration be given by $a(t) = \left\{\frac{2}{10} + \frac{3}{40}t^2 - \frac{1}{80}t^3\right\}g$, where g is the acceleration due to gravity ($g = 9.8$ m/s^2).

(a) Find the velocity at $t = 1$ s if $v(0) = 0$.

Answer: ______________________________

(b) Find the distance traveled between 0 and 4 s if s(0) = 0.

Answer: ______________________________

Maple Lab

Each of the examples in this lab relates to a particular topic chosen from various disciplines. You are urged to try all the exercises, even though the particular topic is not your major area of interest. They all relate to integration, but the particular integral is not given at the outset. The need for integration arises from an analysis of the situation, and setting up the problem is all-important.

The material in the lab in this chapter continues the exploration of how integration can be applied to practical problems. Your task is to follow along in the example, supply the missing steps, and compute the missing output.

ML6–1: Volumes by Integration

Refer to the section on volumes by integration at the beginning of the chapter. This time, you will be asked to calculate the volume of a stadium. The circumference of the stadium is

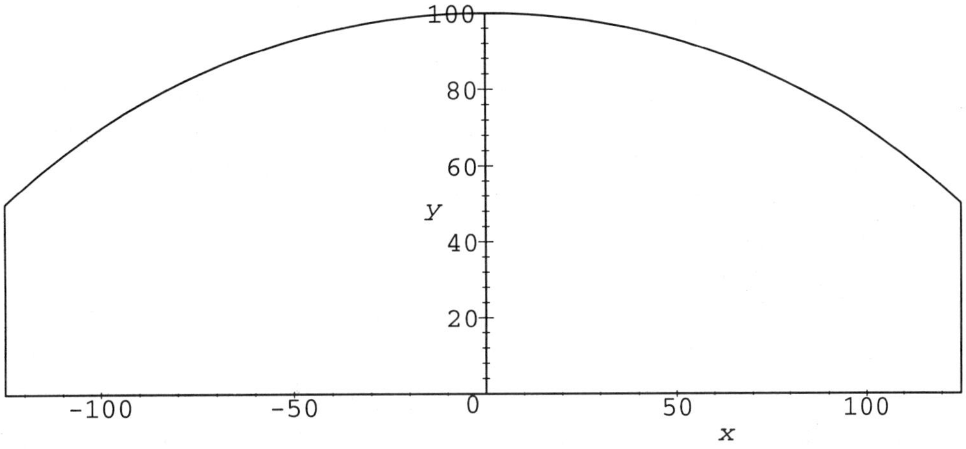

Figure 6.10 A Domed Stadium

a cylindrical wall 50 feet high. The dome of the stadium is in the form of a spherical cap, which is formed by revolving the curve

$$z = -\frac{325}{4} + \frac{1}{4}\sqrt{525625 - 16x^2}$$

around the z axis. The shape of the stadium is shown in Figure 6.10.

(a) Derive the equation for the dome, which is in the form of a spherical cap passing through the points (0, 100) and (125, 50) and having a radius of 725 ft. Use Maple! Write the *solve* command here.

Answer: ______________________________

(b) Find the total volume of the stadium by the cylindrical shell method. ______________

Answer: Maple command: ______________; Volume =______________

(c) Find the total volume of the stadium by the disk method. Check that you get the same answer as in (b). To apply the method of disks, you must separate the calculation into two parts. The stadium is viewed as a cylinder of base radius 125 feet and height 50 feet, plus a spherical cap. Apply the formula for the volume of a cylinder to the first part and find x as a function of the height z. (Really, you need x^2, since the volume of the disk is $2\pi x^2$.

Answer: ______________________________

ML6–2: Work and Energy

How much work does it take to ionize a hydrogen atom? To ionize the atom, its electron must be removed from its position near the nucleus and moved far away from it. To simplify the calculation, we will move the electron infinitely far away from the nucleus. Assume that Coulomb's law holds for all distances. Coulomb's law states that the attractive force on the electron is

$$F = -\frac{1}{4\pi\varepsilon_0}\frac{q_1 q_2}{r^2}$$

where $q_1 = q_2 = 1.602 \times 10^{-19}$ coulombs (C), $\varepsilon_0 = 8.854 \times 10^{-12}$ farads per meter (F/m), and r is the electron-nucleus separation distance in meters (m). The negative sign indicates that the force, F, in newtons (N), is one of attraction between the two particles.

(a) Calculate the work done in moving the electron from its lowest orbit, where it is 5×10^{-11} m away from the nucleus, to a point infinitely far away.

Answer: ____

(b) Convert the answer to electron volts by dividing the answer in (a) by 1.602×10^{-19}. (This unit of energy is more appropriate for the microscopic world.)

Answer: ____

ML6–3: Center of Mass, Centroids, and Moments of Inertia

The notion of center of mass (CM) arises from the science of mechanics. It is effectively the balance point of an object. If an object is suspended at its CM, it will not tend to rotate about this point. Consider the case of a light ruler to which three weights are attached. The weights are 2 kg, 5 kg, and 3 kg, and are attached at 15 cm, 25 cm, and 35 cm, respectively (Figure 6.11).

If a string is attached to the ruler at the CM, it will hang suspended without turning. If the string is attached at any other point, the ruler will topple, because one side will be heavier than the other. To find the CM, choose any point as a possible place to attach the string: we will choose the left-hand side of the ruler, at $x = 0$. The turning effect of a mass is defined as the product of the mass and the distance from the chosen point. A theorem in mechanics states that the sum of the turning effects of all three masses is equal to the product of the CM and total mass. Thus, the CM is found using the command

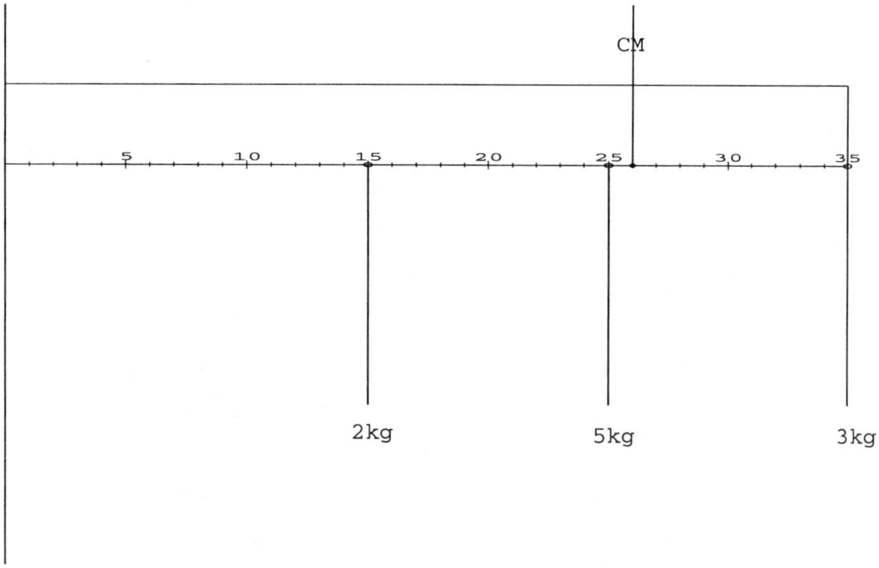

Figure 6.11 Weights Suspended from a Ruler

> solve(15*2+25*5+35*3 = x*10, x);

$$26$$

The center of mass is at $x = 26$ cm. This formula can be generalized to include two-dimensional shapes. The principle is the same, add the turning effects and equate it to the product of the CM and the total mass. Here is the formula for the x coordinate of the center of mass:

> solve(Int(x*y(x), x) = CM*Int(y(x),x), {CM});

$$\left\{ CM = \frac{\int x\,y(x)\,dx}{\int y(x)\,dx} \right\}$$

Note that we had to use $y(x)$ rather than just y in the Maple command line. If we had used y, Maple would treat it as a constant, which would be factored outside the integral sign in both the numerator and in the denominator. Try using y and observe the result. It is certainly not the correct formula for the center of mass.

(a) Find the x coordinate of the center of mass (cm) of the 3, 4, 5 triangle (Figure 6.12).

(i) Write the equation for the hypotenuse and state the domain of x.

Answer: ______________________________

(ii) Using the equation in (i), substitute for y in the integral formula for the CM.

Answer: ______________________________

(iii) Perform the integration to find the x coordinate of the CM.

Answer: ______________________________

(iv) Write the equation for the median line joining the origin to the side of length 4. Specify the domain for x.

Answer: ______________________________

(v) Write the equation for the median line joining the top of the triangle at (3, 4) to its base, which lies on the x axis. Use either the point-slope or the two-point formula for the straight line and state the domain for x.

Answer: ______________________________

(vi) Solve for the intersection of the medians. Is the x coordinate the same as the x coordinate of the CM?

Answer: ______________________________

(b) Find the x coordinate of the center of mass of the triangle defined by the coordinates (0,0), (10, 0), and (18,8) (Figure 6.13).

(i) Write the equation for the longest side and state the domain of x.

Answer: ______________________________

(ii) Using the equation in (i), substitute for y in the integral formula for the CM.

Answer: ______________________________

(iii) Perform the integration to find the x coordinate of the CM.

Answer: ______________________________

(iv) Write the equation for the median line joining the origin to the side of length 4. Specify the domain for x.

Answer: ______________________________

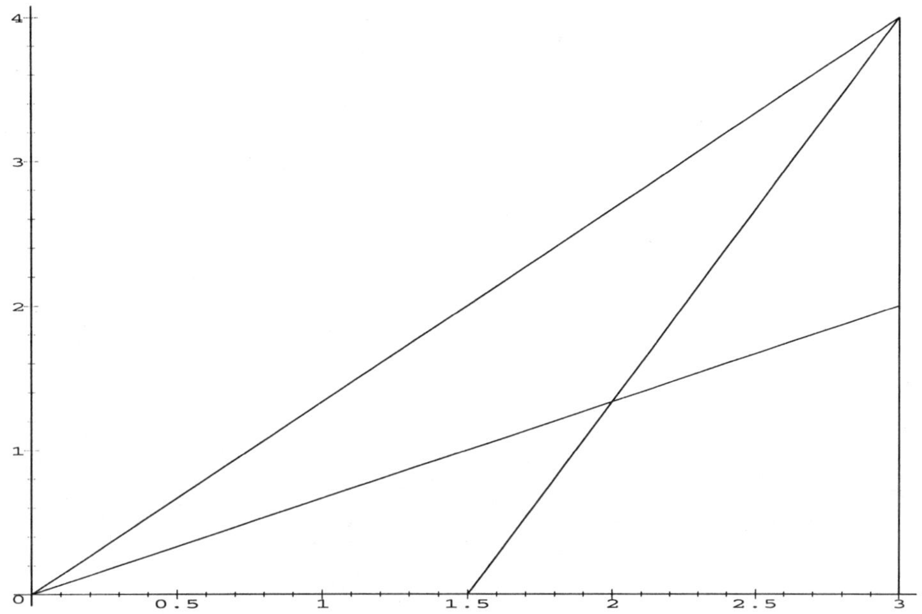

Figure 6.12 The 3, 4, 5 Triangle and Two of Its Medians

(v) Write the equation for the median line joining the top of the triangle at (3, 4) to its base, which lies on the x axis. Use either the point-slope or the two-point formula for the straight line and state the domain for x.

Answer:____

(vi) Solve for the intersection of the medians. Is the x coordinate the same as the x coordinate of the CM?

Answer:____

(c) Find the CM of the figure bounded by the curves $y = x^2$ and $y = \sqrt{x}$ between $x = 0$ and $x = 1$. For this problem, find both the x and the y coordinate values. Note that CM_y is given by the formula

$$CM_y = \frac{\int yx(y)dy}{\int x(y)dy}$$

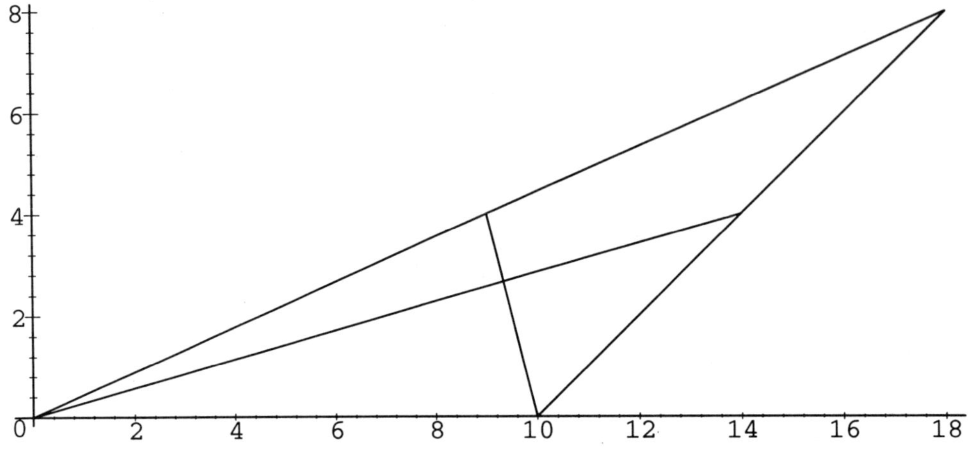

Figure 6.13 The Triangle (0, 0), (10, 0), (18, 8) and Two of Its Medians

However, you should examine the symmetry of the problem to see if you can avoid doing the integrations to find CM_y.

Answer: ____

ML6–4: The Root Mean Square of a Trig Function

Electrical engineers are taught to do calculations in alternating current (AC) theory by dividing the maximum value of the AC current by the square root of 2. This problem attempts to justify that practice.

Definitions:

1. An AC current is given by $i(t) = i_m \sin(\omega t)$, where $i(t)$ is the current at any time, t, and i_m is the maximum value of the current.
2. Ohm's law for resistors ($V = Ri$) holds even if the current through the resistor is AC.
3. The power dissipated in a resistor is $P = Ri^2$, whether direct current or alternating current is used.

A consequence of these definitions is that the *instantaneous power* dissipated by a resistor is $P(t) = R_i(t)^2$. The *energy* dissipated by a resistor is Pt if direct current is involved. In the case of alternating current, the instantaneous power must be integrated: $E = \int P(t)dt$, where the integral is taken over the time interval during which the energy is to be calculated.

(a) Write a formula for the energy dissipated during time dt in a resistor under AC conditions as a function of the instantaneous current.

Answer: ____

(b) Let $t = 2\pi/\omega$. Show that this time represents one full cycle of the AC current.

Answer: ____

(c) Integrate the formula you developed in (a) from $t = 0$ to $t = 2\pi/\omega$ to find the total energy dissipated in one cycle.

Answer: ____

(d) Your answer in (c) is the "average" energy dissipated over one cycle. Compare the answer to $P2\pi/\omega$, the energy dissipated in the same amount of time but by a DC current. What relationship emerges?

Answer: ____

Your work should lead you to the realization that you can use the DC formula for power in AC, so long as you choose the effective AC current (and voltage) to be the maximum value divided by the square root of 2.

ML6–5: Supply Curves and Demand Curves

In economics, the demand curve, $D(x)$, gives the demand price per unit that a consumer is willing to pay for x units of a commodity. We all know that we insist on paying less per unit when we buy in quantity, so $D(x)$ is a decreasing function of x. The supply curve $S(x)$ gives the price per unit that the supplier is willing to sell S units.

Economic equilibrium is reached at the intersection point (x_E, P_E), of these two curves, where x_E is the number of units and P_E is the price of each unit in dollars.

Consumer's surplus is defined as $\int_0^{x_E} D(x)dx - x_E P_E$, and *producer's surplus* is defined as

$x_E P_E - \int_0^{x_E} S(x)dx$.

Given $D(x) = (x - 64)^2$ and $S(x) = 10(1 + \ln(x + 1))$,

(a) Solve for (x_E, P_E).

Answer: ______________________________

(b) Calculate the consumer's surplus.

Answer: ______________________________

(c) Calculate the producer's surplus.

Answer: ______________________________

ML6–6: Exponential Growth

A statistician claims that the number of scientists born in year t, from antiquity to the present day, is given by the formula $S(t) = S_0 e^{kt}$.

(a) Assume, for the sake of argument, that there was 1 scientist born every 10 years at the beginning of our era, that is, in year 0 A.D. Further assume that there were 1,000 scientists born in the year 1900. Determine the constants S_0 and k from these data.

Answer: ______________________________

(b) Assume that all scientists live exactly 70 years. Use integration to approximate all the scientists alive in the world in the year 2000.

Answer: ______________________________

(c) Compare this number to all the scientists who have ever lived. (Use integration again!) What percentage of all scientists who ever lived are alive in the year 2000?

Answer: ______________________________

ML6–7: Depletion of Natural Resources

A more realistic application of the same exponential growth formula of ML6–6 is that of human use of a natural resource, such as the stock of a certain type of fish in the ocean. Due to an expanding world population, the amount of a natural resource, such as fish, being caught during time dt can be estimated by the expression $F_0 e^{kt} dt$, where F_0 and k must be determined. Assume t is in years.

(a) Write an integral formula for the total amount of fish taken during the time interval 0 .. T.

(b) Assume that there were one million of this type of fish caught in 1960, so $F_0 = 10^6$ and the rate of increase in fishing activity is roughly 20% per year, so $k = 0.2$. There are one billion (10^9) of this type of fish in the ocean. Estimate how long will it take to deplete this resource,

(i) Assuming no new fish are born after 1960. *Answer*: ____________________

(ii) Assuming the population breeds very quickly, so that there are always 10^9 fish around until a catastrophe happens, such as all the fish being caught in the same year!

Answer: ____

ML6–8: Annuities: How Much Is Your House Worth?

How much is a house worth? Economists tell us that the capital value of a rental property over T years is approximated by $A_T = \int_0^T R^{-kt} dt$, where R is the annual rent and k is the current continuous interest rate. The worth of a house can be estimated by determining how much it could be rented for, which we are calling R in this problem, and by allowing T to become very large (like infinity). Derive a simple relationship for the capital value by evaluating the integral for this case. Assume that both the rent, R, and k remain constant over the years.

Answer: ____

ML6–9: Hydrostatic Pressure

The pressure, p, in pascals, in a fluid at depth h is given by $p = dgh$, where d is the density of the liquid in kg/m^3 and g is 9.8 m/s^2.

(a) An *incompressible* fluid, like water, has essentially the same density at any depth. Find the depth at which the water pressure equals atmospheric pressure at sea level, which is 14.7 lb/in^2. You must convert lb/in^2 to pascals to do part (b), so do this conversion now.

Answer: ____

(b) Find the total force on the sides of the trough in ML4–8 (Figure 4.13) if $h = 10$ m, $\theta = 45°$, and the trough is filled with water of density $d = 1000$ kg/m^3. Since pressure is defined as force per unit area, calculate the area dA of the side of a slab of water of height dh and depth h, and then use $p(h) = \frac{dF}{dA}$. Integrate to find the total force.

Answer: ____

ML6–10: Probability

In probability theory, the *standard deviation*, σ, of a continuous probability distribution $f(x)$ is defined as $\int_a^b (x-\mu)^2 f(x)dx$, where $\mu = \int_a^b xf(x)dx$ is the *mean* (also called *expectation value* and *average*) of the distribution. The meaning of $f(x)$ is that the probability of a positive outcome in the interval dx is $f(x)dx$. If the probability density $f(x)$ is a constant, than all outcomes are equally probable.

Let the probability of finding a car along a stretch of highway between mile markers a and b be $f(x) = \dfrac{1}{b-a}$, where $b = 20$ mi and $a = 10$ mi.

(a) Calculate the mean of the probability distribution.

Answer: ______________________________

(b) Calculate the standard distribution.

Answer: ______________________________

(c) Change the probability distribution to $f(x) = \dfrac{1}{\sqrt{2\pi}} e^{\frac{\left(x - \frac{b-a}{2}\right)^2}{2}}$ and recalculate the mean and standard deviation. What is the change in the mean and in the standard deviation from the first case?

Answer: ______________________________

ML6–11: Specific Heat

Most substances have a fairly constant specific heat around room temperature. Thus, the energy required to raise the temperature of a given amount of substance is $Q = C\Delta T$, where Q is the amount of energy required in joules, C is the specific heat for the given amount of substance ($C = mc$), and ΔT is the change in temperature. One of the difficulties of producing very low temperatures is the fact that the specific heat of any substance rapidly decreases as the temperature nears absolute zero, which causes any small (and unwanted) input of energy to make the temperature shoot up. In part (a) of this problem, you will calculate the energy to heat a material when the temperature is close to absolute zero.

(a) The specific heat of a given amount of potassium chloride at low temperature is given by the formula $C = 0.16T^3$ mJ/K, $0 < T \leq 4.5$, where K is the Kelvin scale. Zero on this scale is a temperature of absolute zero. The temperature of a warm room is about 300 K.

The unit mJ stands for millijoules, or 10^{-3} joules. How much energy does it take to raise the temperature of the sample of potassium chloride from 1 K to 3 K?

(b) Einstein suggested that a possible formula for the specific heat of a substance is

$$C(T) = \frac{200\left(\frac{10}{T}\right)^2 e^{\frac{10}{T}}}{\left(e^{\frac{10}{T}} - 1\right)^2},$$ with T in degrees kelvin.

(i) Plot $C(T)$ for $T = 0 .. 300$. Sketch the plot here.

(ii) Find the amount of energy required to raise the temperature of this substance from 4 K to 300 K.

Answer: ______________________________

ML6–12: The Area between Two Curves

Find the area of the "double arches" defined by the equations $y = 16 - x^2$ and $y = 14 - x^2$. The area in question is formed by the positive parts of these equations and the x axis (Figure 6.14).

Answer: ______________________________

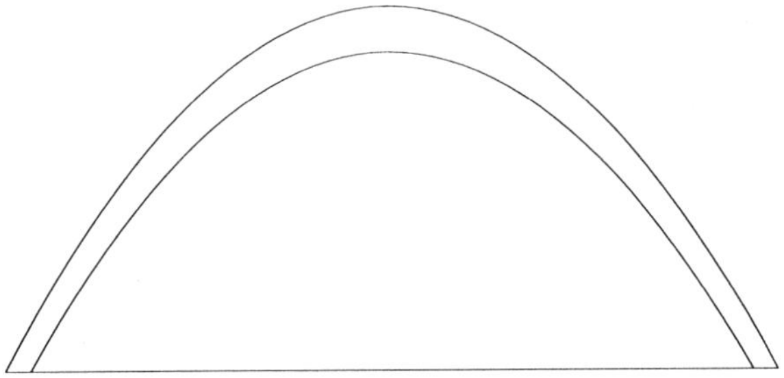

Figure 6.14 Diagram for ML 6–12

CHAPTER 7

Techniques of Integration Using Maple

Objectives for This Chapter

Learn the following manual methods for integration:

1. Change of variable
2. Integration by parts
3. Trigonometric substitution
4. Partial fractions
5. Numerical integration

Maple Commands Used in This Chapter

changevar(x = 10*tan(theta), c, theta);	Part of the student package. The command performs a change of variables in an integration.
expand(c2);	The by-now familiar command to distribute products over sums.
isolate(eq, Int(exp(x)*sin(x), x));	Part of the student package. Similar to the *solve* command.
op(1, c1);	Extract subexpressions from an expression.
simplify(c1, symbolic);	Use this form of the *simplify* command when dealing with roots.

solve(eq, Int(exp(x)*sin(x), x)); The *solve* command can solve for an inert integral form, effectively isolating it to the left-hand side of an equation.

The Need for Integration Techniques

If you completed the Maple lab in the last chapter, you have already reached the conclusion that the important thing is to be able to set up an integral and then let Maple solve it. You can let Maple take care of two things: finding the antiderivative, and doing the mechanical computations, such as substituting for the limits of integration. You are able to attempt a broader class of problems when you use a computer algebra system. The reason is that you do not have to concern yourself with mastering a sufficiently long list of formulas for antiderivatives before you can attempt to solve integral problems.

There are times, however, when Maple needs help. It is possible that an expression can be integrated, yet Maple cannot find the answer. If you know some of the techniques of integration, you may be able to help Maple along, thereby getting the answer. It is also useful to know some of the ways in which mathematicians have integrated expressions before the advent of computer algebra. (After all, someone had to program Maple's algorithms!)

This chapter displays the most common techniques for integrating expressions one or two levels of complication up from the basic formulas. These methods help you understand the mechanics of finding the antiderivative. Many of the examples require commands that are found in the student package. The student package is a collection of routines that help you visualize parts of the calculus, like the *middlebox* command you have already seen, and other commands to help you perform step-by-step solutions to integration problems. Issue the command

```
> with(student);
```

to make sure that all the routines used in this chapter are available in your Maple session.

Change of Variable

Consider the task of finding the antiderivative of $\int \frac{\ln(x)}{x}dx$. The integrand is composed of two elementary functions, $\ln(x)$, and $1/x$. The antiderivative of $\ln(x)$ is $x\ln(x) - x$, and the antiderivative of $1/x$ is $\ln(x)$, but what is the antiderivative of $\ln(x)/x$? How does knowing the elementary antiderivatives help us to solve a problem in integration when the two functions are multiplied together?

The key observation is that this integral can be transformed into an elementary problem by a change of variable. Let $u = \ln(x)$. Then, the differential of u is $du = 1/x\,dx$. Thus, the change of variable, $u = \ln(x)$, transforms $\int \frac{\ln(x)}{x}dx$ to $\int u\,du$. (The latter integral is an elementary one.) We apply the power law formula, since $u = u^1$, to obtain $\int u\,du = \frac{1}{2}\ln(x)^2 + C$. Finally, we use back-substitution to replace u with $\ln(x)$ and du with $1/x\ dx$ to obtain $\int \frac{\ln(x)}{x}dx = \frac{1}{2}\ln(x)^2 + C$. We have found the answer to the integration problem by a change of variable.

Example 7-1

Find $\int_1^e \frac{\ln(x)}{x}dx$.

Solution: This is a definite integral. In the case of definite integrals, we can avoid the back-substitution step so long as we remember to change the limits of integration when we make the substitution $\ln(x) = u$. Since $\ln(1) = 0$ and $\ln(e) = 1$, the change of variable produces $\int_1^e \frac{\ln(x)}{x}dx = \int_0^1 u\,du$. We know that the solution to the latter integral is $u^2/2$. Evaluating this expression at the upper limit yields ½, and at the lower limit, 0; therefore, the value of this definite integral is ½. This result is easily checked using Maple.

> int(ln(x)/x, x = 1 ..exp(1)) = int(u, u = 0 .. 1);

$$\frac{1}{2} = \frac{1}{2}$$

We have reached an important conclusion. In the case of definite integrals, if the problem can be solved by a change of variables, there is no need for back-substitution provided the limits of integration are recalculated in terms of the new variable.

Your Turn. Calculate $\int_0^{\frac{1}{2}\pi} \sin(x)\cos(x)dx$. Hint: Let $\sin(x) = u$ and calculate du.

Answer: ______________________________

The student package contains a command called *changevar* to help you check your work when you perform a change of variable. The command is a little bit picky about how the change of variable relation is written. We use it here to check our transformation in Example 7-1:

> Int(ln(x)/x, x = 1 .. exp(1)) = changevar(ln(x) = u, Int(ln(x)/x, x = 1 .. exp(1)));

$$\int_1^e \frac{\ln(x)}{x}\,dx = \int_0^1 u\;du$$

The command has done its job. The integral has been transformed into the elementary integral and has confirmed our paper and pencil work in Example 7-1. Note, however, the effect of writing the change-of-variable equation the other way around:

> Int(ln(x)/x, x = 1 .. exp(1)) = changevar(u = ln(x), Int(ln(x)/x, x = 1 .. exp(1)));

```
Error, (in changevar) usage: changevar(g(a) = f(b),F(..), b)
```

Writing the equation this way has confused Maple. You can clarify your meaning by telling Maple the name of the transformed variable, in this case u, by listing it as the third parameter in the command:

> Int(ln(x)/x, x = 1 .. exp(1)) = changevar(u = ln(x), Int(ln(x)/x, x = 1 .. exp(1)), u);

$$\int_1^e \frac{\ln(x)}{x}\,dx = \int_0^1 u\;du$$

Now Maple understands! When you are using the *changevar* command, remember what to do if you get an error message about usage: interchange the right- and left-hand sides of the change-of-variable equation or tell Maple the name of the new variable by using a third parameter in the command. The safest way is to always use the third parameter form of the command.

Example 7-2

It may not be obvious which change-of-variable relation will work. The *changevar* command makes the job of testing different equations much simpler. Use it to simplify $c := \int \frac{\arcsin(5x)}{\sqrt{1-25x^2}}\,dx$.

Solution: The choice of a change-of-variable is not obvious to anyone who has not done a similar type of problem. Since the arcsine function appears to be the most troublesome, we might try to eliminate it by the transformation $x = \sin(u)$:

> changevar(x = sin(u), c, u);

$$\int \frac{\arcsin(5\sin(u))\cos(u)}{\sqrt{1-25\sin(u)^2}}$$

Things look more complicated, but before you try something completely different, observe that the expression would be simplified if we had arcsin(sin(u)) instead of arcsin(5 sin(u)). Try the substitution, $5x = \sin(u)$:

> changevar(5*x = sin(u), c, u);

$$\int \frac{1}{5}\frac{\arcsin(\sin(u))\cos(u)}{\sqrt{1-\sin(u)^2}}du$$

Now we do have arcsin(sin(u)). Why didn't Maple simplify it? Arcsin(x) is not a function. It has infinitely many solutions for any given u. To instruct Maple to assume it is a function, use *simplify ... symbolic*:

> simplify(changevar(5*x = sin(u), c, u), symbolic);

$$\frac{1}{5}\int u\ du$$

We arrive at the same integral as Example 7-1, except for the factor 1/5. Its value is obviously $u^2/10 + C$.

Is there any other transformation that simplifies the integral? Try $u = \sqrt{1-25x^2}$. The result is

> ex1 := simplify(changevar(u = sqrt(1 - 25*x^2), c, u), symbolic);

$$ex1 := -\frac{1}{5}\int \frac{\arcsin(\sqrt{1-u^2})}{\sqrt{1-u^2}}du$$

Since the square root term in the denominator is the derivative of the square root term in the numerator, except for a constant term, we might try the further substitution $v = \sqrt{1-u^2}$.

> simplify(changevar(v = sqrt(1 - u^2), ex1, v), symbolic);

$$\frac{1}{5}\int \frac{\arcsin(v)}{\sqrt{1-v^2}}dv$$

Unfortunately, all we have succeeded in doing is to apply the transformation

> changevar(v = 5*x, c, v);

$$\int \frac{1}{5}\frac{\arcsin(v)}{\sqrt{1-v^2}}dv$$

We hope that this example has shown that there is a danger of going around in circles when looking for an appropriate change of variable. Nonetheless, it is only by such trial and error attempts that you gain experience in making substitutions that work.

Your Turn. Find a suitable change of variable for $c := \int_{\frac{1}{3}\pi}^{\frac{1}{2}\pi} \frac{\sin(\theta)}{\sqrt{1+\cos(\theta)}}d\theta$. Hint: Try $x^2 = \cos(\theta)$, and then look for another simple transformation.

Answer: ____

Integration by Parts

The formula for integration by parts may be arrived at by expanding the differential of the product of two functions, *u* and *v*.

```
> D(u*v);
```

$$D(u)v + u\ D(v)$$

The antiderivative of this differential must be *uv*, while the antiderivative of the expanded form is $\int v\,du + \int u\,dv$. Traditionally, the formula is written $\int u\,dv = uv - \int v\,du$.

Integration by parts is a powerful tool. Surprisingly, it can be used to advantage even when it is not obvious that there are two product functions.

When doing integration by parts, you must be able to integrate the part you choose as *v*. Make up a table and fill in all values (Table 7.1).

Making up the table is an important first step. It helps you keep track of the substitutions you need to make as you work out the formula.

Naturally, Maple has a command in the student package to assist you. It is called *intparts*. The following example shows you how to work out a simple integration by hand and then by getting assistance from Maple.

Example 7-3

Use the technique of integration by parts to find $c := \int x\sqrt{1-x}\,dx$.

Solution: First, fill Table 7.2

From the table, substitute in the formula for integration by parts

```
> u*v - Int(v, u) = x*(-2/3*(1-x)^(3/2)) - Int(-2/3*(1-x)^(3/2), x);
```

$$uv - \int v\ du = -\frac{2}{3}x(1 - x)^{3/2} - \int -\frac{2}{3}(1 - x)^{3/2}dx$$

Table 7.1

Integration of v

u =	*dv* =
du =	*v* =

Table 7.2

Table for Example 7-3

$u = x$	$dv = \sqrt{1-x} \cdot dv$
$du = dx$	$v = -\frac{2}{3}(1-x)^{3/2}$

Once you realize that $d(1-x) = -dx$, the integral on the right reduces to $\int v^n dv$. The result is

$$-\frac{2}{3}x(1-x)^{3/2} - \frac{4}{15}(1-x)^{5/2}$$

Now we will do the same problem with Maple's help. The *intparts* command requires two quantities, the integral and the part to be differentiated, *dv*.

> **intparts(c, x);**

$$-\frac{2}{3}x(1 - x)^{3/2} - \int -\frac{2}{3}(1 - x)^{3/2}dx$$

This is the identical result to our manual method. Maple has done the mechanical work of determining *u*, *du*, and *dv*, given *v*. It has performed the substitutions and it presents its results in terms of a fully evaluated term and a new integral. If integration by parts is to work, the transformed integral must be easier to find than the original integral.

Your Turn. Find, using integration by parts, $c := \int \frac{\ln(x)}{\sqrt{x}} dx$.

Answer: ____

Example 7-4

It may be necessary to apply integration by parts more than once to fully evaluate the integral. Here, we will use integration by parts to evaluate $c := \int e^x \sin(x)dx$.

Solution:

> **c1 := intparts(c, sin(x));**

$$c1 := e^x\sin(x) - \int \cos(x)e^x dx$$

We have succeeded in transforming an integral containing a sine term into an integral containing a cosine term. It does not appear that any progress has been made, but we go ahead and apply integration by parts one more time.

> c2 := intparts(Int(exp(x)*cos(x), x), cos(x));

$$c2 := \cos(x)\mathbf{e}^x - \int -\mathbf{e}^x\sin(x)dx$$

We see that what has happened is that after two applications of integration by parts, the original integral reappears. Equating these results yields

$$\int e^x\sin(x)dx = e^x\sin(x) - \cos(x)e^x + \int - e^x\sin(x)dx$$

We can construct this equation from our previously named quantities by using the *op* command to pick up the first term in *c*2:

> eq := c = op(1, c1) - expand(c2);

$$eq := \int \mathbf{e}^x\sin(x)dx = \mathbf{e}^x\sin(x) - \cos(x)\mathbf{e}^x - \int \mathbf{e}^x\sin(x)dx$$

Solving this equation for the integral, we obtain

> Int(exp(x)*sin(x), x) = solve(eq, Int(exp(x)*sin(x), x));

$$\int \mathbf{e}^x\sin(x)dx = \frac{1}{2}\mathbf{e}^x\sin(x) - \frac{1}{2}\cos(x)\mathbf{e}^x$$

Two points should be noted. First, we used the *expand* command to move the negative sign from within the integral sign. Until we do this, Maple will treat the integral in *c*2 as different from the original, and the *solve* command won't work. Also, note that we used the *solve* command to solve for a complex expression, namely the integral itself, instead of solving for a single variable, as we have done in most instances of the solve command.

Since we have loaded the student package, the *isolate* command is available, which provides an even easier way of solving for the integral:

> isolate(eq, Int(exp(x)*sin(x), x));

$$\int \mathbf{e}^x\sin(x)dx = \frac{1}{2}\mathbf{e}^x\sin(x) - \frac{1}{2}\cos(x)\mathbf{e}^x$$

Isolate outputs its result as an equation, by moving the quantity to be isolated to the left-hand side of the equation.

Your Turn. Find, using integration by parts, $c := \int \frac{\cos(x)}{e^x}dx$.

Answer: ______________________________

You may have wondered about the fact that we have ignored the constant of integration when we integrate dv to fill in that part of the table at the start of the integration-by-parts process. This issue is addressed in PP7–2.

Trigonometric Substitution

Trigonometric substitution is usually treated as a separate topic in textbooks on calculus. The only thing different about this type of substitution is that a change of variable is used involving a trigonometric function. The method can be used with certain cases of algebraic functions as well as with purely trigonometric integrals.

Example 7-5

Evaluate $c := \int \frac{1}{\sqrt{x^2 + 100}} dx$ using the method of trigonometric substitution.

Solution: Let $x = 10\tan(\theta)$. The integral becomes

> c1 := changevar(x = 10*tan(theta), c, theta);

$$c1 := \int \frac{10 + 10\tan(\theta)^2}{\sqrt{100\tan(\theta)^2 + 100}} d\theta$$

This integral looks quite complicated: can we simplify the radical?

> c2 := simplify(c1, symbolic);

$$c2 := \int \sqrt{1 + \tan(\theta)^2}\, d\theta$$

Your turn: Complete the problem, given that $\int \sec(\theta) d\theta = \ln(\sec(\theta)\tan(\theta))$

Partial Fractions

The integral $\int \frac{x^2 - 10x + 21}{x^3 - 7x + 6} dx$ looks impressive, but it can be rewritten as the sum of three fractions. The technique is called the *method of partial fractions*.

If you are given three fractions and asked for the sum, you form a common denominator and add. The method of partial fractions undoes this process, attempting to find the original three fractions. In so doing, the integral is reduced to three simpler problems.

The first step is to factor the integrand, in which Maple can be a big help:

> **e1 := (x^2-10*x+21)/(x^3-7*x+6);**

$$e1 := \frac{x^2 - 10x + 21}{x^3 - 7x + 6}$$

> **e2 := factor(e1);**

$$e2 := \frac{(x - 3)(x - 7)}{(x - 1)(x - 2)(x + 3)}$$

The fact that the numerator of $e2$ has been factored is not as important as factoring the denominator. Using the factors in the denominator, we invent three fractions.

> **e3 := A/(x - 1) + B/(x - 2) + C/(x + 3);**

$$e3 := \frac{A}{x - 1} + \frac{B}{x - 2} + \frac{C}{x + 3}$$

Since $e3$ is supposed to be equal to $e2$, we bring $e3$ under a common denominator once more:

> **e4 := normal(e3);**

$$e4 := \frac{Ax^2 + Ax - 6A + Bx^2 + 2Bx - 3B + Cx^2 - 3Cx + 2C}{(x - 1)\ (x - 2)\ (x + 3)}$$

The numerator of $e4$ should be the same as the numerator of $e2$. Therefore, it should be the same as the numerator of $e1$. We can collect the terms in the numerator of $e4$ as a preliminary to equating it to the numerator of $e1$:

> **e5 := collect(numer(e4), x);**

$$e5 := (A + B + C)\,x^2 + (-3C + 2B + A)\,x + 2C - 3B - 6A$$

If the two fractions are to be equivalent, expression $e5$ must be identical to the numerator of $e1$, so we form the equation

> **e6 := e5 = numer(e1);**

$$e6 := (A + B + C)\,x^2 + (-3C + 2B + A)\,x + 2C - 3B - 6A = x^2 - 10x + 21$$

This equation must be an identity, because it must be true for every possible choice of x. This being the case, we can form three separate equations by equating the coefficients of x^2, x, and the constant term:

> eq1 := A + B + C = 1; eq2 := -3*C + 2*B + A = -10; eq3 := 2*C - 3*B -6*A = 21;

$$eq1 := A + B + C = 1$$

$$eq2 := -3C + 2B\ A = -10$$

$$eq3 := 2C - 3B - 6A = 21$$

Check these equations carefully against equation $e6$ because we reentered the data and it is important to guard against transcription errors.

There are three equations and three unknowns. Their solution is

> s1 := solve({eq1, eq2, eq3}, {A, B, C});

$$s1 : = \{A = -3,\ C = 3,\ B = 1\}$$

Since we want to use these solutions, we ask Maple to assign the values to A, B, and C.

> assign(s1);

Now that values have been assigned to A, B, and C, expression $e3$ is:

> e3;

$$-\frac{3}{x-1} + \frac{1}{x-2} + \frac{3}{x+3}$$

Just to be sure, we can ask Maple to add the three fractions:

> normal(e3);

$$\frac{x^2 - 10x + 21}{(x-1)\ (x-2)\ (x+3)}$$

In this way we can verify that $e3$ is identical to $e1$.

The original integrand has been replaced by three integrands because the integral of a sum is equal to the sum of the integrals:

> e7 := Int(e3, x) = Int(op(1, e3), x) + Int(op(2, e3), x) + Int(op(3, e3), x);

$$e7 := \int -\frac{3}{x-1} + \frac{1}{x-2} + \frac{3}{x+3}dx = \int -\frac{3}{x-1}dx +$$

$$\int \frac{1}{x-2}dx + \int \frac{3}{x+3}dx$$

This time, we have let Maple perform the expansion rather than recopying terms.

The integrals in $e7$ are all of the form $\int\frac{1}{x}dx$ and are therefore of the elementary kind. The integral is

> e8 := Int(e3, x) = int(e3, x);

$$e8 := \int -\frac{3}{x-1} + \frac{1}{x-2} + \frac{3}{x+3}\,dx =$$
$$-3\ln(x-1) + \ln(x-2) + 3\ln(x+3)$$

We can check the result by differentiation:

> normal(diff(rhs(e8), x));

$$\frac{x^2 - 10x + 21}{(x-1)(x-2)(x+3)}$$

This is just $e1$ with the denominator factored, so the solution is verified.

We have included the method of partial fractions and shown the solution in a step-by-step fashion not because of the immediate usefulness of the technique, but because it presents such a good opportunity to use Maple for a variety of algebraic manipulations. We should point out that someone experienced enough in the technique can use Maple's conversion to partial fraction command

> convert(e1, parfrac, x);

$$-\frac{3}{x-1} + \frac{1}{x-2} + \frac{3}{x+3}$$

This completes the conversion in one step.

Example 7-6

Find the antiderivative of

> e10 := (4*x + 4)/((x^3 + 4*x));

$$e10 := \frac{4x+4}{x^3+4x}$$

Solution: The difficulty in this example is in trying to find acceptable factors of the denominator without entering into the realm of complex numbers. You are encouraged to attempt a paper and pencil solution before looking at the solution, which follows.

The solution to the stumbling block of having an irreducible quadratic in the denominator is to write the two fractions as $\frac{A}{x} + \frac{Bx+C}{x^2+4}$; that is, generalize the numerator, but always have it of a degree one less than the denominator. To save time and space, now that we are aware of the method, we can use *convert*:

> e11 := convert(e10, parfrac, x);

$$e11 := \frac{1}{x} - \frac{x-4}{x^2+4}$$

The second integrand is easily transformed into two additional integrands. The first is in the form $1/u$ (by the transformation $u = x^2$, and the second is of the form that surrenders to a trigonometric substitution).

The integral is

> Int(e10, x) = int(e11, x);

$$\int \frac{4x+4}{x^3+4x}dx = \ln(x) - \frac{1}{2}\ln(x^2+4) + 2\arctan\left(\frac{1}{2}x\right)$$

You can see how simple integrands turn into complicated integrals.

Your Turn. Find the antiderivative of $e1yt := \dfrac{3x+2}{x(x^2+2)^2}$ by the method of partial fractions.

Answer: ____

Note: This is a difficult problem. Can Maple do the integration without help?

Numerical Integration

You have seen in Chapter 5 how rectangles are used to approximate the area under a curve $f(x)$. This forms the basis for the simplest approach to finding a numerical approximation to the area under a curve. If we build the rectangles whose heights are determined by $f(x)$ at the midpoint of each interval, we are using the *midpoint rule* for numerical integration. Recall that the Maple command that applies the midpoint rule is > *value(middlesum(f(x), x = a .. b));*. This is the most direct approach to numerical integration.

A slightly more elaborate rule for numerical integration is based on forming trapezoids rather than rectangles. The interval $x = a .. b$ is divided into n equal regions as before, although this time a trapezoid is formed by joining the points where the vertical lines intersect $f(x)$ with a straight line. This means that $f(x)$ is approximated by n straight-line segments and in the process, trapezoids are formed.

Example 7-7

Derive the formula for the approximation to the area under a curve by the trapezoidal rule.

Solution: Divide the interval $x = a .. b$ into n equal regions. Let $h = b - \frac{a}{n}$ be the length of any of these subintervals, and construct the vertical lines up to the function $f(x)$. The area of a trapezoid is $\frac{1}{2}\left(h\left(y_i + y_{i+1}\right)\right)$, where y_i and y_{i+1} are the heights corresponding to $f(x_i)$ and $f(x_{i+1})$ (any two adjacent vertical lines that are formed by subdividing the interval $x = a .. b$).

Adding all the trapezoids, $\int_a^b f(x)dx \approx h\left(\frac{y_0+y_1}{2} + \frac{y_1+y_2}{2} + \frac{y_2+y_3}{2} + \ldots + \frac{y_{n-1}+y_n}{2}\right)$.

All but the first and the last y terms, y_0 and y_n, appear twice. Collecting these terms, the sum is

$$\int_a^b f(x)dx \approx h\left(\frac{y_0}{2} + y_1 + y_2 + y_3 + \ldots + y_{n-2} + y_{n-1} + \frac{y_n}{2}\right) = \frac{h}{2}\left(y_0 + 2\sum_{i=1}^{n-1} y_i + y_n\right). \quad (7\text{-}1)$$

Example 7-8

Approximate the integral $\int_0^1 e^{-x^2}dx$ using the trapezoid rule and subdividing the interval of integration into 100 parts.

Solution: Apply the trapezoid rule, Equation 7-1, with $h = 0.01$. We have to express the constants in the exponentials as decimal numbers by adding decimal points, or alternatively we can wrap an *evalf* around the entire command. Either way, this forces the exponentials to be evaluated as decimal numbers.

```
> evalf(0.01/2*(exp(-0^2) + 2*sum(exp(-(x/100)^2), x = 1 .. 99) + exp(-1^2)));

                              .7468180010
```

None of the methods of integration discussed so far in this book seem to help us solve $\int_0^1 e^{-x^2}dx$ analytically. Let us see if Maple knows how to integrate it.

> int(exp(-x^2), x = 0 .. 1);

$$\frac{1}{2}\mathrm{erf}(1)\sqrt{\pi}$$

This is a case where Maple reports a result, but the function is not one of the elementary functions. A reference is given if you look up the Help page for this function (*?erf*). We will not discuss this function further here, but we can ask Maple for the value of the integral by using *evalf*.

> evalf(int(exp(-x^2), x = 0 .. 1));

$$.7468241330$$

The numerical approximation to the integral found by using the trapezoid rule with $h = 100$ is accurate to five decimal places. You might want to check the accuracy of using *middlesum* with 100 rectangles on this same integral. All it takes is a little typing!

Your Turn. Use the trapezoid rule with $h = 100$ to approximate $\int_0^{\frac{\pi}{4}} \frac{\sin(x)}{x} dx$. What is the accuracy of the trapezoid rule in this case?

Answer: ______________________________

Simpson's Rule

Instead of approximating the graph of $f(x)$ with straight line-segments, we could use little bits of parabolas that are sequentially fitted to groupings of three points (because it takes three points to define a parabola). A representative grouping is the set of three points $f(x)$, $f(x+h)$, and $f(x+2h)$. Simpson's rule uses this grouping to fit one parabola to the curve $f(x)$, and $f(x+2h)$, $f(x+3h)$, and $f(x+4h)$ to fit another parabola to the next part of the curve $f(x)$. Simpson's rule results from finding a formula for the sum of the areas of all these parabola-capped strips. It is similar to the trapezoid rule and is only slightly more complicated.

Simpson's Rule:

$$\int_a^b f(x)dx \approx \frac{h}{3}(y_0 + 4y_1 + 2y_2 + 4y_3 + 2y_4 + \ldots + 2y_{n-2} + 4y_{n-1} + y_n) \tag{7-2}$$

The y values have the same meaning as in the case of the trapezoid formula. In words, the formula tells you to divide the interval into an even number of intervals; add the end values, $f(a)$ and $f(b)$, to four times the function evaluated at the odd-numbered y indices and twice the function evaluated at the even-numbered y indices; and then multiply the whole result by $h/3$. You see why formulas are so much more useful than words in most cases! It takes many more words to describe the same idea, but here, the key idea is to take four times the odd-numbered functional values and two times the even-numbered functional values. Amazingly, weighting the data points in this way produces a more accurate result.

Example 7-9

Approximate the integral $\int_0^1 e^{-x^2}dx$ using Simpson's rule and subdividing the interval of integration into 100 parts.

Solution: This is the same as the previous example except that this time, Simpson's rule is to be used.

> 0.01/3*(exp(-0^2) + 4*sum(exp(-((2*x + 1)/100.)^2), x = 0 .. 49) + 2*sum (exp(-((2*x)/100.)^2), x = 1 .. 49) + exp(-1.^2));

```
.7468241324
```

The result is accurate to over eight decimal places, almost double that of the trapezoid rule, yet requires only a few more calculations.

Your Turn. Use Simpson's rule with $h = 100$ to approximate $\int_0^{\frac{\pi}{4}} \frac{\sin(x)}{x}dx$. What is the accuracy of the numerical integration in this case?

Answer: ____

Maple uses a combination of analysis and numerical techniques when you ask it to provide a numerical answer to an integral. You should do the same. If you need to evaluate an integral, first ask whether the integral can be expressed symbolically. The symbolic form gives you good insight into the character of the solution in general. If such a symbolic answer is not forthcoming, you should plot the integrand, again to get as much insight into the nature of the problem as you can. Then you can apply numerical integration with confidence.

Paper and Pencil Exercises

PP7–1

Since this chapter is on using Maple to do the integrations, there isn't much need for paper and pencil work here, so these exercises are short. To give you an idea of the amount of numerical work that has to be done, carry out Simpson's rule by hand on the integral

Table 7.3

Integration by Parts with No Constant of Integration

$u = \arctan(x)$	$dv = x$
$du = \dfrac{dx}{1+x^2}$	$v =$

$\int_0^4 4x - x^2 + 3$ using just 5 points ($h = 1$). Compare your numerical answer with the value you calculate by doing the integration symbolically and evaluating the antiderivative at $x = 0$ and $x = 4$.

Answer: ______

PP7–2

When applying the formula for integration by parts, the constant of integration is usually ignored when finding the antiderivative of dv. Try integrating $c := \int x \arctan(x)dx$ by parts, with $u = \arctan(x)$ and $dv = x$. Fill in Table 7.3 and see how far you can get with the solution.

Now try integration by parts using a constant of integration. Use Table 7.4

Do you see that you can factor the expression for v and thereby simplify the problem? What is the integral?

Answer: ______

Table 7.4

Integration by Parts with a Constant of Integration

$u = \arctan(x)$	$dv = x$
$du = \dfrac{dx}{1+x^2}$	$v = \dfrac{1}{2}x^2 + \dfrac{1}{2}$

PP7–3

Reduce the integral to a simpler form by the method indicated. You need not perform the integration step by hand.

(a) $\int \frac{3x}{x^2 + 1} dx$, substitution: $u = x^2 + 1$ *Answer*: ____________

(b) $\int e^x \sqrt{4 + e^x}\, dx$, substitution: $u = 2 + e^x$ *Answer*: ____________

(c) $\int \ln(x) dx$, integration by parts *Answer*: ____________

(d) $\int x \arctan(x) dx$, integration by parts *Answer*: ____________

(e) $\int \tan(x)^2 dx$, trigonometric substitution: *Answer*: ____________

(f) $\int \frac{1}{\sqrt{1 - x^2}} dx$, trigonometric substitution: $x = sin(\theta)$

Answer: ____

Maple Lab

ML7–1

Use *intparts* to integrate $c := \int x \arctan(x) dx$. You may have to solve an equation containing the integral. Provide a full solution and display all the steps here:

Answer: ____

____

____

____

____

____

ML7–2: The Trapezoid Rule

(a) Derive the formula of a trapezoid whose base, h, lies on the x axis and whose heights are y_1 and y_2.

Answer: ____

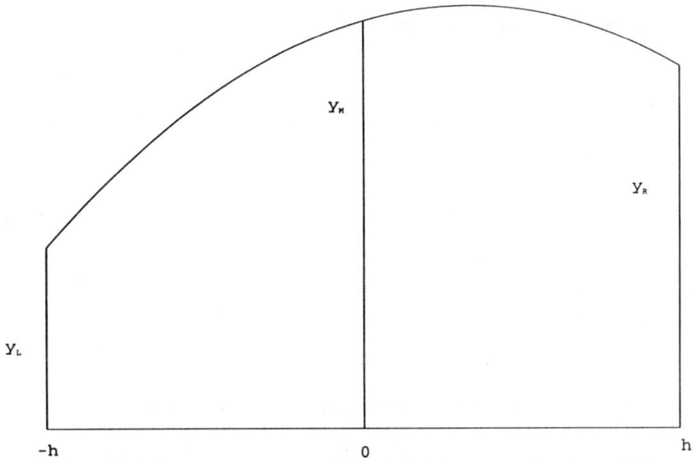

Figure 7.1 Fitting a Quadratic to the Tops of the Three Heights y_L, y_M, y_R

(b) Use the trapezoid rule with 100 intervals to integrate ln(ln(x)), x = 10 .. 20.

Answer: ______________________________

(c) Use Maple to perform the same integration. What is the percent error using (b)?

Answer: ______________________________

ML7–3: Simpson's Rule

(a) Fit a quadratic $Ax^2 + Bx + C$ to the three points $(-h, y_L)$, $(0, y_M)$, and (h, y_R), where y_L, y_M, and y_R are the heights of the function $f(x)$ at three equally spaced points (Figure 7.1). You determine the coefficients A, B, and C by substituting the three known points into the quadratic form and solving the resulting equations for A, B, and C. Summarize your method of solution here:

Answer: ______________________________

____

____

____

____

(b) Find the area under the curve defined in (a).

Answer: ______________________________

(c) Join a number of areas like the one in (b) together to derive the formula for Simpson's rule.

Answer: ______________________________

(d) Construct the Maple command to approximate $\int_{10}^{20} \ln(\ln(x))dx$.

Answer: ______________________________

(e) Evaluate the command in (d) to provide the numerical result.

Answer: ______________________________

(f) Compare the accuracy of Simpson's rule to the trapezoid rule for this integral.

Answer: ______________________________

CHAPTER 8

Sequences and Series

Objectives for This Chapter

1. Define sequences and series

2. Derive the formulas for Maclaurin series and Taylor series

3. Construct series approximations from the definitions of Maclaurin series and Taylor series

4. Use Maple commands to construct series approximations to functions

Maple Commands Used in This Chapter

series(arctan(x), x, 10);	This command and the one following are the two central commands of this chapter. The *series* command expands a function about $x = 0$.
taylor(cos(x), x = -Pi/2, 10);	The *taylor* command expands the function about $x = a$ other than 0.
(D@@n)(p)(0)	The *n*th derivative of the function *p,* evaluated at 0.
[S];	List notation. The object [*S*] is a *list*.
about(x);	Find out information about the variable *x*.
assume(x, real); additionally(x > 1);	The *assume* command gives Maple information about a variable, the *additionally* command adds extra information.

`assume(x, RealRange(0, Open(1)));`	Another way of giving Maple information about a variable. Specifies the domain $0 \le x < 1$
`convert([S], list);`	Convert object *S* to a list.
`convert([S], set);`	Convert object *S* to a set.
`eval(subs(x = 0 , [seq( diff(sin(x), x$n), n = 1 .. 4)]));`	An example of previously defined commands all used together.
`Int(1/x^2, x = 1 .. infinity)`	The inert form of the integral command, where the range of integration is infinite.
`limit(n!/n^n, n = infinity);`	A limit at infinity.
`map(convert, L2, list);`	Apply the *convert* procedure to each operand of *L2*, turning each operand into a list.
`plot({pts}, style = point);`	The style of the *plot* command to produce points rather than a smooth line.
`S2[10];`	Pick out the 10th element of a list.
`seq(n^2, n = 1 .. 20);`	Form a *sequence* of 20 squares.
`Sum(1/2^n, n = 1.. infinity) = sum(1/2^n, n = 1.. infinity);`	The inert form of the *sum* command, where the number of terms is infinite.
`unapply(t813a, x);`	Create a function from an expression.
`value(SP);`	Evaluate *SP*, where *SP* is an *Int*, a *Diff*, or some other inert form of a command.

Sequences and Series

The idea of a series, a sum of terms, is important because it is possible to approximate a function using polynomial. Let the function $f(x)$ be approximated by the polynomial

$$f(x) = a_0 + a_1 x + a_2 x^2 + a_3 x^3 \tag{8-1}$$

If we add more terms to the polynomial, choosing the coefficients with care, the approximation gets better and better. This leads to the idea that the correspondence between the function $f(x)$ and the polynomial $pn(x)$ might be exact if we "added" an infinite number of terms.

We have not yet defined what we mean by the statement "adding an infinite number of terms," since a step-by-step addition process is clearly impossible. To make sense of the idea, we will need to start with a simpler concept, that of an infinite sequence. Taking one more step backward, we will define what is meant by a finite sequence, and from there move to infinite sequences and then on to infinite sums. A sequence is best defined as a function whose domain is the positive integers. More informally, a sequence is any ordered listing of numbers, such as the squares of the natural numbers or their cubes, taken in order. As long as there is a formula for the sequence, there is a Maple command, called *seq*, to generate the terms. For instance, to generate the sequence of squares from 1 to 20, the command is

> S := seq(n^2, n = 1 .. 20);

```
S := 1, 4, 9, 16, 25, 36, 49, 64, 81, 100, 121, 144, 169,
            196, 225, 256, 289, 324, 361, 400
```

There are two closely related concepts in Maple relating to sequences. Objects like the above are called expression sequences, and the same thing in square brackets is called a list. Thus,

> [S];

```
[1, 4, 9, 16, 25, 36, 49, 64, 81, 100, 121, 144, 169, 196,
            225, 256, 289, 324, 361, 400]
```

is a list. Sequences and lists are almost the same. They both preserve the order of their elements. On the other hand, a set is an unordered collection of unique elements. If we use the *convert* command to change *S* to a list and a set, as in

> convert([S], list);

```
[1, 4, 9, 16, 25, 36, 49, 64, 81, 100, 121, 144, 169, 196,
            225, 256, 289, 324, 361, 400]
```

> convert([S], set);

```
{1, 4, 9, 16, 25, 36, 49, 64, 81, 100, 121, 144, 169, 196,
            225, 256, 289, 324, 361, 400}
```

very little seems to have been accomplished. Now observe the difference when we choose the sequence *cos((n Pi)/2):*

> S1 := seq(cos(n*Pi/2), n = 0 .. 20);

```
S1 := 1, 0, -1, 0, 1, 0, -1, 0, 1, 0, -1, 0, 1, 0, -1, 0,
            1, 0, -1, 0, 1
```

The list is

> **[S1];**

$$[1, 0, -1, 0, 1, 0, -1, 0, 1, 0, -1, 0, 1, 0, -1, 0, 1, 0, -1, 0, 1]$$

whereas the set is

> **convert([S1], set);**

$$\{0, -1, 1\}$$

Converting S1 to a list preserves the order of the sequence, and the list retains all of the terms of the sequence. The set is a much different object, however. It contains only the unique elements 0, +1, and −1. Thus, sequences and lists are almost interchangeable in Maple, but neither is closely related to a set. All three have their uses, and the *convert* command permits you to choose the data structure you want.

One situation requiring a list is in the *plot* command. To produce the plot of the sequence of squares, we need to form a list of points, then use the plot option *style* = *point* to plot points rather than lines. Here are the commands (see also Figure 8.1).

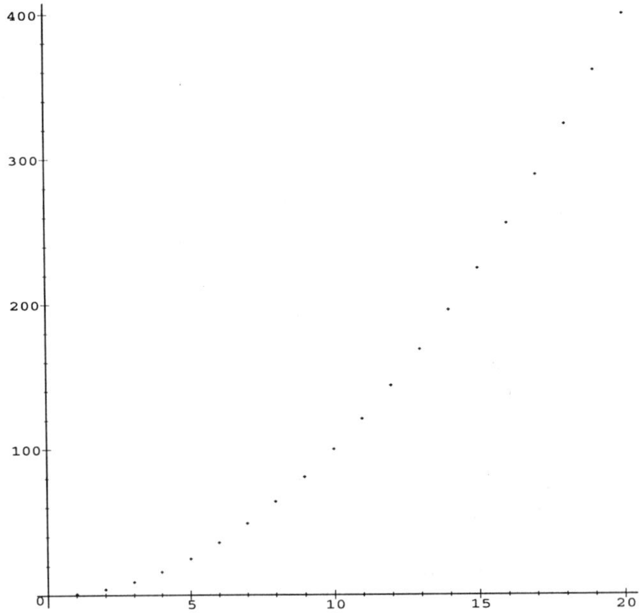

Figure 8.1 A Plot Generated from a List of Points

> **pts := seq([n, n^2], n = 0 .. 20);**

```
pts := [ 0, 0] , [ 1, 1] , [ 2, 4] , [ 3, 9] , [ 4, 16] , [ 5, 25] ,
[ 6, 36] , [ 7, 49] , [ 8, 64] , [ 9, 81] , [ 10, 100] , [ 11, 121] ,
[ 12, 144] , [ 13, 169] , [ 14, 196] , [ 15, 225] , [ 16, 256] , [ 17,
289] , [ 18, 324] , [ 19, 361] , [ 20, 400]
```

> **plot({pts}, style = point);**

Note that there are a number of points to be plotted, so we should enclose the name *pts* in curly braces to denote a set of points. In this case, enclosing the points in square brackets, to denote a list of points, would also work.

Infinite Sequences

A function whose domain is *N* (the set of natural numbers, which here may include, at times, 0) has a value for every whole number, so an infinite sequence results. The infinite sequence

$$f(0), f(1), f(2) \ldots$$

is usually written *a*0, *a*1, *a*2, ..., in order to simplify the notation. The idea is that there is a rule by which every number in the sequence may be determined.

It is easy to pick out the *i*th member of a sequence in Maple. Let a particular sequence *S*2 be defined by:

> **S2 := seq(n*(n^2 - 1)*(3*n + 2)/24, n = 0 .. 20);**

```
S2 := 0, 0, 2, 11, 35, 85, 175, 322, 546, 870, 1320, 1925,
2717, 3731, 5005, 6580, 8500, 10812, 13566, 16815, 20615
```

The 10th member of sequence *S*2 is

> **S2[10];**

```
870
```

Sequences start at *S*2[1] rather than *S*2[0]:

> **S2[0];**

```
Error, invalid subscript selector
```

To make the sequence into a list, simply enclose it in square brackets,

> L2 := [S2];

```
L2 := [0, 0, 2, 11, 35, 85, 175, 322, 546, 870, 1320, 1925,
 2717, 3731, 5005, 6580, 8500, 10812, 13566, 16815, 20615]
```

and to make it into a list of lists, use the *map* command:

> LL2 := map(convert, L2, list);

```
LL2 := [[0], [0], [2], [11], [35], [85], [175], [322],
[546], [870], [1320], [1925], [2717], [3731], [5005],
 [6580], [8500], [10812], [13566], [16815], [20615]]
```

Note that you must convert *S2* to a list before you can convert it to a list of lists. The following command, which operates on the sequence *S2*, produces an error:

> LL2 := map(convert, S2, list);

```
Error, wrong number (or type) of parameters in
              function convert
```

You can make an attempt at plotting an infinite sequence. After some adjustments to the *y*-axis scale, the following command:

> plot(1/(10 + n) + 2, n = 0 .. infinity, y = 2 .. 2.1, style = point);

gives you a picture of the behavior of the function over its full domain (Figure 8.2). Of course, the points are not measurable on this plot, but it still indicates that the sequence has values near 2 as *n* gets large. (*Note:* The scale of the *y* axis is 2 to 2.1; therefore, the bottom of the *y* axis is at $y = 2$.)

Sequences may be defined from expressions, functions, and procedures. An example of a sequence built from a procedure is

> f := n -> n*(n + 1)*(2*n + 1)/6;

$$f := n \to \frac{1}{6}n(n+1)(2n+1)$$

> S3 := seq(f(n), n = 0 .. 15);

```
S3 := 0, 1, 5, 14, 30, 55, 91, 140, 204, 285, 385, 506,
              650, 819, 1015, 1240
```

Maple's procedures can be as complex as you like, because you can use Maple's built-in programming language to build procedures for specific tasks that you would like to automate. We won't delve into the topic of building user-defined procedures in this book, because it is about using the extensive prebuilt commands and procedures. As you have seen, there are

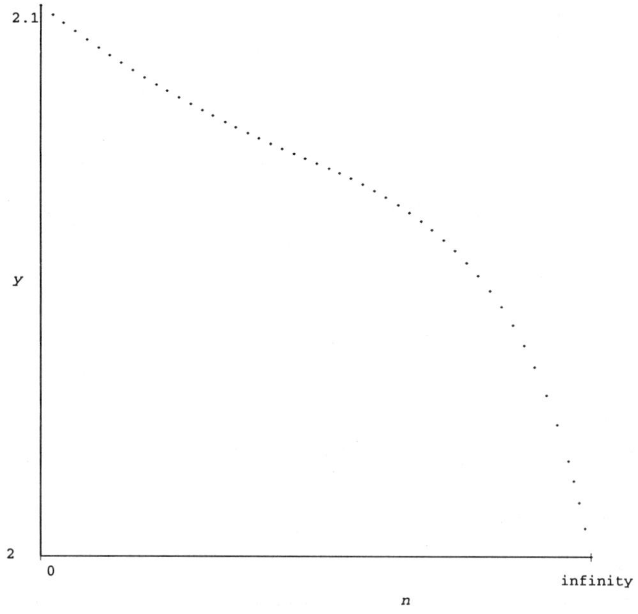

Figure 8.2 A Plot Whose Horizontal Axis Range Is Infinity

Maple commands for all the elementary operations. In the next chapter, you will see that there is even a Maple command for solving differential equations.

The tools for studying sequences are basically the same as the ones we have seen earlier for studying continuous functions:

(a) *Plotting;* However, using the plot option *style* = *point* may be more instructive.

(b) *Limits:* An infinite sequence tends to a limit if all the terms except the first few remain as close as you please to the limit point, which you estimate by looking at the graph you plot in (a). Call this limit point *L*. You can choose some small number, like 10^-100, and all the terms after the *n*th (you will have to calculate the value of *n*) will be within 10^-100 of *L*.

Example 8-1

Examine the sequence a_n = *n*!/*n*^*n* to find the limit as *n* approaches infinity.

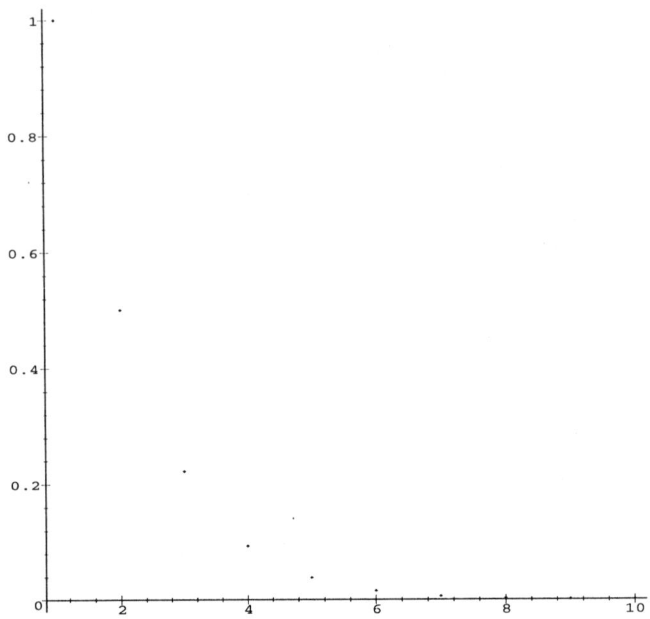

Figure 8.3 Inferring the Limit of $\frac{n!}{n^n}$ from the Graph of the Points in the Sequence

Solution:

(a) Plot the sequence

> plot([seq([n, n!/n^n], n = 1 .. 10)], style = point);

The sequence is decreasing and the terms are all positive. These two facts plus the shape of the graph suggest that the sequence has limit zero (Figure 8.3). To confirm this supposition, use the limit command:

> limit(n!/n^n, n = infinity);

$$0$$

Your Turn. Find the limit of

> a[n] = a^(1/n), a > 0;

$$a_n = a^{\left(\frac{1}{n}\right)}, \quad 0 < a$$

Hint: Since a is a variable, you must choose representative values for it before plotting the sequence. Try plotting the sequence if $a = 0$ to see why. Next, try $a = 1/100$, then $a = 100$. Finally, apply Maple's limit command.

> limit(a^(1/n), n = infinity);

$$1$$

Series

The sum of a sequence of numbers is called a *series*. There are two common types of series that we will address first.

Arithmetic Series

The series

$$s_n = a + (a + d) + (a + 2d) + (a + 3d) + \ldots + (a + nd) = \sum_{k=0}^{n} (a + kd) \tag{8-2}$$

is called an arithmetic series. The sum of this series is easily found after observing that the sum of the natural numbers $1 + 2 + 3 + 4 + 5 + \ldots + n = \frac{n(n+1)}{2}$. Rearranging the terms in Equation 8-2 produces

$$s_n = na + d\sum_{k=0}^{n} k = na + d\frac{n(n+1)}{2} = \frac{(n+1)(2a+nd)}{2} \tag{8-3}$$

Maple finds this result easy to derive. Just issue the commands

> a1 := (sum(a+k*d, k = 0 .. n));

$$a1 := a(n+1) + \frac{1}{2}d(n+1)^2 - \frac{1}{2}(n+1)d$$

> a2 := simplify(a1);

$$a2 := an + a + \frac{1}{2}dn^2 + \frac{1}{2}dn$$

You might want to expand the right-hand side of Equation 8-3 to show that it reduces to $a2$.

Geometric Series

In a geometric series, each term is a constant multiple of the one preceding it. The general form of a geometric series is

$$g_n = a + ar + ar^2 + ar^3 + \ldots ar^{n-1} + ar^n = \sum_{k=0}^{n} ar^k \quad (8\text{-}4)$$

The sum g_n is expressible as a compact formula, which is derived by multiplying each term of the series by r and subtracting these terms from g_n.

$$g_n = a + ar + ar^2 + ar^3 + \ldots + ar^{n-1} + ar^n$$

$$rg_n = ar + ar^2 + ar^3 + \ldots + ar^{n-1} + ar^n + ar^{n+1}$$

$$g_n - rg_n = a - ar^{n+1}$$

$$g_n = \frac{a(1 - r^{n+1})}{(1 - r)} \quad (8\text{-}5)$$

The formulas for the sums of the arithmetic series and geometric series should be familiar to you from your precalculus course.

Infinite Series

The importance of infinite sequences lies in the fact that they permit us to put meaning to the ideal of an infinite sum. These objects are called *infinite series*. If you are asked to sum the infinite series

$$\frac{1}{2} + \frac{1}{4} + \frac{1}{8} + \frac{1}{16} ..?$$

you might be tempted to reply, as Zeno did 2,500 years ago, that there is no answer because each addition takes a finite time and there are an infinite number of additions required. Ignoring Zeno's objection for the moment, if we plot the finite sums,

```
> plot([seq([n, sum(2^(-r), r = 1 .. n)], n = 1 .. 10)], style = point);
```

we see that the limit seems to be 1 (Figure 8.4).

Now to address Zeno's objection: we remove the extraneous notion of time from the problem by defining the sum of an infinite series as the limit of the sequence of partial sums as n approaches infinity. We do not need any more than that to say what we mean by the sum of infinitely many terms. The need for infinite sequences is established: they permit us to define what is meant by an infinite sum.

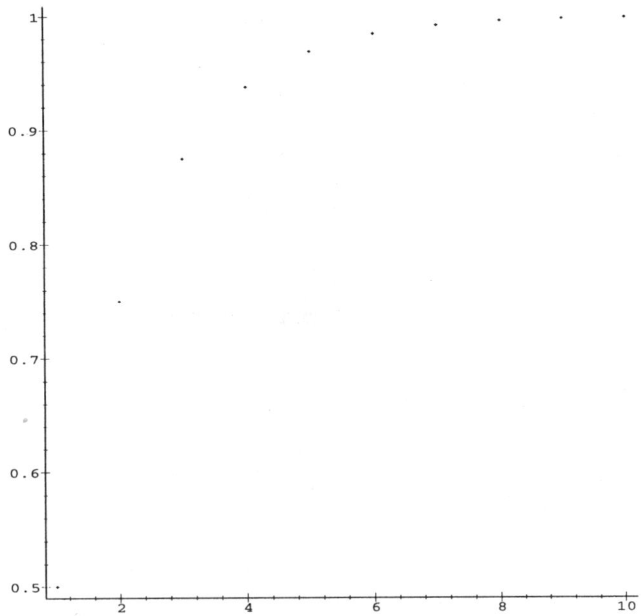

Figure 8.4 Graphical Approach to Finding the Limit of a Sum

Example 8-2

Find the sum of the infinite series $\sum_{n=1}^{\infty} \frac{1}{2^n}$

Solution: We have seen that the points representing the partial sums seem to be converging to 1. Let Maple perform the sum:

> Sum(1/2^n, n = 1.. infinity) = sum(1/2^n, n = 1.. infinity);

$$\sum_{n=1}^{\infty} \frac{1}{2^n} = 1$$

Note the use of both the capitalized (inert form) and the lower-case versions of the *sum* command. It gives you a nice way of writing both the summation notation and the result. In fact, you could have Maple calculate the partial sums for you. Copy the command and replace the word *infinity* by *r*.

> **Sum(1/2^n, n = 1.. r) = sum(1/2^n, n = 1.. r);**

$$\sum_{n=1}^{r} \frac{1}{2^n} = -2\left(\frac{1}{2}\right)^{(r+1)} + 1$$

The sum is 1, less a term containing $\frac{1}{2^r}$. In the limit, this latter term is 0, so the sum to infinity is indeed 1.

Your Turn. Find the sum $1 + x + x^2 + x^3 + ?$ and show that it diverges (becomes infinitely large) if $x >= 1$.

Hint: Begin with the following Maple commands;

> **assume(x, RealRange(0, Open(1)));** (This is one way of saying 0 <= x < 1)

> **SP := Sum(x^n, n = 0 .. k);** (Form the inert sum command)

> **S := value(SP);** (Evaluate the sum)

> **Limit(S, k = infinity) = limit(S, k = infinity);** (Find the limit of the sum as the number of terms go to infinity)

Answer: ____

Example 8-3: Convergence

Find the sum of the series $1 + \frac{1}{2} + \frac{1}{3} + \frac{1}{4} + ?$

Solution: The terms in this series have limit 0, so you might be led to conclude that terms after a certain number contribute little. This is not the case. No matter how far out in the series you go, you can always take enough terms so that the accumulated sum is increased by 1.

> **SP := Sum(1./n, n = r .. k);**

$$SP := \sum_{n=r}^{k} \left(\frac{1.}{n}\right)$$

Take r to be 100 and experiment to find the number of terms that are required to add to a number greater than 1.

> value(subs(r = 100, k = 270, SP));

$$1.000110815$$

Your Turn. How many terms do you need to take to get a number bigger than 1, starting with $a_{1,000} = 1/1,000$? Use Maple to experiment.

Answer: ______________________________

Example 8-3 shows that it is not enough that the terms in the sum have limit 0 for an infinite series to have a finite sum. Not only that, remember that these series have been defined in terms of an infinite sequence, and sequences are ordered objects. We will see shortly that the order of the terms may be important in some series.

Rearranging Terms

Consider the series $1 - \frac{1}{2} + \frac{1}{3} - \frac{1}{4} + \frac{1}{5} - \frac{1}{6} + ?$ defined by

> Sk := Sum((-1)^(n + 1)/n, n = 1 .. k);

$$Sk := \sum_{n=1}^{k} \frac{(-1)^{(n+1)}}{n}$$

By forming the sum 1 + (–1/2 + 1/3) + (–1/4 + 1/5) ..., you see that all the terms after the first are negative, so the sum must be less than 1. But by writing the sum as (1 – 1/2) + (1/3 – 1/4) ..., all the terms in brackets are positive, so the sum must be greater than 0. However, regrouping the terms into positive and negative halves, we obtain 1 + 1/3 + 1/5 + 1/7 + ... –(1/2 + 1/4 + 1/6 + ...). The two series are both divergent, resulting in infinity – infinity, a meaningless result. Thus, the order in which the terms are taken in this case is very important. Perhaps because of this Maple is reluctant to assign a value to the sum:

> Limit(Sk, k = infinity);

$$\lim_{k \to \infty} \sum_{n=1}^{k} \frac{(-1)^{(n+1)}}{n}$$

> value(Limit(Sk, k = infinity));

$$undefined$$

Nonetheless, Maple is able to compute the answer we are looking for if we make the problem into an infinite sum from the outset:

```
> Sk := Sum((-1)^(n + 1)/n, n = 1 .. infinity);
```

$$Sk := \sum_{n=1}^{\infty} \frac{(-1)^{(n+1)}}{n}$$

```
> value(Sk);
```

If the terms are summed in the order specified by n, without rearrangement, the series does converge to ln(2). Such series are called conditionally convergent. Conditionally convergent series must be treated with great care. A course in advanced calculus would allow us to prove that by rearranging the terms in a conditionally convergent series, the newly rearranged series can be made to converge to any number you choose!

In contrast, the series $\sum_{n=1}^{\infty} a_n$ is *absolutely convergent* if the series $\sum_{n=1}^{\infty} |a_n|$ is convergent. Any convergent series that consists solely of positive terms is *absolutely convergent*. Since it is not enough that the individual terms in a series approach zero for the series to converge, tests of convergence must be devised. We will list some of these tests here.

Convergence Tests

A Comparison Test: Suppose that for sequences a_n and b_n, 0<= a_n and a_n <= b_n for all n. If $\sum b_n$ converges, then so does $\sum a_n$.

A Second Comparison Test: If a_n >= 0 and b_n >= 0, and if $\lim_{n \to \infty} \frac{a_n}{b_n}$ = c and c $\neq$ 0, then $\sum a_n$ converges if, and only if, $\sum b_n$ converges.

The Ratio Test: If a sequence a_n consists of all positive terms, then form the limit of the ratio $\lim_{n \to \infty} \frac{a_n + 1}{a_n} = r$. If $r < 1$, then $\sum a_n$ converges, whereas it diverges if $r > 1$. If $r = 1$ the test is inconclusive: the series may converge or diverge.

The Integral Test: The sequence a_n has been defined as the values of $f(n)$, where n is a whole number. If it is possible to make $f(n)$ into a function of the real valued variable x, such that $f(x) = f(n)$ whenever $x = n$, then $\sum a_n$ converges if, and only if, $\int_1^{\infty} f(x)dx$ exists.

Example 8-4

Show that $\sum_{n=1}^{\infty} \frac{1}{n^2}$ is convergent by the integral test.

Solution: The continuous function corresponding to $\frac{1}{n^2}$ is $f(x) = \frac{1}{x^2}$. The integral of this function is

> Int(1/x^2, x = 1 .. infinity) = int(1/x^2, x = 1 .. infinity);

$$\int_1^{\infty} \frac{1}{x^2} dx = 1$$

The integral exists (is finite), so the corresponding series must be convergent. Is the value of the infinite series also 1? It's easy to find out by evaluating the sum directly.

> S := sum(1/n^2, n = 1 .. infinity); evalf(S);

$$S := \frac{1}{6}\pi^2$$

$$1.644934068$$

The sum of the series is a little more than 1, and surprisingly, it involves the number π. The proof of Maple's result is beyond the scope of this text.

Your Turn.

(a) Show that $\sum_{n=1}^{\infty} \frac{1}{n^2 + 1}$ is convergent using the integral test, and then find its sum.

Answer: ____

(b) Compare the values of the integral and the sum. *Hint:* In the case of the sum, you may need to convert the answer to decimal form to interpret it.

Answer: ____

(c) The sum is bigger than the integral. Show why this is so graphically, using the *leftbox* command.

Answer: ____

Example 8-5

State when $\sum_{n=1}^{\infty} \frac{1}{n^p}$ is convergent.

Solution: By the integral test,

> **Int(1/x^p, x = 1 .. infinity) = int(1/x^p, x = 1 .. infinity);**

$$\int_1^{\infty} \frac{1}{x^p}\,dx = \lim_{x \to \infty} -\frac{x^{(-p+1)}}{p-1} + \frac{1}{p-1}$$

The term $-\frac{x^{(-p+1)}}{p-1}$ gives the clue we need. As x gets large, p must be greater than 1 so that the whole term will tend to zero.

> **assume(p > 1);**

> **limit(-1/(p-1)*x^(-p+1)+1/(p-1),x = infinity);**

$$\lim_{x \to \infty} -\frac{x^{(-p\sim+1)}}{p\sim-1} + \frac{1}{p\sim-1}$$

Why aren't we getting the answer we expect? We try putting more restrictions on x:

> **assume(x, real); additionally(x > 1);**

> **limit(-1/(p-1)*x^(-p+1)+1/(p-1), x = infinity);**

$$\lim_{x\sim \to \infty} -\frac{x\sim^{(-p\sim+1)}}{p\sim-1} + \frac{1}{p\sim-1}$$

> **limit(1/(p-1)*x^(-p+1), x = infinity);**

$$\lim_{x\sim \to \infty} \frac{x\sim^{(-p\sim+1)}}{p\sim-1}$$

> **10000000^(-.03);**

$$.6165950019$$

```
> about(x); about(p);
```

```
                Originally x, renamed x~:
     is assumed to be: RealRange(Open(1),infinity)
                Originally p, renamed p~:
     is assumed to be: RealRange(Open(1),infinity)
```

No approach seems to work, other than the first. Here we had to work without Maple's help by observing that p must be greater than 1.

The Root Test: If $\lim_{n \to \infty} |a_n|^{\left(\frac{1}{n}\right)} = l$ and $l < 1$ then $\sum a_n$ is absolutely convergent. If $l > 1$, the series diverges, and if $l = 1$, the series may either converge or diverge.

Example 8-6

Test the series $\sum_{n=1}^{\infty} x^n$ for convergence by applying the root test.

Solution: If x = 1, then $(x^n)^{\left(\frac{1}{n}\right)} = 1$ for all n, so $\lim_{n \to \infty} (x^n)^{\left(\frac{1}{n}\right)} = x$. Therefore, the series diverges.

Using Maple to verify that $(x^n)^{\left(\frac{1}{n}\right)} = 1$

```
> limit((1^n)^(1/n), n = infinity);
```

```
                    1
```

If $x \neq 1$,

```
> limit((x^n)^(1/n), n = infinity);
```

```
                    x
```

Therefore, the series converges if $x < 1$ and diverges if $x \geq 1$.

Your Turn. Using the root test, show that the series $\sum_{n=0}^{\infty} \frac{1}{x^{(2n+1)}}$ converges if $x > 1$.

Answer: ______________________________

Expansion of Functions in Series Form

There are many times when it is advantageous to represent a function defined by a complicated formula as a polynomial approximation. As long as the function does not contain too many twists and turns over the interval of interest, it can be approximated by a low-order polynomial. You can understand the possibility of this happening by observing that if the function looks almost like a straight line, it can be approximated by a linear polynomial; if it looks like a parabola, it can be approximated by a quadratic; and so on. As the degree of the approximating polynomial increases, so does the number of changes of curvature it can approximate. We will begin the discussion by approximating a function near $x = 0$.

Maclaurin Series. Let the power series be given by the polynomial whose first nine terms are given by

> p := x -> a[0] + a[1]*x + a[2]*x^2 + a[3]*x^3 + a[4]*x^4 + a[5]*x^5 + a[6]*x^6 + a[7]*x^7 + a[8]*x^8;

$$p := x \rightarrow a_0 + a_1 x + a_2 x^2 + a_3 x^3 + a_4 x^4 + a_5 x^5 + a_6 x^6 + a_7 x^7 + a_8 x^8$$

The value of the polynomial p at $x = 0$ is $p(0) = a_0$. Stated another way, the value of the function at $x = 0$ is given by the constant coefficient, a_0. The function and its power series approximation coincide at $x = 0$. Now take the first derivative of the power series and evaluate it at $x = 0$.

> D(p)(0);

$$a_1$$

Similarly, the higher derivatives of the series are

> (D@@2)(p)(0);

$$2 a_2$$

> (D@@3)(p)(0);

$$6 a_3$$

> (D@@4)(p)(0);

$$24 a_4$$

> (D@@5)(p)(0);

$$120 a_5$$

While Maple does not give the general *n*th derivative, it is easy to see by working out the first few terms by hand that it must be

> **(D@@n)(p)(0) = n!*a[n];**

$$(D^{(n)})(p)(0) = n!a_n$$

Solving for the coefficient a_n, we find

> **a[n] = (D@@n)(p)(0)/n!;**

$$a_n = \frac{(D^{(n)}(p)(0)}{n!}$$

We argue that, just as the series matches the function at $x = 0$ by our choice of a_0, the best choice for a_1, and in fact all the other terms, should be given by the last equation. Thus the rule: to find the coefficients of the power series, find the *n*th derivatives for the function you are trying to approximate, evaluate the results at $x = 0$, and divide by $n!$. This series matches the function best near $x = 0$, because there, not only do the function and the series have the same value, but the first derivative is the same, and so on. Thus, if we can compute the various derivatives of a given function, we can determine the coefficients of its corresponding power series. How many coefficients do we need to compute? Obviously, computing more coefficients will make it possible to still have a good correspondence between the function and its series for values of x farther and farther away from $x = 0$.

Example 8-7

Compute the coefficients in the power series for $\mathbf{e}^x$.

Solution: Since

> **Diff(exp(x), x$n) = exp(x);**

$$\text{Diff}(\mathbf{e}^x, x\$n) = \mathbf{e}^x$$

and $\mathbf{e}^0 = 1$, we have $a_n = \frac{1}{n!}$. Therefore,

> **exp(x) = Sum(x^n/n!, n = 0 .. infinity);**

$$\mathbf{e}^x = \sum_{n=0}^{\infty} \frac{x^n}{n!}$$

Your Turn. Compare e^2 with its power series approximation $1 + 2 + \frac{2^2}{2} + \frac{2^3}{3!} + \frac{2^4}{4!} + \ldots$.

(a) How many terms do you need for four-figure accuracy?

Answer: ____________________

(b) How many terms do you need if $x = 0.003$?

Answer: ____________________

Example 8-8

Find the power series expansion for $\sin(x)$.

Solution: The four distinct derivatives of $\sin(x)$ are

```
> eval(subs(x = 0 , [seq( diff(sin(x), x$n), n = 1 .. 4)]));
```

$$[1, 0, -1, 0]$$

after which the cycle starts again:

```
> eval(subs(x = 0 , [seq( diff(sin(x), x$n), n = 5 .. 8)]));
```

$$[1, 0, -1, 0]$$

Therefore, the power series expansion of $\sin(x)$ is

```
> sin(x) = Sum((-1)^(n+1)*x^(2*n + 1)/(2*n + 1)!, n = 0 .. infinity);
```

$$\sin(x) = \sum_{n=0}^{\infty} \frac{(-1)^{(n+1)} x^{(2n+1)}}{(2n+1)!}$$

The first few terms are

```
> expand(sum((-1)^(n+1)*x^(2*n + 1)/(2*n + 1)!, n = 0 .. 10));
```

$$-x + \frac{1}{6}x^3 - \frac{1}{120}x^5 + \frac{1}{5040}x^7 - \frac{1}{362880}x^9 + \frac{1}{39916800}x^{11} - \frac{1}{6227020800}x^{13} + \frac{1}{1307674368000}x^{15} - \frac{1}{355687428096000}x^{17} + \frac{1}{121645100408832000}x^{19} - \frac{1}{51090942171709440000}x^{21}$$

Your Turn. Estimate the sine of 30 degrees using terms of the power series up to x^9. Compare your answer to the exact answer of 1.2. What is the percent error? (Remember to convert 30° to radians!)

Answer: ____________________

Example 8-9

Find the power series expansion for ln(1 + x).

Solution: The form (1 + x) is used instead of x since we can choose $x = 0$ and still have ln(1 + x) defined. Remember: ln(0) is undefined, but ln(1) = 0, which is a perfectly respectable number.

The first 10 derivatives of ln(1 + x) at $x = 0$ are

> subs(x = 0, [seq(diff(log(1 + x), x$n)/n!, n = 1 .. 10)]);

$$\left[1, \frac{-1}{2}, \frac{1}{3}, \frac{-1}{4}, \frac{1}{5}, \frac{-1}{6}, \frac{1}{7}, \frac{-1}{8}, \frac{1}{9}, \frac{-1}{10}\right]$$

Thus $\ln(1 + x) \approx x - \frac{x^2}{2} + \frac{x^3}{3} - \frac{x^4}{4} + \frac{x^5}{5} - \frac{x^6}{6}$ .. ?

Your Turn. Compute ln(e) using the power series expansion for e^x. Compare your answer to ln(e) = 1. State the number of terms you used in the power series expansion and the percent error of the approximation.

Answer: Number of terms:_______________; percent error:_______________

Because power series are so useful in approximating other functions, Maple has a command for their construction. The *series* command calculates the first few polynomial terms of any combination of standard functrions.

Example 8-10

Calculate the first ten terms of arctan(x) using the *series* command.

Solution: Since arctan is one of the functions known to Maple, all you have to do is to formulate the *series* command, giving it the arctan function and specifying the number of terms.

> series(arctan(x), x, 10);

$$x - \frac{1}{3}x^3 + \frac{1}{5}x^5 - \frac{1}{7}x^7 + \frac{1}{9}x^9 + O(x^{10})$$

The third parameter, *10*, specifies the number of terms you want, while the second term specifies the variable for the expansion, in case there are other parameters.

Your Turn. State the command that finds the first seven terms of cos(x).

Answer: ____

Taylor Series

The expansion of a function need not begin at $x = 0$. Some other value may be taken as the starting point for the expansion. This time we assume that the function can be approximated by a polynomial of the form

```
>p(x) = a[0] + a[1]*(x - a) + a[2]*(x - a)^2 + a[3]*(x - a)^3 + a[4]*(x - a);
```

$$p(x) = a_0 + a_1(x - a) + a_2(x - a)^2 + a_3(x - a)^3 + a_4(x - a)$$

It is not too difficult to show that the kth coefficient, a_k, is given by

$$a_k = \frac{(D^{(k)})(p)(a)}{k!} \tag{8-6}$$

where $D^{(k)}(p)(a)$ represents the kth derivative of p, evaluated at a. To illustrate this formula, we will use it to recompute the series expansion of the natural logarithmic function of Example 8-8.

Example 8-11

Find the expansion of $\ln(x)$ about $x = 1$ by direct calculation. Note that this is equivalent to finding the expansion of $\ln(1 + x)$ about $x = 0$.

Solution: First, identify the parameter a in the previous expansion of $p(x)$. Since the expansion is to be about $x = 1$, then parameter $a = 1$. Next, compute the derivatives, as we did in Example 8-8, using Formula 8-2.

```
> subs(x = 1, [seq(diff(log(x), x$n)/n!, n = 1 .. 10)]);
```

$$\left[1, \frac{-1}{2}, \frac{1}{3}, \frac{-1}{4}, \frac{1}{5}, \frac{-1}{6}, \frac{1}{7}, \frac{-1}{8}, \frac{1}{9}, \frac{-1}{10}\right]$$

The list of coefficients is the same because we are using essentially the same command. This time, though, we multiply these coefficients by the appropriate power of $(x - 1)$.

$$\ln(x) = x - 1 - \frac{(x-1)^2}{2} + \frac{(x-1)^3}{3} - \frac{(x-1)^4}{4} \cdots \tag{8-7}$$

Your Turn. Calculate ln(1) using formula 8-7. Is the answer exact?

Answer: ______________________________

When a power series is obtained by expanding about some value other than 0, it is called a *Taylor series*. The Maple command is *taylor*($f(x), x = a, n$), where $f(x)$ is the function (written as an *expression*) whose Taylor expansion is desired, a is the value about which the expansion takes place, and n is the desired number of terms.

Example 8-12

Find the Taylor expansion of cos(x) about $x = -\pi/2$ and compare it to the power series for sin(x) about $x = 0$.

Solution: The Maple command for the Taylor series for cos(x) about $x = -\pi/2$ is

> ex812 := taylor(cos(x), x = -Pi/2, 10);

$$ex812 := x + \frac{1}{2}\pi - \frac{1}{6}\left(x + \frac{1}{2}\pi\right)^3 + \frac{1}{120}\left(x + \frac{1}{2}\pi\right)^5 - \frac{1}{5040}\left(x + \frac{1}{2}\pi\right)^7 + \frac{1}{362880}\left(x + \frac{1}{2}\pi\right)^9 + O\left(\left(x + \frac{1}{2}\pi\right)^{10}\right)$$

You know that the cosine is an even function, yet the terms in *ex812* all have odd powers. If we shift the y axis to $-\pi/2$ by the following substitution:

> ex812a := subs(x = x-Pi/2, ex812);

$$ex812a := x - \frac{1}{6}x^3 + \frac{1}{120}x^5 - \frac{1}{5040}x^7 + \frac{1}{362880}x^9 + O(x^{10})$$

and compare this to the series expansion of *sin*(x),

> series(sin(x), x, 10);

$$x - \frac{1}{6}x^3 + \frac{1}{120}x^5 - \frac{1}{5040}x^7 + \frac{1}{362880}x^9 + O(x^{10})$$

we see that the expressions are identical. Since there is a well-known trigonometric identity stating that $\cos(x - \pi/2) = \sin(x)$, we should not be surprised that the two expansions turn out to be identical. Another way of thinking about the result is to think of shifting the cosine curve a distance of $\pi/2$ to the right, where it becomes a sine curve.

Your Turn. Even polynomials can be approximated by polynomials. You can approximate a polynomial of degree 11 with a polynomial of lesser degree.

(a) Use Maple to find the Taylor series to degree 5 for

> ex812yt := x + x^3 + x^5 + x^7 + x^9 + x^11;

about $x = 1$.

Answer: ______________________________

(b) How close is the approximation when $x = 1.05000$?

Answer: ______________________________

When either the *series* or the *taylor* commands are used, Maple responds with a polynomial and an additional term that gives some indications of the order of the remaining terms. Often, a pure polynomial is required, without the order term to get in the way. One of the options in Maple's *convert* command allows you to strip the order term from the series. Example 8-13 shows how.

Example 8-13

Find the Taylor expansion, with terms up to degree 6, of

> ex813 := (x + 2)/(x^2 - 1);

$$ex813 := \frac{x + 2}{x^2 - 1}$$

about $x = -2$ and express it as a pure polynomial.

Solution:

> t813 := taylor(ex813, x = -2, 7);

$$t813 := \frac{1}{3}(x + 2) + \frac{4}{9}(x + 2)^2 + \frac{13}{27}(x + 2)^3 + \frac{40}{81}(x + 2)^4 + \frac{121}{243}(x + 2)^5 + \frac{364}{729}(x + 2)^6 + O((x + 2)^7)$$

We use the number 7 in the *taylor* command because we want the order term (the O term) to be at least one greater than the degree of the highest required term in the polynomial. Now convert *t813* to a pure polynomial:

> t813a := convert(t813, polynom);

$$t813a := \frac{1}{3}x + \frac{2}{3} + \frac{4}{9}(x + 2)^2 + \frac{13}{27}(x + 2)^3 + \frac{40}{81}(x + 2)^4 + \frac{121}{243}(x + 2)^5 + \frac{364}{729}(x + 2)^6$$

The only difference in the last two Maple outputs is the absence of the order term in the latter expression, *t813a*. As a matter of interest, you could convert *t813a* into a function if you so desired. This is a case of an "irregular verb" in Maple, because the command is *not, convert(t813a, function)*, but is instead

> t813b := unapply(t813a, x);

$$t813b := x \rightarrow \frac{1}{3}x + \frac{2}{3} + \frac{4}{9}(x+2)^2 + \frac{13}{27}(x+2)^3 + \frac{40}{81}(x+2)^4 + \frac{121}{243}(x+2)^5 + \frac{364}{729}(x+2)^6$$

The *unapply* command is used to convert from an expression (such as *t813a*) to a function (such as *t813b*).

Your Turn. How do you turn a *function* into an *expression*? If

> f813yt := x -> 1/(1 - x);

state the command that produces the expression $1/(1-x)$ from the function whose name is *f813yt*.

Answer: ______________________________

Paper and Pencil Exercises

PP8–1

Compute the limits of the following sequences if they exist. Otherwise, describe the behavior as oscillating, or diverging, to $+\infty$ or $-\infty$.

(a) $\lim_{x \to \infty} \dfrac{3n+4}{2n-7}$ *Answer*: ______________

(b) $\lim_{x \to \infty} \sqrt{n+1} - \sqrt{n}$ *Answer*: ______________

(c) $\lim_{x \to \infty} \sqrt[n]{n}$ *Answer*: ______________

(d) $\lim_{x \to \infty} \dfrac{\sum_{r=1}^{n} r^x}{n^{x+1}}$ *Answer*: ______________

(e) $\lim_{x \to \infty} \dfrac{n+1}{n} - \dfrac{n}{n+1}$ *Answer*: ______________

(f) $\lim_{x \to \infty} \dfrac{a^n + (-1)^n/n}{a^{n-1} + (-1)^{n+1}/(n+1)}$ *Answer*: ______________

(g) $\lim_{x \to \infty} \dfrac{\sin(nx)}{n}$ *Answer*: ______________

(h) $\lim_{x \to \infty} \sin(\sqrt{n}x)$ *Answer*: ______________

(i) $\lim_{x \to \infty} \dfrac{\sin(x/\sqrt{n})}{n}$ *Answer*: ______________

(i) $\lim_{x \to \infty} na^n$, $|a|<1$ *Answer*: ______________

Maple Lab

ML8–1

Examine $\sum_{n=1}^{\infty} \dfrac{1}{n(n+1)}$ to determine whether or not it converges to a finite value. Try the following Maple commands.

(a):

> S := k -> Sum(((n*(n+1))^(-1), n = 1 .. k)); *Answer*: ______________

(b):

> Sk := value(S(k)); *Answer*: ______________

(c):

> limit(Sk, k = infinity); *Answer*: ______________

ML8–2

Find the value of the infinite series

$$\sum_{n=1}^{\infty} \frac{1}{n(n+1)(n+2)}$$

Answer: ______________

ML8-3

Even though the *harmonic series* $\sum_{n=1}^{\infty} \frac{1}{n}$ is divergent, you can evaluate the partial sums. Plot the behavior of the harmonic series as follows:

> Sn := Sum(1/r, r = 1 .. n);

> plot([seq([n, sum(1/r, r = 1 .. n)], n = 1 .. 20)], style = point);

Does the plot show a diverging function? If you have difficulty deciding, try more points. What is your conclusion? Write a detailed explanation of how you came to your decision.

Answer: ____

____

____

____

ML8-4

Consider the problem of finding an approximation for $n!$ when n becomes very large. It is fairly obvious that, if x has X digits before the decimal place, the function x^n has about nX digits. Similarly, the number e^x has roughly the same number of digits as 10^x, so once more, the number of digits is fairly easy to count. The number $e^{1,000,000}$ has about 4,000,000 decimal places, since $e^{1,000,000} \cong 10^{434,294}$. How many digits are there in the number $(10^6)!$? To estimate the number of digits, do the following.

(a) Try to bring $n!$ down to manageable size by taking its logarithm, and then express $\ln(n!)$ as a series.

Answer: ____

(b) Approximate this series as an integral. What are the limits of integration?

Answer: ____

(c) Perform the integration. You should now have an estimate for $\ln(n!)$. From what we have said about the number of digits in e^x, $\ln(n!)$ is an estimate of the number of digits in $n!$. How many digits does 100! have, according to this estimate?

Answer: ____

ML8–5

Some series have a pattern in their coefficients that is easy to describe. For example, the pattern of the coefficients in the series expansion of e^x is just $1/n!$ Use Maple to find the pattern of the coefficients in the expansion $\frac{1}{2}\ln\left(\frac{1+x}{1-x}\right)$.

Answer: ______________________________

ML8–6

Use Maple to find the limit, as $n \to$ of the sequences a_n, whose general term is given below:

(a) $\dfrac{e^n + e^{-n}}{e^n - e^{-n}}$ *Answer*: ______________

(b) $\sqrt{n+2} - \sqrt{n}$ *Answer*: ______________

(c) $n \sin\left(\dfrac{1}{n}\right)$ *Answer*: ______________

(d) $\dfrac{n!}{\sqrt{2\pi n}\, n^n e^{-n}}$ *Answer*: ______________

(e) $n \cos\left(\dfrac{n\pi}{2}\right)$ *Answer*: ______________

ML8–7: Present Value

The present value of an IOU (or other promissory note) for \$100 that comes due a year from now is currently worth $\frac{100}{1+i}$, with i being the interest rate on invested money. The reason that the IOU is worth less than \$100 today is that you could invest $\frac{100}{1+i}$ in a savings account and it would be worth \$400 in a year, when the bank adds the accumulated interest to your deposit. Thus, money that is owed to you with payment due in the future is worth less than the face value of the note today.

(a) Suppose you have an IOU from a very reliable friend, who promises to pay you \$100 in ten years. The interest rate is 7%, compounded yearly. What is the present value of the IOU?

Answer: ______________________________

(b) Payments that you owe to someone else are just like IOUs. If you are charged \$1,000 per month rent on your apartment, how much should you offer to pay "up-front" so that you do not have to pay rent again for one year? Assume the interest is compounded monthly and the monthly interest rate is 1%.

Hint: Your first payment is due one month from now, so its present value is $\frac{1{,}000}{1+0.01}$. Your next payment is due in two months, so its present value is $\frac{1{,}000}{(1+0.01)^2}$, and so on. Write a formula for the sum using Maple.

Answer: ____

(c) Say that you win the lottery and you want to pay your rent *infinitely* far into the future, so you can live rent free the rest of your life. How much should you write the check for, assuming a constant interest rate of 1% per month, compounded monthly?

Answer: ____

(d) The landlord offers to sell you the place for \$80,000. Is this a fair price?

Answer: ____

ML8–8: Differentiating a Power Series

Recall that $\frac{1}{1-x} = \sum_{n=0}^{\infty} x^n = 1 + x + x^2 + x^3 + \ldots, -1 < x < 1.$

(a) Use Maple to find the power series representation of $\frac{1}{(1-x)^2}$.

Answer: ____

(b) Differentiate $\frac{1}{1-x}$ with respect to x and differentiate its power series expansion term by term. Compare your answer with (a).

Answer: ____

(c) Use Maple to find the power series representation of $\frac{1}{(1-x)^3}$. Compare this with the derivative of (b).

Answer: ____

(d) What hypothesis can you formulate about integrating term by term? Test your hypothesis by integrating $\ln(1 + x)$.

Answer: ______________________________

(e) Without using Maple, what is the sum $x + 2x^2 + 3x^3 + \ldots$

Answer: ______________________________

ML8–9

In Example 8-11, "Your Turn," you found that the series approximation gave the correct answer for $\ln(x)$ when $x = 1$, the point about which the expansion is taken. Show that the series expansion of *any* function evaluated at the point about which the expansion is taken gives an exact answer.

Answer: ______________________________

____

____

CHAPTER 9

Differential Equations

Objectives for This Chapter

1. State the characteristic of a differential equation
2. Understand what is meant by a particular and a general solution to a differential equation
3. Define the terms *order* and *degree* as they apply to differential equations
4. Apply the method of change of variable
5. Define and solve exact differential equations and define the term, *integrating factor*
6. Find the general solution for the first-order linear differential equation
7. Solve differential equations of order two and solve simultaneous differential equations
8. Learn the Maple commands to solve differential equations analytically and numerically

Maple Commands Used in This Chapter

This chapter contains more commands than usual, but most have been used before. The only new commands are *dsove*, *Deplot*, *odeplot*, and *laplace*. Other, less common, commands are included for ease of reference.

(D@@3)(y)	The third derivative, in operator form.
assume(k >0);	Place restrictions on a variable.
collect(deq99b, du);	Collect terms containing *du* together.
convert((D@@3)(y)(t), diff) ;	Convert from *D* form to *diff.*

convert(student[isolate] (expr, radical);	Convert a *RootOf* form to radical form.
D[1](f);	The partial derivative of *f* with respect to its first argument.
DEtools[DEplot](deq915, y(x), x= 1 .. 10, [[ic915]], y= 0 .. 12, dirgrid=[10,10], color=black, linecolor=COLOR(RGB, 0, 0, 0), scaling=constrained, stepsize=0.1);	The *Deplot* command with some of its options.
diff(y(x), x$2) + diff(y(x), x) - 6*y = 18;	A second-order linear differential equation.
dsolve(deq96, y(x));	The *dsolve* command which is central to the solution of differential equations.
dsolve({deqn, IC}, y(x), type = numeric);	The form of the *dsolve* command when a numeric solution to the differential equation is desired.
eliminate({sol94a, d1}, c);	Eliminate the variable *c* from two equations.
eval(subs(ps, deq2));	The by-now familiar combination of commands for substituting a value into an expression and then converting the result to decimal form.
expand(dx*subs(diff(y(x), x) = dy/dx, deq96));	The *expand* command applied to a term containing *diff*.
inttrans[laplace](ex922, t, s);	Maple's command for taking the Laplace transform of an expression.
op(1, op(2, d1)) = 0;	The *op* command is useful for picking terms out of lists or sets, when just part of Maple's output is needed.
plots[display]({p1, p2});	Plots may have different options. This command permits two plots, each with different options, to be displayed together.
plots[implicitplot]({family}, x = 0 .. 10, y = -10 .. 10, color = black);	Plot a set of implicit functions together.

plots[odeplot](nsol920, [t, v(t)], 0 .. 10, labels=['t','v(t)'], color = black);	Plot the solution to a differential equation that was solved numerically.
readlib(eliminate):	The *eliminate* command needs to be loaded using this command before it can be used.
simplify(diff(lhs(sol93), x) - lhs(deq93));	Another collection of useful commands.
student[isolate](sol912, y(x));	The command, similar to *solve*, that isolates a variable, in this case $y(x)$.

What Is a Differential Equation?

A *differential equation* is an equation containing derivatives or differentials. Every time we wrote a formula for a derivative in the previous chapters, we were writing a differential equation. Don't get the idea that the subject of differential equations is trivial just because every formula for a derivative is technically a differential equation. There are much more challenging types of differential equations than the ones that can be solved simply by taking the antiderivative of an elementary function. While Equation 9-1 is a differential equation, so is Equation 9-2.

$$\frac{\partial}{\partial x}y(x) = \cos(x) \tag{9-1}$$

$$\left(\frac{\partial}{\partial x}y(x)\right) + y(x) = 2\cos(x) \tag{9-2}$$

In the first case, a solution to Equation 9-1 is obviously $y(x) = \sin(x)$, while the solution to Equation 9-2 is not obvious without doing a little guesswork. We will merely scratch the surface of this vast but very useful subject in this chapter. Like integrals, differential equations may not have solutions in terms of elementary functions. Even if a solution in terms of these functions exists, Maple may not be able to find it. Moreover, Maple may provide a solution that you find difficult to interpret. Such are the hazards of the subject, but there is some comfort in knowing that your field of study likely involves only a few types and these can be studied intensively.

Before we wade into the techniques of solution, we need some definitions. These will help us talk about the various types of differential equations. We have seen that a differential equation contains derivatives or differentials. Some standard techniques require that the equation be written in differential form, but Maple does not provide a mechanism for writing differentials. We will have to write our equations using derivatives when we ask Maple to help us solve them. There are two broad classes of differential equations. This chapter will discuss only *ordinary differential equations*, which are those containing only one independent variable. If a differential equation contains partial derivatives, then it is called a *partial differential equation*. Such equations must of necessity contain more than one independent variable. If there is a single independent variable, such as x in $y = f(x)$, then the partial derivative $\frac{\partial y}{\partial x}$ is identical to the ordinary derivative $\frac{dy}{dx}$. Since we have discussed mainly examples of a single independent variable in this book, we have not made much of an issue of the way Maple always writes its derivative symbol as a partial derivative.

The *order* of a differential equation is that of the highest-order derivative in the equation. The *degree* of a differential equation is the algebraic degree of its highest-order term. A *linear differential equation* of order n is of the first degree in the dependent variable and its derivatives, and its highest derivative is of order n. Consider the following examples:

$$M(x, y)dx + N(x, y)dy = 0 \tag{9-3a}$$

$$\frac{dy}{dx} + P(x)y = Q(x) \tag{9-3b}$$

$$a_3(x)\frac{d^3y}{dx^3} + a_2(x)\frac{d^2y}{dx^2} + a_1(x)\frac{dy}{dx} + a_0(x)y = f(x) \tag{9-3c}$$

$$\frac{dy}{dx} + P(x)y = Q(x)y^n \tag{9-3d}$$

$$\left(\frac{dy}{dx}\right)^2 + P(x)y = Q(x) \tag{9-3e}$$

$$r\left(\frac{d^2y}{dx^2}\right)^2 + \left[1 + \left(\frac{dy}{dx}\right)^3\right](y) = y^3(3x^4 + 2x^2 + x + 1) \tag{9-3f}$$

Equation 9-3a is a differential equation expressed in the form of differentials. It is equivalent to $M(x, y) + N(x, y)\frac{dy}{dx} = 0$; thus, it is first-order and linear, and therefore, of degree 1. The dependent variable, $y(x)$, does not appear at all, but the equation could be forced into linear form by adding the term y, which, upon rearrangement, yields $\frac{dy}{dx} + 0y = -\frac{M(x, y)}{N(x, y)}$, showing that it is indeed a first-order linear differential equation, as is Equation 9-3b. Equation 9-3c is third-order linear, and Equation 9-3d is first order and nonlinear (but still of degree 1).

Equation 9-3e is first order but of degree 2, and therefore nonlinear. Finally, Equation 9-3f is second order and of degree 2 (not 3, even though the first derivative term is raised to the third power). It is decidedly nonlinear. The appearance of any one of the terms $\left(\frac{d^2y}{dx^2}\right)^2$, $\left(\frac{dy}{dx}\right)^3$, or y^3 makes the differential equation nonlinear.

A more general way of describing a first-order differential equation is through the relation $F(x, y, y') = 0$, where $y' = dy/dx$. Similarly, a second-order equation can be described by the relation $F(x, y, y', y'')$, where $y'' = d^2y/dx^2$. A *solution* to a differential equation is the relation $f(x, y) = 0$ such that it satisfies the original differential equation. Unfortunately, this notation cannot be used in Maple, because the single quote sign causes a syntax error if used in this context.

A Note on Notation

Maple writes all its derivatives as partial derivatives, even in the case of the derivative of a function of only one independent variable. Therefore, we will use both notations interchangeably:

$$\frac{dy}{dx} = \frac{\partial y}{\partial x}, \frac{d^2y}{dx^2} = \frac{\partial^2 y}{\partial x^2}, \dots$$

Example 9-1

State the order and degree of this differential equation.

$$2\left(\frac{\partial^3}{\partial x^3}y\right)^3 + \left(\frac{\partial^2}{\partial x^2}y\right)\left(\frac{\partial}{\partial x}y\right) + x^2\left(\frac{\partial}{\partial x}y\right) = A\,\sin(x)$$

Solution: The order is 3 and the degree is also 3, since the highest-order derivative is $\frac{\partial^3}{\partial x^3}y$ and it appears to the third power (thus degree 3).

Your Turn. State the order and degree of

$$2\left(\frac{\partial^4}{\partial x^4}y\right)^2 + \left(\frac{\partial^2}{\partial x^2}y\right)^5\left(\frac{\partial}{\partial x}y\right)^6 + x^2\left(\frac{\partial}{\partial x}y\right) = A\,\sin(x)$$

Answer: ____

Particular and General Solutions for Differential Equations

By definition, a solution to a differential equation satisfies the original equation. Just because one solution has been found, it does not mean that there are no others. The *general solution* encompasses all possible solutions. Any of the restricted solutions are called *particular solutions.* We have stated that $y = \sin(x)$ is a solution to the differential equation $\frac{dy}{dx} = \cos(x)$, but it is a particular solution. The most general solution is $y = \sin(x) + C$, where C is a constant. Any value of C satisfies the differential equation, since its derivative with respect to x is 0.

Initial Conditions and Boundary Conditions

Since a constant of integration is introduced every time an integration is performed in the solution, a first-order differential equation contains one arbitrary constant, a second-order differential equation contains two independent constants, and so on. The order of the differential equation determines the number of independent constants. The extra information that is needed to determine the values of these constants is called the *initial condition* (or *conditions*). In more complicated cases, the information may not be an initial value but a value at some other time or place. In this case, they are called *boundary conditions.* Initial conditions and boundary conditions are known values of the solution to the differential equation that can be used to determine the values of the arbitrary constants of integration in the general solution.

Example 9-2

Show that

$$y = -3 + \_C1\ e^{(2x)} = \_C2\ e^{(-3x)}$$

is a solution to the differential equation

$$\left(\frac{\partial^2}{\partial x^2} y\right) + \left(\frac{\partial}{\partial x} y\right) - 6y = 18$$

Solution: Define the equation and its proposed solution in Maple.

> deq2:= diff(y(x), x$2) + diff(y(x), x) - 6*y = 18;

$$deq2 := \left(\frac{\partial^2}{\partial x^2}y(x)\right) + \left(\frac{\partial}{\partial x}y(x)\right) - 6y = 18$$

> ps := y(x) = -3 + _C1*exp(2*x) + _C2*exp(-3*x);

$$ps := y(x) = -3 + \_C1e^{(2x)} + \_C2e^{(-3x)}$$

Substitute the solution into the differential equation.

> subs(ps, deq2);

$$\left(\frac{\partial^2}{\partial x^2}(-3 + \_C1e^{(2x)} + \_C1e^{(-3x)})\right) +$$

$$\left(\frac{\partial}{\partial x}(-3 + \_C1e^{(2x)} + \_C2e^{(-3x)})\right) + 18 - 6\_C1e^{(2x)} - 6\_C22^{(-3x)} = 18$$

If you look at the last equation, you see that all that Maple has done is replace y with the proposed function of x, but it has not done any evaluation of the derivatives. We use *eval* to do the evaluation.

> eval(subs(ps, deq2));

$$18 = 18$$

The differential equation is satisfied, and therefore the proposed solution is indeed at least a particular solution. In fact, it is the general solution. By varying the two constants, $\_C1$ and $\_C2$, all possible solutions of this equation are obtained.

Some differential equations may be better solved using a relation of the form $F(x, y) = 0$ rather than a function. In this case, there are multiple solutions to the differential equation.

Example 9-3

Show that $xy(x) - x^2 + \ln(y(x)) = c$ is a solution to the following differential equation

> deq93 := (x + 1/y(x))*diff(y(x), x) + y(x) - 2*x = 0;

$$deq93 := \left(x + \frac{1}{y(x)}\right)\left(\frac{\partial}{\partial x}y(x)\right) + y(x) - 2x = 0$$

Solution: Define the supposed solution as

> sol93 := x*y(x) - x^2+ln(y(x)) = c:

We will take the derivative of the left-hand side of *sol*93 implicitly and compare it to the differential equation. Obviously, the derivative of the right-hand side of *sol*93 is 0, just like the differential equation. After simplification, the difference between the left-hand side of *sol*93 and *deq*93 is 0, showing that they are identical, therefore *sol*93 satisfies *deq*93.

```
> simplify(diff(lhs(sol93), x) - lhs(deq93));
```

$$0$$

It was possible to check that the proposed function of x in Example 9-3 was indeed the solution to the differential equation *deq*93. Generally, things are not so easy. To check the solution, you must take its derivative and compare it to the differential equation. The difficulty arises in eliminating the arbitrary constant, which may be buried deeply in the equation $F(x, y, y', c) = 0$ obtained by implicit differentiation. We will show two cases. The first yields to our manipulations, whereas the second shows the difficulties one can encounter.

Example 9-4

(a) Show that the differential equation

```
> deq94a := x*diff(y(x), x) - 2*y(x) = 0;
```

$$deq94a := x\left(\frac{\partial}{\partial x}y(x)\right) - 2y(x) = 0$$

has the solution

```
> sol94a := y(x) = c*x^2;
```

$$sol94a := y(x) = x^2 c$$

(b) Apply the previous procedure to

```
> deq94b := diff(y(x), x) -((y(x) - 2)/x)^2 = 0;
```

$$deq94b := \left(\frac{\partial}{\partial x}y(x)\right) - \frac{(y(x) - 2)^2}{x^2} = 0$$

to show that a solution is

```
> sol94b := y(x) - 2 - x/(1 + c*x) = 0;
```

$$sol94b := y(x) - 2 - \frac{x}{1 + xc} = 0$$

Solution:

(a) Find the derivative of the supposed solution. This time we will take the derivative of the whole equation at once, since the *diff* command supports this option.

> d1 := diff(sol94a, x);

$$d1 := \frac{\partial}{\partial x}y(x) = 2xc$$

The derivative also contains the constant *c*. We have two equations containing *c*, namely *d*1 and *sol*94*a*. The library function *eliminate* can be used to remove *c*.

> readlib(eliminate):

> eliminate({sol94a, d1}, c);

$$\{c = \frac{y(x)}{x^2}\},\ \{-x\left(\frac{\partial}{\partial x}y(x)\right) + 2y(x)\}$$

The second expression is the required expression with *c* eliminated. The expression is equated to 0 and is seen to be identical to *deq*94*a*.

(b) Apply the same steps as in part (a) to obtain

> d2 := diff(sol94b, x);

$$d2 := \left(\frac{\partial}{\partial x}y(x)\right) - \frac{1}{1 + xc} + \frac{xc}{(1 + xc)^2} = 0$$

> a := eliminate({sol94b, d2}, c);

$$a := \{c = \frac{-y(x) + 2 + x}{x(y(x) - 2)}\},\ \{\left(\frac{\partial}{\partial x}y(x)\right)x^2 - y(x)^2 + 4y(x) - 4\}$$

The second expression, equated to 0, should be the same as the differential equation *deq*94*b*, but it looks different. Subtracting this expression from the left-hand side of *deq*94*b* will not help either, because by looking carefully at the expression in *a* you can discover that it is a multiple of the left-hand side of *deq*94*b*. This makes no essential difference, because multiplying 0 by anything is still 0. Selecting the second expression in *a* requires the use of the *op* command because the expression is given as a set with one element. We must select the first element of the second set in *a* using the command *op*(1, *a*[2]. With this realization, we can confirm the equality of the expression in *a* and the left-hand side of *deq*94*b* as follows:

> simplify(op(1, a[2])/x^2 - lhs(deq94b));

$$0$$

Equation *sol*94*b* is a solution of *deq*94*b*, but it took some work to prove it. This is really the more representative case. It generally takes some careful thought to show that a proposed solution is truly a solution to a differential equation. In other words, even when you have a solution to a differential equation, it may be difficult to prove it.

There are a number of examples in PP9–3 to help you practice verifying that a proposed solution really does satisfy its differential equation. Since verification is such an important step, make sure you know how to apply the paper-and-pencil method as well as the Maple method.

Converting an Algebraic Equation to a Differential Equation

You can always form a differential equation from any given equation. The order of the differential equation will be the same as the number of arbitrary constants in the given equation.

Example 9-5

Find a differential equation whose solution is

> sol95 := (1 - x) - c*(1 - y(x)) = 1;

$$sol95 := 1 - x - c(1 - y(x)) = 1$$

Solution: Take the derivative of equation *sol*95:

> d := diff(sol95, x);

$$d := -1 + \left(\frac{\partial}{\partial x}y(x)\right)c = 0$$

Eliminate *c*.

> d1 := eliminate({sol95, d}, c);

$$d1 := \left\{c = \frac{x}{-1 + y(x)}\right\}, \left\{1 - y(x) + x\left(\frac{\partial}{\partial x}y(x)\right)\right\}$$

The required differential equation is

> op(1, op(2, d1)) = 0;

$$1 - y(x) + x\left(\frac{\partial}{\partial x}y(x)\right) = 0$$

Your Turn. Find a differential equation whose solution is

> sol95a := (1 + x) - c*(1 + y(x)) = 1;

$$sol95a := 1 + x - c(1 + y(x)) = 1$$

Answer: ____

Equations that contain an adjustable parameter, such as *sol*95 in Example 9-5, define what is known as a *family of curves*. The family may be visualized by plotting. Generate a sequence of ten curves by giving *c* ten equally spaced values and then plot them all on the same graph (Figure 9.1):

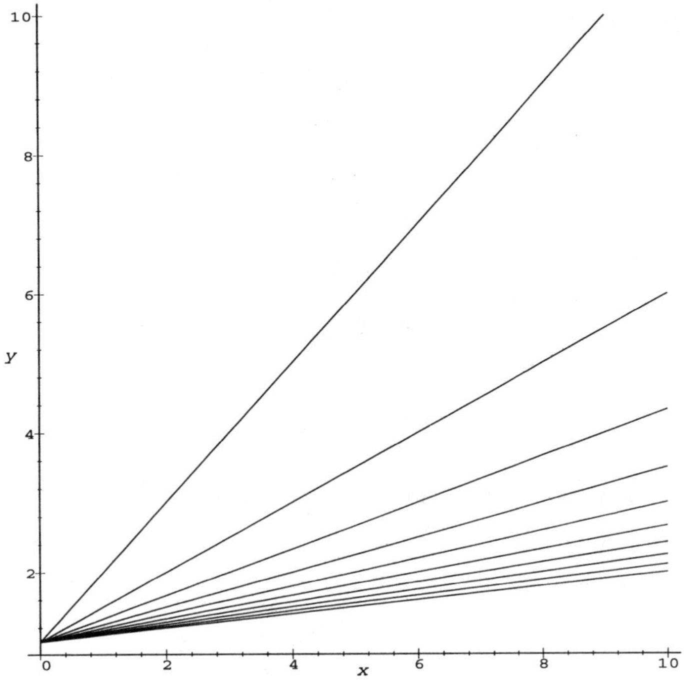

Figure 9.1 The Family of Curves Formed by Giving c Different Values in $1 + x - c^*(1 + y(x)) = 1$

> family := seq(sol95, c = 1 .. 10):

> plots[implicitplot]({family}, x = 0 .. 10, y = -10 .. 10, color = black);

One of the great benefits of using Maple is the ability to generate such graphs. You can change the c parameter values to something different, say negative values from –10 to –1, and see what change this makes in the picture. You can easily see what effect a parameter change has on your solution once you have that solution.

Variables Separable

The first-order differential equation

$$Q(x, y)\frac{dy}{dx} = -P(x, y) \tag{9-4a}$$

can be written in differential form

$$P(x, y)dx + Q(x, y)dy = 0 \tag{9-4b}$$

If both P and Q are expressible as products of a function of x only and a function of y only, the differential equation is said to have *variables separable*. If P and Q are factorable in this manner, we have

$$f(x)g(y)dx + h(x)k(y)dy = 0 \tag{9-5a}$$

Dividing through by $g(y)h(x)$, we obtain, assuming $g(y)h(x) \neq 0$,

$$\frac{f(x)}{h(x)}dx + \frac{k(y)}{g(y)}dy = 0 \tag{9-5b}$$

Each term in Equation 9-5b is the differential of a single variable; therefore, we can integrate term by term to obtain the formula for the solution:

$$\int \frac{f(x)}{h(x)}dx + \int \frac{k(y)}{g(y)}dy = c \tag{9-5c}$$

The differential equation is considered to be solved when it has been reduced to the form of Equation 9-5c, even if the integrals do not exist in terms of a finite number of elementary functions, as long as these integrals can be shown to exist.

Example 9-6

Use Maple and the method of separation of variables to find the solution to the differential equation:[1]

> deq96 := a*(x*diff(y(x), x) - 2*y(x)) = x*y(x)*diff(y(x), x);

$$deq96 := a\left(x\left(\frac{\partial}{\partial x}y(x)\right) - 2y(x)\right) = xy(x)\left(\frac{\partial}{\partial x}y(x)\right)$$

Then use Maple's built-in command for solving differential equations and verify both results.

Solution: To apply the technique of separation of variables, we must somehow get around the fact that Maple does not understand differentials. The trick is to substitute the ratio of two unknowns, *dy* and *dx*, for the derivative and then multiply through by *dx*. Lastly, we expand the result so that we can collect terms later.

1. Remember to issue the command *a* := *'a'*; to insure that *a* is undefined, since it was defined in the previous example.

> deq96a := expand(dx*subs(diff(y(x), x) = dy/dx, deq96));

$$deq96a := a\ x\ dy - 2dx\ a\ y(x) = x\ y(x)\, dy$$

To collect terms in *dx* on one side of the equation and in *dy* on the other, we use the following syntax, which is nothing more than the familiar algebraic rule of subtracting equals from equals:

> deq96b := ((a*x*dy = a*x*dy) - deq96a)/(y(x)*x) ;

$$deq96b := 2dx\,\frac{a}{x} = \frac{a\ x\ dy - x\ y(x)\ dy}{x\ y(x)}$$

This result can be simplified, but this time we need to apply the *simplify* command to each side separately.

> deq96c := lhs(deq96b) = simplify(rhs(deq96b));

$$deq96c := 2\frac{dx\ a}{x} = \frac{dy\ (a - y(x))}{y(x)}$$

Observe that we have accomplished the separation of variables. The left-hand side of Equation *deq96c* contains only terms in *x* and *dx*, while the right-hand side contains only terms in *y*(*x*) and *dy*. Thus, the two sides can be integrated. We remember to supply the constant of integration. This is a step that requires concentration, since it was copied rather than letting Maple generate it from *deq96c*. The reason was stated before: Maple does not understand differentials, so we must rewrite the equation in Maple's standard form for integrals.

> sol96a := int(2*a/x, x) = subs(y = y(x), int((a - y)/y, y)) + C;

$$sol96a := 2a\ \ln(x) = -y(x) + a\ \ln(y(x)) + C$$

We have obtained a solution by the application of one of the standard methods, namely, separation of variables (see Figure 9.2). Maple has its own differential equation solver, called *dsolve*. To use it, you supply the command with the differential equation and you name the function to be solved for, in this case, *y*(*x*).

> sol96b := dsolve(deq96, y(x));

$$sol96b := -\frac{1}{2}\ln(y(x)) + \frac{1}{2}\frac{y(x)}{a} + \ln(x) = \_C1$$

Again, the two solutions, *sol96a* and *sol96b*, look different. We can take the derivatives of both and compare them to the given equation, *deq96*.

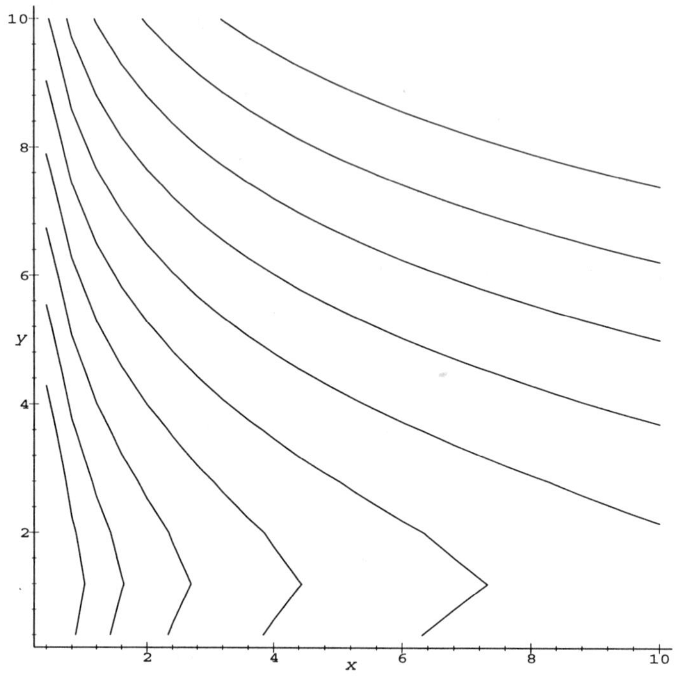

Figure 9.2 The Family of Curves for *sol*96a

> simplify(diff(sol96a, x));

$$2\frac{a}{x} + \frac{\left(\frac{\partial}{\partial x}y(x)\right)(a - y(x))}{y(x)}$$

> simplify(diff(sol96b, x));

$$-\frac{1}{2}\frac{\left(\frac{\partial}{\partial x}y(x)\right)ax - xy(x)\left(\frac{\partial}{\partial x}y(x)\right) - 2y(x)a}{y(x)ax} = 0$$

It is easier to see that the first of these equations is equivalent to *deq*96, but a little cross-multiplication and further simplification shows that the second is also a solution. There is another way to check that the two solutions are the same, and that is by plotting. If the plots look the same, it does not prove that the two equations are identical, but it does provide strong evidence that they are. As such, it forms a valuable extra check on your work, and it gives you a picture of the behavior of the solutions as well. Form the family of curves, just as was done in Example 9-5 (Figure 9.3).

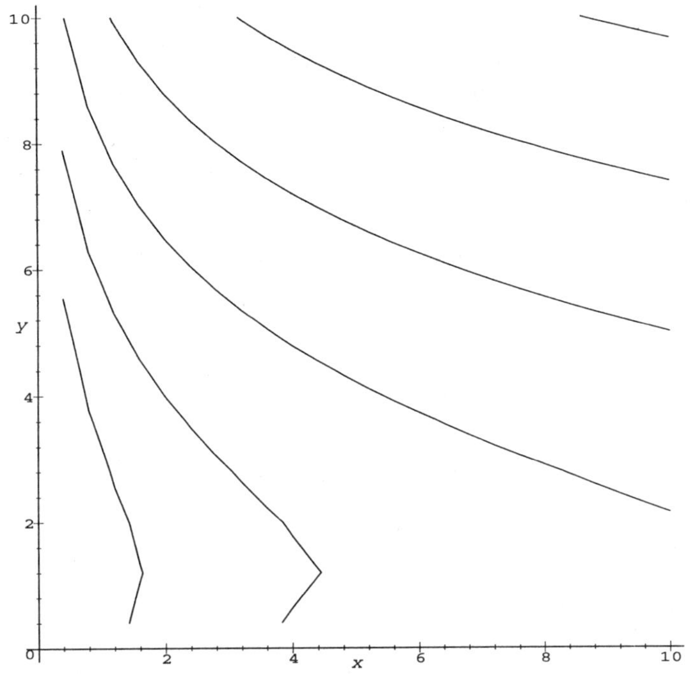

Figure 9.3 The Family of Curves for *sol96*

> **family := seq(subs(a = 1, sol96a), C = 1 .. 10):**

> **plots[implicitplot]({family}, x = 0 .. 10, y = -10 .. 10, color = black);**

> **family := seq(subs(a = 1, sol96b), C = 1 .. 10):**

> **plots[implicitplot]({family}, x = 0 .. 10, y = -10 .. 10, color = black);**

There is an important observation to make before we attempt to compare the two plots. The constants in the two equations are not the same! There is no need for $\_C1$ to be equal to C. In other words, one value of $\_C1$ may correspond to a different value of C, yet may give the same graph. In fact, it looks as if a value of $\_C1 = 1$ produces roughly the same curve as $C = 2$. Nonetheless, the families of curves look the same, even though different values of the constants would have to be used in order to have Figure 9.3 look exactly like Figure 9.2.

Elementary Change of Variable

Two of the most powerful ideas in mathematics are embodied in the phrases, "Let x be a variable that denotes a quantity," and "Substitute a function of x for x in a given expression."

It is the second of these phrases that concerns us here. Given a differential equation that appears difficult to solve, a properly chosen change of variable may produce an equation that is easier to solve. Once the solution to the transformed equation has been made available, the change-of-variable equation can be used to recover the solution to the original differential equation. The technique of change of variable requires an elementary fluency with partial differentiation.

The *partial derivative* of a function $z = f(x, y)$ with respect to x is written as

$$f_x = \frac{\partial f(x, y)}{\partial x} = \frac{\partial z}{\partial x} \tag{9-6}$$

Maple uses a variation of the middle expression in Equation 9-6, namely $\frac{\partial}{\partial x}f(x, y)$. The partial derivative of $f(x, y)$ with respect to x is understood as the ratio of the change in the function to a small change in x, with y held constant. Similarly, $\frac{\partial f}{\partial y}$ is the ratio of the change in the function to a small change in y, with x held constant. To be more precise, the definition of the partial derivatives $\frac{\partial f}{\partial x}$ and $\frac{\partial f}{\partial y}$ are

$$\frac{\partial f}{\partial x} = Lim_{\Delta x \to 0} \frac{f(x + \Delta x, y) - f(x, y)}{\Delta x} \tag{9-7a}$$

$$\frac{\partial f}{\partial y} = Lim_{\Delta y \to 0} \frac{f(x, y + \Delta y) - f(x, y)}{\Delta y} \tag{9-7b}$$

Partial derivatives are calculated in exactly the same way as ordinary derivatives. If the partial derivative is to be taken with respect to the independent variable x, then all other terms in the expression, whether they are constants or other independent variables, are treated as constants for the purposes of carrying out the differentiation.

Example 9-7

Find the following partial derivatives:

(a) $\frac{\partial}{\partial x}f(x, y)$, if $f(x, y) = x^2 + 2xy + y^2$

(b) $\frac{\partial}{\partial y}f(x, y)$, if $f(x, y) = x^2 + 2xy + y^2$

(c) $\frac{\partial}{\partial x}f(x, y)$, if $f(x, y) = x^2 + y^2$

(d) $\frac{\partial}{\partial x} f(x, y)$, if $f(x, y) = \frac{x}{y} + \ln(y)$

(e) $\frac{\partial}{\partial y} f(x, y)$, if $f(x, y) = x^2 - y^2$ and $x = 1 - y$

Solution:

(a) Type the following equation in Maple:

> diff(x^2+2*x*y +y^2, x);

$$2x + 2y$$

Since y is independent of x, you do *not* write $y(x)$ in the Maple input line, but write simply y. This informs Maple that y is an independent variable. Maple's differentiation process treats y as a constant, producing the correct result.

(b) The identical answer is arrived at if this expression is differentiated with respect to y:

> diff(x^2+2*x*y +y^2, y);

$$2x + 2y$$

Because of the symmetry of the function, a small change in x produces the same change in the function as a small change in y.

(c) Here we show the operator method of finding the partial derivative. Define $x^2 + y^2$ as a function:

> f := (x, y) -> x^2 + y^2;

$$f := (x, y) \rightarrow x^2 + y^2$$

When you use Maple's *D* operator, you need to specify which variable. This is done by noting the order of appearance of x and y in the defining equation. The variable x appears first, so the way to specify the derivative of the function with respect to x is

> D[1](f);

$$(x, y) \rightarrow 2x$$

The answer appears as a function. Note that the derivative with respect to y, $D[2](f)$, is $(x, y) \rightarrow 2y$, a different result. This time the two partial derivatives are the same only if $x = y$.

(d) Returning to the *diff* notation for this problem, the Maple command is

> diff(x/y + ln(y), x);

$$\frac{1}{y}$$

Since y is treated as a constant, it does not matter that it appears in the second term as ln(y), because the derivative of ln(y) with respect to x is still zero.

(e) In this example, x is *not* an independent variable, but is itself a function of y. Thus, once the substitution $x = 1 - y$ is made, $f(x, y)$ becomes

> f(x, y) = subs(x = 1 - y, x^2 - y^2);

$$f(x, y) = (1 - y)^2 - y^2$$

and is no longer a function of x. The derivative of this function with respect to y is

> diff((1-y)^2-y^2, y);

$$-2$$

Why? Well, the expression $(1 - y)^2 - y^2$ can be simplified to $1 - 2y$, so the derivative with respect to y is just -2.

Your Turn. Find the indicated derivatives of $f(x, y)$.

(a) $\frac{\partial}{\partial x}(f(x, y))$, if $f(x, y) = e^{(xy)} + y^2 + 3$

Answer: ____

(b) $\frac{\partial}{\partial y}f(x, y)$, if $f(x, y) = e^{(xy)} + y^2 + 3$

Answer: ____

(c) Calculate $\frac{\partial}{\partial x}f(x, y)$ at $x = 1, y = 1$, and then compare it to $\frac{\Delta f}{\Delta x}$, by setting $\Delta x = 0.001$ in Equation 9-7a. (Do not take the limit; simply work out the value of the ratio for $\Delta x = 0.001$.)

Answer: ____

The Total Differential

The *total differential* of a function $f(x, y)$ is defined as

$$df(x, y) = \frac{\partial f}{\partial x}dx + \frac{\partial f}{\partial y}dy \qquad (9\text{-}8)$$

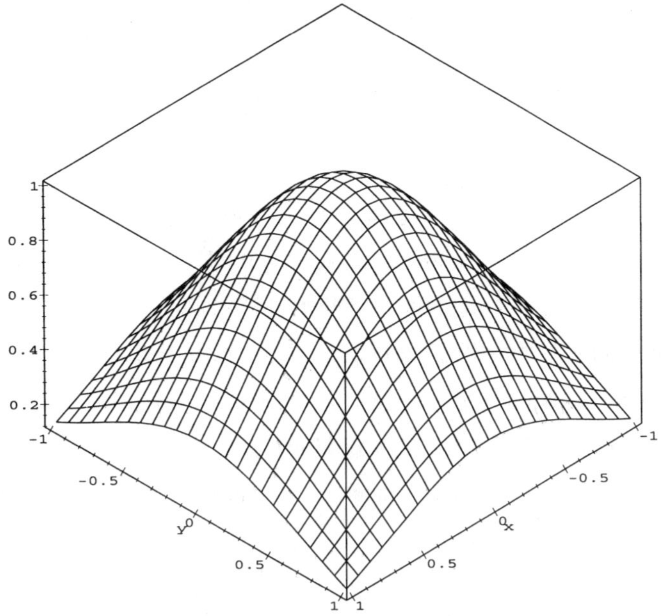

Figure 9.4 A Function $f(x, y)$, Showing Contour Lines in the x and y Directions

It is the sum of the partial derivatives with respect to both x and y, multiplied by their respective differentials. Imagine a surface generated by the plot of $f(x, y)$, such as the one in Figure 9.4. As the contour lines intersect, they form little rectangles with sides that can be thought of as having lengths dx and dy. Pick any place on the surface. The height of the point on the surface is $f(x, y)$. If you imagine going along a contour line lying in the direction of the x axis for a distance dx and then turning and going for a distance dy along the y contour line, you will arrive at a new place on the surface. The new height is $f(x + dx, y + dy)$, and the difference in height is $d(f(x, y) = f(x + dx, y + dy) - f(x, y)$, which is also the total differential.

Example 9-8

(a) Find the total differential of $f(x, y) = 3y^3 - x^2y$ and then evaluate it at $x = 1 + 2\sqrt{2}\sqrt{5}$, $y = 1$, with $\Delta x = 0.001$ and $\Delta y = 0.001$.

(b) Calculate $f(x + \Delta x, y + \Delta y) - f(x, y)$, where x, y, Δx, and Δy have the same values as in part (a). Do you get the same result?

Solution:

(a) Apply the definition of the total differential

```
> ex93 := diff(3*y^3 - x^2*y, x)*dx + diff(3*y^3 - x^2*y, y)*dy;
```

$$ex93 := -2xydx + (9y^2 - x^2)\,dy$$

Then substitute the given values for *x*, *y*, *dx*, and *dy* and convert to a decimal number.

```
> evalf(subs(x = 1 + 2*sqrt(2)*sqrt(5), y = 1, dx = 0.001, dy = 0.001, ex93));
```

$$.05929822128$$

(b) Use the function without taking derivatives and substitute the given values.

```
> evalf(subs(x = 1 + 2*sqrt(2)*sqrt(5), y = 1, dx = 0.001, dy = 0.001, ex93b));
```

$$.05930487$$

Again the results agree to nearly four places. This is strong evidence that the results would be even closer if *dx* and *dy* were made smaller. You can check this supposition quite easily by modifying the command to do so.

Your Turn.

(a) Find the total differential of $f(x,y) = 2x - 3y$ if $\Delta x = 0.003$ and $\Delta y = 0.002$.

(b) Calculate $f(x + \Delta x, y + \Delta y) - f(x, y)$, if Δx and Δy have the same values as in part (a). Do you get the same result? Explain and state why it is not necessary to know definite values of *x* and *y* here.

The Method of Change of Variables

Armed with this amount of knowledge of partial derivatives and total differentials, we are now ready to apply the method of change of variables to problems in first-order differential equations. To apply the method, follow these steps:

1. Write the equation in differential form.
2. Make a guess at the transformation equation. A good choice is to pick out a common sub-expression or to pick the inverse function of some function appearing in the original equation.
3. Write the differentials *dy* or *dx*, or both, in terms of the new variable(s).
4. Solve the new equation in terms of the new variable(s).
5. Recover the original variables. This means that the equations for the changes of variable must be invertible.

Example 9-9

Solve $(2x + 3y + 4)dx + (2x + 3y + 5)dy = 0$ by the method of change of variables.

Solution: This equation is easily solved by Maple's *dsolve* command, but we will provide a step-by-step solution using the change-of-variables method. We use the indeterminates *dx* and *dy*, which Maple treats like any other variable. They are not differentials as far as Maple is concerned.

> deq99 := (2*x+3*y+4)*dx + (2*x+3*y+5)*dy = 0;

$$deq99 := (2x + 3y + 4)\,dx + (2x + 3y + 5)\,dy = 0$$

The expression $2x + 3y$ occurs twice, so a possible set of change of variable equations is

> cv1 := u = 2*x + 3*y: cv2 := v = 3*y:

Solving these for x and y, we show that the equations are invertible:

> solve({cv1, cv2}, {x, y});

$$\{ y = \frac{1}{3}v,\ x = \frac{1}{2}u - \frac{1}{2}v \}$$

From these equations, it is obvious that the total differentials are

> td1 := dx = 1/2*du-1/2*dv; td2 := dy = 1/3*dv;

$$td1 := dx = \frac{1}{2}du - \frac{1}{2}dv$$

$$td2 := dy = \frac{1}{3}dv$$

The transformed differential equation is

> deq99a := subs(x = 1/2*u-1/2*v, y = 1/3*v, td1,td2, deq99);

$$deq99a := (u + 4)\left(\frac{1}{2}du - \frac{1}{2}dv\right) + \frac{1}{3}(u + 5)dv = 0$$

We simplify and collect terms as follows:

> deq99b := simplify(deq99a);

$$deq99b := \frac{1}{2}u\ du - \frac{1}{6}u\ dv + 2du - \frac{1}{3}dv = 0$$

> deq99c := collect(deq99b, du);

$$dec99c: = \left(\frac{1}{2}u + 2\right)du - \frac{1}{6}u\ dv - \frac{1}{3}dv = 0$$

> **deq99d := collect(deq99c, dv);**

$$deq99d := (-\frac{1}{6}u - \frac{1}{3})dv + \left(\frac{1}{2}u + 2\right)du = 0$$

> **deq99e := op(1, lhs(deq99d))/(-1/6*u-1/3) + simplify(op(2, lhs(deq99d))/(-1/6*u-1/3)) = 0 ;**

$$deq99e := dv - 3\frac{(u + 4)du}{u + 2} = 0$$

Remember that most of the commands in this sequence were generated by copying and pasting. Very little typing was required. In addition, the last command was edited by putting in the *simplify* command after viewing Maple's output. It became obvious that an extra *simplify* command would make the final equation simpler.

In *deq99e*, the variables are separated, and we can integrate term by term. Since Maple does not understand that *dv* and *du* are differentials, we must pick out the integrands, either manually or by using the *op* command. In this case, it's far easier to copy the second integrand, the one for *du*, rather than trying to pick out the factors using the *op* command. Even so, we will show you how it's done using the *op* command.

> **I99 := Int(1, v) + Int(op(1, op(2, op(1, deq99e))) * op(2, op(2, op(1, deq99e)))*op(4, op(2, op(1, deq99e))), u) = C;**

$$I99 := \int 1 \, dv + \int - 3\frac{u + 4}{u + 2}du = C$$

By performing these integrals, we find the solution to the differential equation in terms of the transformed variables.

> **sol99a := value(I99);**

$$sol99a := v - 3u - 6 \ln(u + 2) = C$$

All that remains is to rewrite the equation in terms of the original variables.

> **sol99 := subs(cv1, cv2, sol99a);**

$$sol99sv := - 6y - 6x - 6 \ln(2x + 3y + 2) = C$$

To show that *sol99sv* is indeed the solution, we take its derivative with respect to *x*, but first we must make sure to inform Maple that *y* is really *y*(*x*).

> d0 := subs(y = y(x), sol99sv);

$$d0 := -6y(x) - 6x - 6\ln(2x + 3y(x) + 2) = C$$

> d1 := diff(lhs(d0), x)=0;

$$d1 := -6\left(\frac{\partial}{\partial x}y(x)\right) - 6 - 6\frac{2 + 3\left(\frac{\partial}{\partial x}y(x)\right)}{2x + 3y(x) + 2} = 0$$

Solving for the derivative, we find

> d2 := solve(d1, diff(y(x), x));

$$d2 := -\frac{3y(x) + 4 + 2x}{2x + 3y(x) + 5}$$

which is the same value for *dy/dx* that we would get by isolating *dy/dx* in *deq99*. We could, as a final check, use *dsolve*.

Your Turn. Use *dsolve* on the following equation and show that the solution is equivalent to *sol99*.

> deq99a := (2*x+3*y(x)+4) + (2*x+3*y(x)+5)*diff(y(x), x) = 0;

$$deq99a := 2x + 3y(x) + 4 + (2x + 3y(x) + 5)\left(\frac{\partial}{\partial x}y(x)\right) = 0$$

Answer: ____

Interpretation of the First-Order Differential Equation

A first-order differential equation may be solved for *dy/dx* and written in the form

$$\frac{dy}{dx} = f(x, y) \tag{9-9}$$

Equation 9-9 states that a slope is associated with each point (x, y) for which the function $f(x, y)$ is defined. The differential equation can be used to find the little pieces of tangent lines all over the plane by evaluating enough x, y pairs. If you begin at any one of these little tangent segments and draw a line from it to the adjacent one, and so on, you will construct an approximation to a solution curve for the differential equation. By taking a segment that is not part of this curve and following it, another solution is found. In short, a family of curves can be constructed in this manner. If you know the initial condition, you can select the one particular curve that satisfies this condition.

MapleV4 has a command for automatically plotting these tangent segments. We will give an example of the command, applied to a very important differential equation. This equation models growth and decay. It is derived from the observation that in some systems, the rate of change of a quantity depends only on the amount of that quantity in existence at any given time. In biology, the quantity can be the number of bacteria in a tank of nutrients. If there are only a few bacteria compared to the number that the tank can hold, and there is plenty of food, the bacteria will multiply without hindrance. In this case, the growth of new bacteria will only depend on the amount of bacteria present at any given time. Let $B(t)$ be the number of bacteria at any time t. The rate of change of this number is $dB(t)/dt$, and our model assumes that this is proportional to the amount present. Writing this information as a differential equation, we have

$$\frac{\partial}{\partial t}B(t) = k\, B(t) \text{ or} \tag{9-10}$$

Rate of Change = Constant of Proportionality × Amount Present.

The same approach can be taken to systems in which some quantity is decaying. The classic example is the decay of a radioactive substance. Let there be $N(t)$ atoms of a radioactive substance at time t. The law of radioactive decay (in most cases) states that there is a constant probability that any given atom will decay during the next small time interval. Each decay event is independent of any other decay event. It is like tossing a very large number of 20-sided dice and observing how many land with the face labeled *1* on top. Those are the atoms that have decayed during that time interval. Those that come up *1* are removed, and the dice are tossed again. The number of dice removed at each toss depends only on two parameters, (a) the number of sides of the dice, in this case 20, and (b) the number of dice used in the initial toss. Substitute radioactive atoms for dice, and you have the case of radioactive decay.

We make the assumption that there are a huge number of radioactive atoms in our sample when we begin the experiment. For example, a microgram of pure radium contains about 10^{15} atoms. The large number of atoms justifies our use of a continuous variable $N(t)$ to approximate them. From our discussion, you can see that the rate of decay $dN(t)/dt$ is proportional to the number of atoms at time, t, multiplied by some constant of proportionality. The differential equation is

$$\frac{\partial}{\partial t}N(t) = -\, k\, N(t) \tag{9-11}$$

We introduce the negative sign explicitly to indicate that there will be fewer atoms as time goes on.

Next, consider an example from electronics. A charged capacitor is connected to a resistor to form a complete circuit. The resistor is characterized by a resistance value, R, in ohms. The voltage across a capacitor is given by $V(t) = q/C$, where q is the amount of charge in coulombs on one of the capacitor's plates and C is a constant, called the *capacitance*, measured in farads. If a charged capacitor is connected to a resistor, a current will flow. Current is defined as the rate of change of charge, dq/dt. The charge flows from one plate of the capacitor, through the resistor, and onto the other plate of the capacitor. *Ohm's law* states that the voltage across a resistor is related to the rate of change of charge by the equation:

$$V(t) = R\frac{dq}{dt} \tag{9-12}$$

The same voltage exists across the capacitor but in the opposite sense, so we have an equation relating the voltages across both resistor and capacitor:

$$V(t) = R\frac{dq}{dt} = -\frac{q}{C} \tag{9-13}$$

Here, the quantity that is decaying is the electric charge. Moving the constant R in Equation 9-13 to the right-hand side of Equation 9-14, we have

$$\frac{dq}{dt} = -\frac{1}{RC}q \tag{9-14}$$

The constant of proportionality in this case is $-1/RC$.

Investments

As our last example, we derive the equation for the growth of money in an investment that has a constant interest rate. Let the amount of money in the investment be called $P(t)$. This is the value of the investment at any given time t, in years. We make the further assumption that the interest is calculated at each instant. Since the money is invested in a fund that bears a constant rate of return, the growth is proportional to the amount currently in the account, and the differential equation is

$$\frac{dP(t)}{dt} = kP(t) \tag{9-15}$$

In all these examples, the common thread is the fact that the rate of change of the quantity is proportional to the amount present.

$$\frac{dy}{dx} = \pm ky \tag{9-16}$$

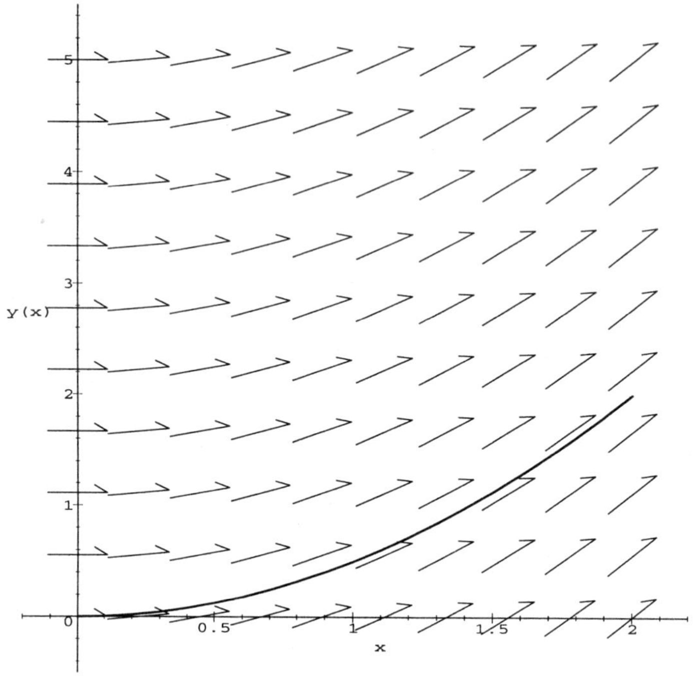

Figure 9.5 Direction Field for $\frac{\partial}{\partial x}y(x) = kx$, with $k = 1$

Plotting the Tangent Segments

Equation 9-16 is one of the simplest, yet it is of great importance. The central objective of the study of differential equations is to find the general solutions to equations involving derivatives. In the case of Equation 9-16, we can plot the direction field for the equation by assigning values to k (Figure 9.5). First, let $k = 1$. The command to plot the little tangent segments for this differential equation contains quite a few parameters. First we will give the command and then we'll explain its parameters. Here it is:

> DEtools[DEplot](diff(y(x),x)= x, y(x), x= 0 .. 2, [[y(0) = 0]], y=0..5, dirgrid=[10,10], color=black, linecolor=COLOR(RGB, 0, 0, 0));

The command itself, *DEtools[DEplot]*, indicates that it is part of the Detools package. The parameters are (a) *diff(y(x),x)= x*, which is the differential equation to be plotted, (b) the dependent variable $y(x)$, (c) the range for the independent variable, $x = 0 .. 2$, (d) the initial condition for the solid black curve that represents a particular solution, $[[y(0) = 0]]$ given as a "list of lists" inside double square brackets, (e) the range for the dependent variable, $y =$

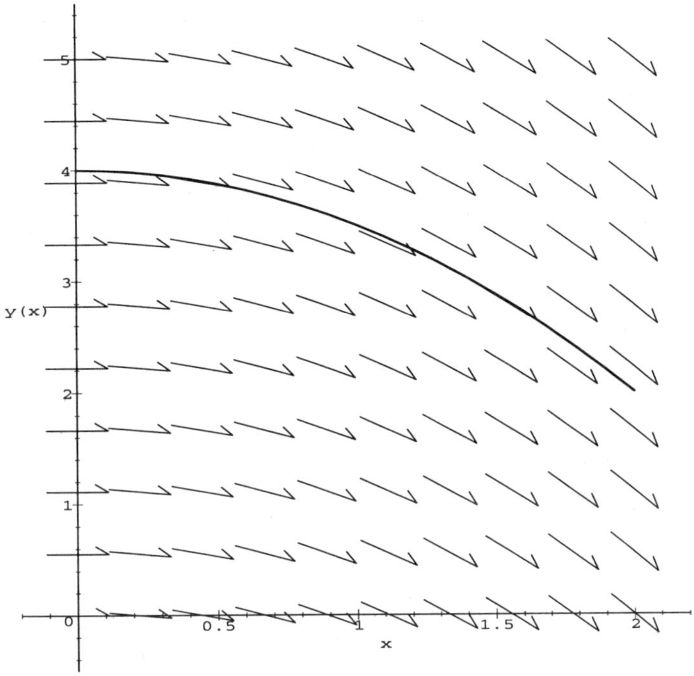

Figure 9.6 Direction Field for $\frac{\partial}{\partial x}y(x)=kx$, with $k=-1$.

0..5, (f) the number of tangent segments to include in the plot, *dirgrid*=[10,10], (g) the color of the tangent segments, *color=black*, and lastly, (h) the color of the line representing the particular solution satisfying the given initial condition, *linecolor=COLOR(RGB, 0, 0, 0))*; The color and grid information may be omitted and Maple will supply default values.

An example of decay exists when $k=-1$ (Figure 9.6). Here is the command and the plot of its direction field, including a solution for the initial condition *[[y*(0) = 4]]:

> DEtools[DEplot](diff(y(x),x)= -x, y(x), x= 0 .. 2, [[y(0) = 4]], y=0..5, dirgrid=[10,10], color=black, linecolor=COLOR(RGB, 0, 0, 0));

Trial-and-Error Solutions

Many of the simpler differential equations can be solved by trial and error. In the case of Equation 9-16, ask yourself, "What function has a derivative that is, except for a multiplicative constant, the same as the function itself?" As you scan the list of derivatives for the elementary functions, you will encounter the function $y=e^x$, whose derivative is also e^x. If the list you are looking at contains $y = e^{kx}$, you see that its derivative is ke^{kx}. Obviously, this is a solution to the differential Equation in 9-16, but is it the most general? Generalizing the function further, we write as our trial solution $y = Ae^{kx} + C$. The derivative of this function is $dy/dx = kAe^{kx}$, which is equal to the trial solution only if $C = 0$. It is known that a general solution to an ordinary differential equation contains as many arbitrary constants as the order of the equation. A first-order differential equation has one arbitrary constant, corresponding to the constant that is introduced as a result of the single integration required to solve the problem.

Example 9-10

Solve the differential equation for an investment where the rate of interest is a constant and is compounded at each instant. Find the value of k if the effective yearly interest rate is 7%.

Solution: The differential equation for this problem is

$$\frac{\partial}{\partial t}P(t) = k\,P(t) \tag{9-17}$$

You have seen that the solution is $P(t) = Ae^{kt}$, the solution having been found by trial and error. Let $P(0)$ be the investment amount at $t = 0$. Then $P(0) = Ae^{k0} = A$. The constant A is identified with the amount of money invested at time $t = 0$.

If the effective yearly interest rate is 7%, then an investment of $A = P(0)$ dollars will grow to $P(1) = 1.07\,P(0)$ dollars. Thus, at $t = 1$ year,

$$P(1) = P(0)e^{k1} = 1.07\,P(0) \tag{9-18}$$

Canceling $P(0)$ from the second two terms of Equation 9-18, we have $e^k = 1.07$, and we can solve for k:

```
> solve(exp(k) = 1.07, {k});
```

$$\{k = .06765864847\}$$

We would leave this number of decimal places in our value for k even though the interest rate is given to only three figures because investors consider 7% an exact number and would expect all calculations based on the derived value of k to be exact to the penny, even if large sums were invested. The formula for the growth of the investment for every dollar invested becomes

```
> p := t -> exp(.06765864847*t);
```

$$p := t \to \mathbf{e}^{(.06765864847)}$$

If $1 million is invested, we require that the amount after 1 year is exactly $1,070,000.00, which is what we get by applying our formula:

> **P = 1e6*p(1);**

$$P = .107000000010^7$$

Your Turn. If $1 billion (1*e*9) is invested, is the formula with $k = 0.06765864847$ still correct?

Answer: ____

We see from these last few examples that there is a way of solving first-order differential equations without the need to apply any method at all. Make up a table of derivatives of the common functions, expressing the derivative in terms of the dependent variable as well as expressing it in the usual form containing only the dependent variable. This table becomes a "dictionary" for trial-and-error solutions.

Example 9-11

Make up the table entry for $y = kx^n + C$.

Solution: Take the derivative of this funciton, then express it in terms of the dependent variable, $y(x)$, as well as the independent variable, x.

> **y := k*x^n + C:**

> **Diff('y', x) = diff(y, x);**

$$\frac{\partial}{\partial x} y = \frac{k x^n n}{x}$$

The right-hand side can be rewritten as $\frac{\partial}{\partial x} y = \frac{n(y - C)}{x}$, or in differential form, $x\, dy = n(y - C)dy$.

Your Turn. Make up the table entry for $y = A \sin(x)$.

Answer: ____

Given a first-order differential equation, see if the form of the equation matches any of those in your dictionary. If there is a match, you are on your way to solving the differential equation! Many of the more common differential equations can be solved this way. Even so, there are disappointingly few differential equations that will yield to this approach. Maple can solve a lot more, but you need to be involved in the solution, and the best way is to experience the solution of common types by the standard methods. That way, you will be better able to interpret Maple's results and to decide on alternative methods when the straightforward approach fails in some way.

The search for some more general methods of solution still relies on classifying the differential equation. In the end, we will let Maple search for a solution based on its own methods. If all these fail, you can fall back on the numerical and graphical approaches. These require that initial conditions be known, but they can provide approximate solutions when analytic solutions are not possible. In the next section, we return to the case where it is possibile to rearrange the differential equation so that each differential term contains only its own variable.

Variables Separable, Continued

Consider the problem of finding the solution to

$$\frac{\partial}{\partial x} y = -ky^2 \qquad (9\text{-}19)$$

This form would appear in a table of differential equations derived from elementary functions, since $y = \frac{1}{kx}$, $\frac{dy}{dx} = -\frac{1}{kx^2} = -\frac{k}{k^2x^2} = -ky^2$. Therefore, we would discover the solution to differential Equation 9-19 by looking at the table entry for the elementary function $y = 1/kx$. The solution we have found in this way cannot be the most general solution because there is no constant of integration in the result. There are only a couple of places to put such a constant, so a little more trial and error will produce the full solution, but we will look for a better method in this section.

If we write the equation in differential form, we have $dy = -ky^2dx$. Moving the term containing dx to the left-hand side and dividing through by y^2, we have $kdx + \frac{dy}{y^2} = 0$. The variables are separated. Each differential is multiplied by its own variable, or a constant. Integrating, we have $\int kdx + \int \frac{dy}{y^2} = C$, $kx - \frac{1}{y} = C$, $y = \frac{1}{kx - C}$. The constant of integration is present, and thus we have the most general solution to Equation 9-19.

Example 9-12

Find the solution to Equation 9-19 using Maple.

Solution:

> sol912 := dsolve(diff(y(x), x) = -k*y(x)^2, y(x));

$$sol912 := \frac{1}{y(x)} = kx + \_C1$$

> sol912a := student[isolate](sol912, y(x));

$$y(x) = \frac{1}{kx + \_C1}$$

We identify $\_C1 = -C$, and the two solutions are identical.

Your Turn: Find the solution to $\frac{dy}{dx} = -ky^3$ by separating variables. Check using the *dsolve* command.

Answer: ______________________________

Example 9-13

Solve differential equation *deq*913 by separating variables. Check the solution using Maple.

> deq913 := diff(y(x), x) + k*x*y(x)^3 = 0;

$$deq913 := \left(\frac{\partial}{\partial x}y(x)\right) + kxy(x)^3 = 0$$

(a) To begin the separation of variables, write the equation in differential form.

> eq913a := dy + k*x*y^3*dx - 0;

$$eq913a := dy + kxy^3dx$$

(b) Divide by y^3 to separate the variables.

> eq913b := expand(eq913a/y^3);

$$eq913b := \frac{dy}{y^3} + kxdx$$

(c) Integrate.

> eq913c := int(1/y^3, y) + int(k*x, x) = C;

$$eq913c := -\frac{1}{2}\frac{1}{y^2} + \frac{1}{2}kx^2 = C$$

(d) Check by using Maple's *dsolve* command.

> eq913d := dsolve(deq913, y(x));

$$eq913d := \frac{1}{y(x)^2} = kx^2 + \_C1$$

You can use a little bit of mental manipulation to show that the two solutions are the same.

(e) If you want to express the solution as a function, you can use the command:

> convert(student[isolate](eq913d, y(x)), radical);

$$y(x) = \sqrt{\frac{1}{kx^2 + \_C1}}$$

If you ask Maple to isolate *y(x)* in *eq913d,* it writes the solution in *RootOf* form. The purpose of *convert(. . ., radical)* is to write *y(x)* in terms of the square root radical sign.

Your Turn. Find, by separating variables and by *dsolve,* the solution to

> deq913yt := k*(x*diff(y(x), x) + 2*y) = x*y(x)*diff(y(x), x);

$$deq913yt := k\left(x\left(\frac{\partial}{\partial x}y(x)\right) + 2y\right) = xy(x)\left(\frac{\partial}{\partial x}y(x)\right)$$

Answer: ____

The constant of integration can be determined if an initial condition is known. The next example shows how.

Example 9-14

Solve

> deq914 := 2*y(x) + x^2*diff(y(x), x) = -x*y(x);

$$deq914 := 2y(x) + x^2\left(\frac{\partial}{\partial x}y(x)\right) = -1$$

given the initial condition

> ic914 := y(7/2) = 1/ln(2);

$$ic914 := y\left(\frac{7}{2}\right) = \frac{1}{\ln(2)}$$

Solution: Note that we have given a name to the differential equation and to the equation describing the initial condition. This will make it easy to formulate the *dsolve* command.

(a) To apply the technique of separation of variables, we rewrite *deq*914 in differential form.

> eq914a := 2*y*dx +x^2*dy = -x*y*dx;

$$eq914a := 2ydx + x^2dy = -xydx$$

(b) Divide through by x^2y.

> eq914b := expand(eq914a/(x^2*y));

$$eq914b := 2\frac{dx}{x^2} + \frac{dy}{y} = -\frac{dx}{x}$$

(c) The variables are separated, so the solution is found by integrating term-by-term. The solution is

> eq914b := int(2/x^2, x) + int(1/y, y) = int(-1/x, x) + C;

$$eq914b := -\frac{2}{x} + \ln(y) = -\ln(x) + C$$

Determine the constant *C* by substituting the initial condition:

> eq914b1 := subs(x = 7/2, y = 1/ln(2), eq914b);

$$eq914b1 := -\frac{4}{7} + \ln\left(\frac{1}{\ln(2)}\right) = -\ln\left(\frac{7}{2}\right) + C$$

> solve(eq914b1, {C});

$$\{ C = -\frac{4}{7} - \ln(\ln(2)) + \ln\left(\frac{7}{2}\right) \}$$

> eq914b2 := subs(C = -4/7-ln(ln(2))+ln(7/2), eq914b);

$$eq914b2 := -\frac{2}{x} + \ln(y) = -\ln(x) - \frac{4}{7} - \ln(\ln(2)) + \ln\left(\frac{7}{2}\right)$$

(d) Using Maple,

> dsolve({deq914, ic914}, y(x));

$$eq914c := = y(x)\ \frac{7}{2}\frac{e^{\left(\frac{2}{x}\right)}}{x e^{\left(\frac{4}{7}\right)}\ln(2)}$$

Are these two solutions the same? They are, but it takes a bit of manual algebra to prove it. This is a case where Maple's simplification routines have trouble showing the equality of the two equations. Nevertheless, you can always verify that the solutions are almost the same by plotting both functions on the same graph. You will see a single line, but you can nudge the lines apart by adding a small offset, like 0.1, to one of the functions. Then you will see two lines.

Your Turn. Solve $\frac{dy}{dx}$, $y(0) = 1$. What is the shape of the solution?

Answer: ____

Review

We have been discussing two methods, the straightforward one, where you define the differential equation and use Maple's *dsolve* command or method using the "fake" differential expression. Once you have separated the variables, you convert to an equation containing integrals. Since you do part of this conversion by hand, you must be especially careful to copy terms correctly, keeping track of minus signs, and the like. The first method is general because it works on a wider class of differential equations than the variables separable type. The second method is a more step-by-step method, and you may find that the variables cannot be separated. If this happens, you will need to try other techniques, such as the ones we discuss in the following sections.

Graphing the Solution to a Differential Equation

Here we introduce the first of the methods that work even when Maple cannot find a symbolic solution to a differential equation. Recall that a first-order differential equation gives information about the slopes of the solution line at points throughout the plane. Using this information, you could plot tiny line segments having the slope of the solution line at various selected points on the plane. To be more specific, you could mark off a rectangle in the x, y plane, say $x = 0 .. 10$ and $y = 0 .. 10$, and construct a table for the 121 grid points (x_i, y_j), $i, j = 0 .. 10$. The entries in the table are the slopes dy/dx at each one of these points. Using these data, you draw small line segments having the calculated slopes, centered on the grid points. You now have a picture of all these slopes. You construct a particular solution by selecting any point on the plane and following its slope line until it joins another slope line. As you continue to follow the slope lines, you construct a possible solution to the differential equation.

Example 9-15

Construct the slope lines for

> deq915 := diff(y(x), x) = -y(x)/x;

$$deq915 := \frac{\partial}{\partial x}y(x) = -\frac{y(x)}{x}$$

with the initial condition

> ic915 := y(1) = 10;

$$ic915 := y(1) = 10$$

Solution: You have seen that the command *Deplot* has many options. Here it is in all its glory.

> DEtools[DEplot](deq915, y(x), x= 1 .. 10, [[ic915]], y= 0 .. 12, dirgrid=[10,10], color=black, linecolor=COLOR(RGB, 0, 0, 0),scaling=constrained, stepsize=0.1);

If Maple can find a solution to the differential equation, as it can in this case, you can find the solution and plot it.

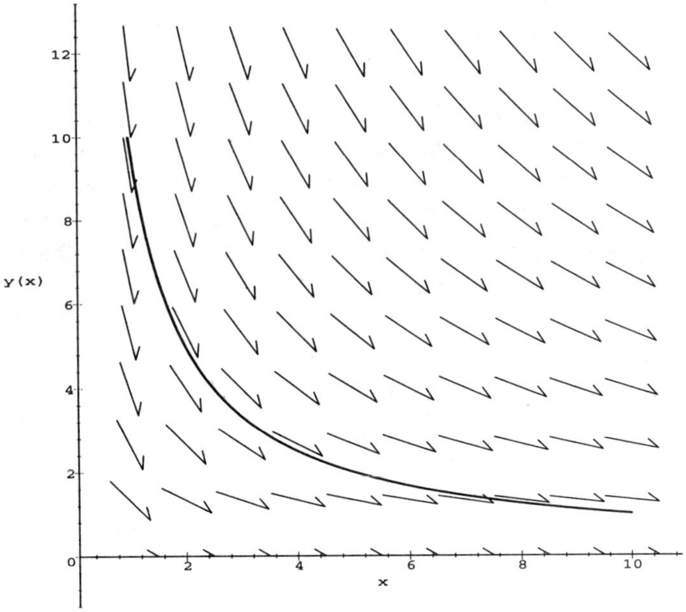

Figure 9.7 Direction Field for Example 9-15

```
> s915 := dsolve({deq915, ic915}, y(x));
```

$$s915 := y(x) = \frac{10}{x}$$

Plotting the right side of *s*915 gives the same line as Figure 9.7.

Your Turn. Construct the slope lines for

```
> deq915yt := diff(y(x), x) = - x/y(x); ic915yt := y(0) = 1;
```

$$deq915yt := \frac{\partial}{\partial x} y(x) = -\frac{x}{y(x)}$$

$$ic915yt := y(0) = 1$$

Describe the shape of the particular solution.

Answer: ____

Example 9-16

Construct the slope lines for

```
> deq916 := x + sqrt(1 + y(x)^2)*diff(y(x), x) = 0; ic916 := y(0) = 4;
```

$$deq916 := x + \sqrt{1 + y(x)^2}\left(\frac{\partial}{\partial x} xy\right) = 0$$

$$ic916 := y(0) = 4$$

The direction lines in this plot cross the particular solution (Figure 9.8). This cannot be! You need to recognize that the curve for a particular solution cannot cross any of the slope segments, and certainly a direction field in which all the segments are horizontal straight lines cannot represent a field whose slope is $-x/y(x)$. We will discuss this further after we ask you to look at one further example:

Your Turn. Construct the slope lines and solution for

```
> deq916yt := diff(y(x), x) = -x/y(x); ic916yt := y(0) = 1;
```

$$deq916yt := \frac{\partial}{\partial x} y(x) = -\frac{x}{y(x)}$$

$$ic916yt := y(0) = 1$$

using the command

```
> DEtools[DEplot](deq916yt, y(x), x= -2 .. 2, [[ic916yt]], y= -.1..1, dirgrid=[20,10],
color=black, linecolor=COLOR(RGB, 0, 0, 0),scaling=constrained, stepsize=0.1);
```

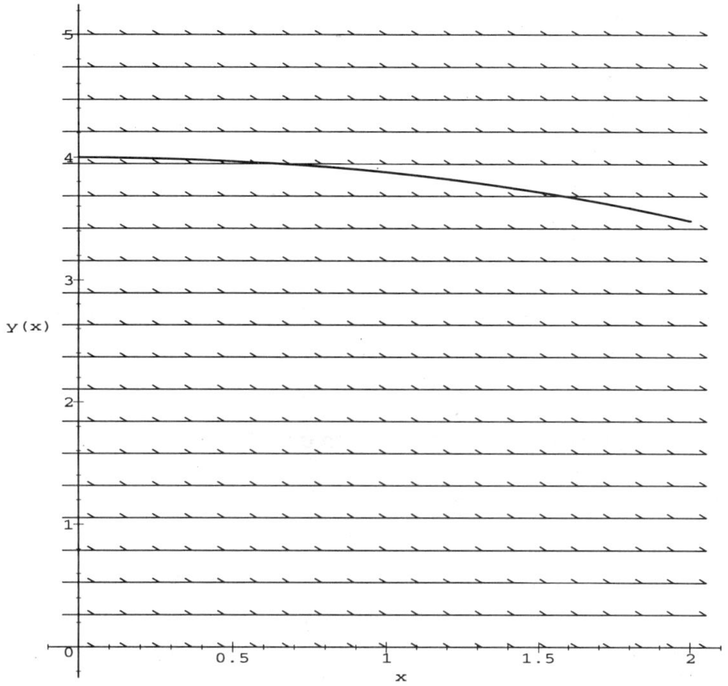

Figure 9.8 An Incorrect Plot for the Direction Field for Example 9-16

What is wrong with the plot? Solve the differential equation for this initial condition and compare the solution to the plot.

Answer: ____

Now review the last two examples. They answer the question, "Why bother to learn techniques for the solution of differential equations?" The last examples show that you cannot blindly accept the graphs that may be produced by the differential equation solver. You need to supply your own assessment about whether the graphs are correct or not. If the slopes of the direction field are all horizontal and don't follow the slope of the solution curve, then you must realize that something is wrong. Example 9-16, in the section called "Your Turn," shows that the solver may add an extra piece to the solution. Not only that, a whole branch of the curve has been missed. (Try *ic916yta := y(0) = –1;* as the initial condition, then you will see the other half of the full solution.) Verification of the solution is most important. Always ask yourself if the solution is reasonable and then check any suspicious behavior.

Exact Differentials and Integrating Factors

The total differential is given by Equation 9-8. The expression on the right-hand side of 9-8 is a differential form like those we have already encountered. If such a form is equated to 0, we have a differential equation expressed in differential form rather than in derivative form. The differential form may or may not be such that it satisfies $df(x, y) = \frac{\partial f}{\partial x}dx + \frac{\partial f}{\partial y}dy$. If it does, the differential equation is called *exact*. The solution of the equation is simply $f(x, y) = C$.

The differential equation

$$M(x, y)dx + N(x, y)dy = 0 \tag{9-20}$$

will be exact if there is some function $f(x, y)$ such that

$$\frac{\partial f}{\partial x} = M(x, y) \text{ and } \frac{\partial f}{\partial y} = N(x, y) \tag{9-21a}$$

From these equations, we see that

$$\frac{\partial^2 f}{\partial y \partial x} = \frac{\partial}{\partial y}M(x, y) \text{ and } \frac{\partial^2 f}{\partial x \partial y} = \frac{\partial}{\partial x}N(x, y) \tag{9-21b}$$

We will assume that the two partial derivatives, $\frac{\partial^2 f}{\partial y \partial x}$ and $\frac{\partial^2 f}{\partial x \partial y}$ are equal. That is, it does not matter in which order the derivatives are taken. Thus,

$$\frac{\partial}{\partial y}M(x, y) = \frac{\partial}{\partial x}N(x, y) \tag{9-21c}$$

Conversely, if the conditions in 9-21b hold, it can be shown that the right-hand side of Equation 9-20 is an exact differential. The solution to differential Equation 9-20 starts by integrating $\frac{\partial f}{\partial x} = M(x, y)$ with respect to x.

$$f = \int^{x} M(x, y)dx + g(y) \tag{9-22}$$

The little x over the integral sign means that y is treated as a constant in the integration. Since y was held constant when computing $\frac{\partial f}{\partial x}$, it is held constant during the integration as well. The integral in Equation 9-22 can be performed in Maple by writing y instead of $y(x)$ in the expression $M(x, y)$. That way, Maple will treat y as a constant, just as we require. Any part of the function f that consists solely of terms in y alone will drop out of the partial derivative of *f* with respect to *x*, so we add an unknown function *g*(*y*), representing these lost terms. Using the value of *f* computed from 9-22, we examine the second part of 9-21a:

$$\frac{\partial f}{\partial y} = \frac{\partial}{\partial y}\left(\int^{x} M dx + g(y)\right) = \frac{\partial}{\partial y}\int^{x} M dx + \frac{d}{dy}g(y) = N \qquad (9\text{-}23)$$

In Equation 9-22 we have written *dg*/*dy* because *g* is a function of *y* only. Solving for *dg*/*dy*, we find

$$\frac{d}{dy}g(y) = N(x, y) - \frac{\partial}{\partial y}\int^{x} M dx \qquad (9\text{-}24)$$

We have said that *g* is a function of *y* only, so the right-hand side of Equation 9-24 should also be a function of *y* only. If we take the derivative of the right-hand side of 9-24 with respect to *x*, we have $\frac{\partial}{\partial x}(N - \frac{\partial}{\partial y}\int^{x} M dx) = \frac{\partial N}{\partial x} - \frac{\partial M}{\partial y} = 0$. This must be true since $\frac{\partial N}{\partial x} = \frac{\partial M}{\partial y}$ from 9-21c.

Integrating 9-24, we have

$$g = \int\left(N - \frac{\partial}{\partial y}\int^{x} M dx\right)dy \qquad (9\text{-}25)$$

Substituting 9-25 in 9-22 gives us

$$f(x, y) = \int^{x} M(x, y)dx + \int\left(N - \frac{\partial}{\partial y}\int^{x} M dx\right)dy \qquad (9\text{-}26)$$

We have spent some time on this derivation because of the importance of exact differential equations. They provide the solution to a most common and useful type of differential equation, the linear differential equation that appears in the next section.

Obviously, an example of the application of Equation 9-26 is required, and here it is.

Example 9-17

Show that

> ex917 := dx/sqrt(x^2 + y^2) + (1/y - x/(y*sqrt(x^2+y^2)))*dy = 0;

$$ex917 := \frac{dx}{\sqrt{x^2 + y^2}} + \left(\frac{1}{y} - \frac{x}{y\sqrt{x^2 + y^2}}\right)dy = 0$$

is exact, then solve it using Equation 9-26.

Solution: The first thing to do is to identify the quantities *M* and *N* in *ex*917.

> M := 1/sqrt(x^2+y^2); N := 1/y - x/(y*sqrt(x^2+y^2));

$$M := \frac{1}{\sqrt{x^2 + y^2}}$$

$$N := \frac{1}{y} - \frac{x}{y\sqrt{x^2 + y^2}}$$

Calculate the derivatives in 9-21b to show that differential equation *ex*917 is exact.

> Diff('M', y) = simplify(diff(M, y));

$$\frac{\partial}{\partial y} M = -\frac{y}{(x^2 + y^2)^{3/2}}$$

> Diff('N', x) = simplify(diff(N, x));

$$\frac{\partial}{\partial x} N = -\frac{y}{(x^2 + y^2)^{3/2}}$$

The partial derivatives are the same; therefore, differential equation *ex*917 is exact. Note the use of single quote marks around *M* and *N* in the last two commands. The single quotes prevent the evaluation of *M* and *N* on the left-hand side of the Maple output, giving us a nice-looking equation.

Write the formula for the solution of an exact differential equation as a Maple command. Use the inert forms for the derivative and the integrals so that the formula itself is displayed. Then show the actual integrals that need to be done using the *eval* command. The evaluation takes place using the *value* command. The problem is being done this way to show each step in the solution. It can be done in one step by replacing the inert forms, *Int* and *Diff*, with *int* and *diff* in the following command.

> s917 := Int('M', x)+Int('N' - Diff(Int('M', x), y), y);

$$s917 := \int M dx + \int N - \left(\frac{\partial}{\partial y}\int M dx\right) dy$$

> s917a := eval(s917);

$$s917a := \int \frac{1}{\sqrt{x^2 + y^2}} dx + \int \frac{1}{y} - \frac{x}{y\sqrt{x^2 + y^2}} - \left(\frac{\partial}{\partial y}\int \frac{1}{\sqrt{x^2 + y^2}} dx\right) dy$$

> s917b := value(s917);

$$s917b := \ln(x + \sqrt{x^2 + y^2})$$

The solution to differential equation *ex*917 is $\ln(x + \sqrt{x^2 + y^2})$. This solution is verified by writing *ex*917 in differential equation form and then using *dsolve*:

```
> deq917 := 1/sqrt(x^2 + y(x)^2) + (1/y(x) - x/(y(x)*sqrt(x^2+y(x)^2)))*diff(y(x), x)
= 0;
```

$$deq917 := \frac{1}{\sqrt{x^2 + y(x)^2}} + \left(\frac{1}{y(x)} - \frac{x}{y(x)\sqrt{x^2 + y(x)^2}}\right)\left(\frac{\partial}{\partial x}y(x)\right) = 0$$

```
> sol917 := dsolve(deq917, y(x));
```

$$sol917 := \ln\left(x + \sqrt{x^2 + y(x)^2}\right) = \_C1$$

Your Turn. Show that

```
> ex917yt := (y/x^2 + y + 1/y)*dx + (x - 1/x - x/y^2)*dy = 0;
```

$$ex917yt := \left(\frac{y}{x^2} + y + \frac{1}{y}\right)dx + \left(x - \frac{1}{x} - \frac{x}{y^2}\right)dy = 0$$

is exact. Then solve it by applying Equation 9-26. Check the solution by writing *ex917yt* as a differential equation and using *dsolve*.

Answer: ______________________________

We are now ready to apply our knowledge about exact differential equations to the first-order linear differential equation.

The First-Order Linear Differential Equation

Integrating Factors

If a differential equation $Mdx + Ndy = 0$ is multiplied by an expression $f(x, y)$ so that $fMdx + fNdy$ is an exact differential, then f is said to be an *integrating factor* for the differential equation. For example, $\frac{1}{x^2}$ is an integrating factor for the differential equation $xdy - ydx = 0$ because $d\left(\frac{x}{y}\right) = \frac{xdy - ydx}{x^2}$. The solution of the differential equation is therefore simply $y/x = C$. The use of an integrating factor to convert a differential equation into an exact differential equation provides the key to the solution of the first-order linear differential equation, whose most general form is either

$$\frac{dy(x)}{dx} + P(x)y(x) = Q(x) \tag{9-27a}$$

or its equivalent differential form

$$dy(x) + P(x(y)x)dx = Q(x)dx \tag{9-27b}$$

After spending so much time on exact differential equations, you must suspect that we will look for an integrating factor to make Equation 9-27a into an exact differential equation. We may discover the form of the integrating factor by solving the simpler differential equation:

$$\frac{dy(x)}{dx} + P(x)y(x) = 0 \qquad \text{(9-28a)}$$

Separating variables by dividing by $y(x)$, we obtain

$$\frac{dy}{y} + P(x)dx = 0 \qquad \text{(9-28b)}$$

whose solution is $\ln(y) + \int P(x)dx = \ln(C)$, where we have chosen $\ln(C)$—after all, $\ln(C)$ is still a constant—as the constant of integration in order to make the final formula simpler. Raise each side of the equation to the power e to obtain $ye^{\int P(x)dx} = C$ and take the total differential of the right-hand side of this equation. The result is

$$d\left(ye^{\int Pdx}\right) = e^{\int Pdx}(Pydx + dy) \qquad \text{(9-29)}$$

The term in brackets on the right-hand side of Equation 9-29 is simply the left-hand side of 9-27a, so multiplying it by the integrating factor $e^{\int Pdx}$ does produce an exact differential, namely $d\left(ye^{\int Pdx}\right)$. We must multiply the right-hand side of 9-27a by the same integrating factor to preserve equality. With this multiplication, equation 9-27a becomes

$$e^{\int pdx}(dy(x) + P(x)y(x)dx) = e^{\int pdx}Q(x)dx \qquad \text{(9-30)}$$

The left-hand side is an exact differential whose integral we know is $ye^{\int Pdx}$, and the right-hand side is a function of x only. The solution to the first-order linear differential equation is therefore

$$ye^{\int pdx} = \int e^{\int pdx}Q(x)dx + C \qquad \text{(9-31)}$$

The equation is important because it models many processes in science, engineering, economics, and other disciplines.

Example 9-18

Suppose a skydiver jumps from an aircraft at a height of 4,572 meters (15,000 feet). Find a differential equation for her vertical speed assuming that air resistance retards her downward motion with a force proportional to her speed in the downward direction.

Derive, then solve, the differential equation and determine the constant in the air resistance term if her mass is 50 kilograms (kg), and her *terminal velocity* is 45 meters per second (m/s), which is nearly 100 miles per hour. An object falling through the atmosphere eventually reaches a constant speed, called its terminal velocity, as long as it is far enough above the ground when it starts its fall.

Solution: The forces acting on the skydiver are gravity (*mg*) downward and air resistance (*kv*) upward. Newton's second law of motion states that the total force on the skydiver produces a change in velocity given by the equation

> deq918 := m*g - k*v(t) = m*diff(v(t), t);

$$deq918 := mg - k\ v(t) = m\left(\frac{\partial}{\partial t}v(t)\right)$$

where *m* is her mass in kilograms, *g* is a constant equal to 9.8 meters per second squared (m/s^2), *k* is an undetermined constant whose units are kilograms per second (kg/s), and *t* is the time in seconds (s).

Dividing *deq*918 through by the constant *m*, we obtain an equation whose form is the same as Equation 9-27. It is

> deq918a := diff(v(t), t) + k/m*v(t)= g;

$$deq918a := \left(\frac{\partial}{\partial t}v(t)\right) + \frac{k\,v(t)}{m} = g$$

We identify the terms in Equation 9-31 for the solution to linear differential Equation 9-27 and apply it to the particular case of *deq*918*a*. The integrating factor is

> exp(Int(P(x), x)) = exp(Int(k/m, t));

$$\mathbf{e}^{\left(\int P(x)dx\right)} = \mathbf{e}^{\left(\int \frac{k}{m}dt\right)}$$

The value of this integral is

> f918 := value(exp(int(k/m,t)));

$$f918 := \mathbf{e}^{\left(\frac{kt}{m}\right)}$$

Substituting this in Equation 9-31, we have

> s918 := v(t)* exp(k/m*t) = Int(g* exp(k/m*t), t)+ C;

$$s918 := v(t)\mathbf{e}^{\left(\frac{kt}{m}\right)} = \int g\mathbf{e}^{\left(\frac{kt}{m}\right)}dt + C$$

We can isolate $v(t)$ with the command

```
> s918a := expand(value(s918)/exp(k/m*t));
```

$$s918a := v(t) = \frac{mg}{k} + \frac{C}{e^{\left(\frac{kt}{m}\right)}}$$

To determine the constant C, we assume that $v(0) = 0$. That is, the skydriver's downward speed when she jumps from the plane is 0, and her stopwatch begins timing then. Substituting known parameters, we have

```
> s918b := subs(m = 50, g = 9.8, v(t) = 0, t = 0, s918a);
```

$$s918b := 0 = \frac{490.0}{k} + \frac{C}{e^0}$$

Note that it is *very important* to make the $t = 0$ substitution after the substitution $v(t) = 0$. If the order of these two substitutions is reversed, Maple will write $v(0)$ for $v(t)$, making sure the substitution $v(t) = 0$ impossible because $v(t)$ is no longer in the formula! If you find this confusing, try the command with the substitutions reversed for yourself and observe the results.

Now solve for C:

```
> c918 := solve(s918b, C);
```

The solution to the differential equation at this point is

```
> s918c := subs(C = c918, m = 50, g = 9.8, s918a);
```

$$s918c := v(t) = \frac{490.0}{k} - \frac{490.0}{e^{\left(\frac{1}{50}kt\right)}k}$$

Now we determine the constant k by observing that in the limit of very long times, the speed $v(\infty) = 45$ is the limit of the right-hand side of $s918c$. To do so, we find that we must assume that k is a real positive number.

```
> assume(k >0);
```

```
> k918 := solve(45 = limit(rhs(s918c), t = infinity), k);
```

$$k918 := 10.88888889$$

We have arrived at the complete solution to our differential equation:

> sol918 := subs(C = c918, k = k918, m = 50, g = 9.8, s918a);

$$sol918 := v(t) = 44.99999999 - \frac{44.99999999}{e^{(.2177777778)}}$$

An inspection of *sol*918 reveals that at $t = 0$, the speed is indeed 0 (since $e^0 = 1$), and at large values of t, the second term will be insignificant, so the terminal velocity is 45 m/s.

Your Turn. If we define "a long time" as 5 m/k, how many seconds is this, how fast is the skydiver going at this time, and how far has she fallen?

Answer: 5*m*/*k* = _______ seconds; v(5m/*k*) = _______ m/s; distance = _______ m

Linear Differential Equations with Constant Coefficients

An Introduction to the Operator Method

When the degree of a linear differential equation is two or higher, the operator method, which is discussed next, simplifies the work. It provides a method of solving many equations of practical importance. You have met Maple's differential operator D in previous chapters. Using this notation, a third-order linear differential equation may be written

> (D@@3)(y) + a*(D@@2)(y) +b*D(y) + c*y = 0 ;

$$(D^{(3)})(y) + a(D^{(2)})(y) + bD(y) + cy = 0$$

This form of a differential equation can be converted to the more familiar form using *convert, diff*:

> convert((D@@3)(y)(t) + a*(D@@2)(y)(t) +b*D(y)(t) + c*y(t) = 0, diff) ;

$$\left(\frac{\partial^3}{\partial t^3}y(t)\right) + a\left(\frac{\partial^2}{\partial t^2}y(t)\right) + b\left(\frac{\partial}{\partial t}y(t)\right) + cy(t) = 0$$

There is an algebra for the D operators that we will not go into here. Consult any text on technical calculus for a treatment of this subject. We will only do enough to give you the flavor of the approach.

Example 9-19

Solve

> **deq919 := diff(y(x), x$2) + 4*diff(y(x), x) - 5*y(x) = 0;**

$$deq919 := \left(\frac{\partial^2}{\partial x^2}y(x)\right) + 4\left(\frac{\partial}{\partial x}y(x)\right) - 5y(x) = 0$$

Solution: Since we want to use the operator method, we convert this equation to operator form.

> **deq919a := convert(deq919, D);**

$$deq919a := (D^{(2)})(y)(x) + 4\ D(y)(x) - 5y(x) = 0$$

If we "factor" $y(x)$, the equation resembles a quadratic equation:

> **eq919 := (D^2 + 4*D - 5)*y = 0;**

$$eq919 := (D^2 + 4D - 5)\,y = 0$$

This quadratic form is factorable:

> **factor(eq919);**

$$(D + 5)\,(D - 1)\,y = 0$$

Let $z = (D - 1)y$. Then $eq919$ becomes $(D + 5)z = 0$, or $Dz + 5z = 0$. In this last form, it is equivalent to the differential equation

> **eq919a := convert(D(z)(x) + 5*z(x) = 0, diff);**

$$eq919a := \left(\frac{\partial}{\partial x}z(x)\right) + 5z(x) = 0$$

Equation $eq919a$ is a first-order linear differential equation whose solution is easily found.

> **s919a := dsolve(eq919a, z(x));**

$$s919a := z(x) = \mathbf{e}^{(-5x)}\ \_C1$$

Once $z(x)$ is known, substitute the solution in $z = (D - 1)y$, and solve for y, since this, too, is a first-order differential equation. Substituting the appropriate terms in $z = (D - 1)y$, we arrive at

> **deq919b := op(2, s919a) = convert(D(y)(x) - y(x), diff);**

$$deq919b := \mathbf{e}^{(-5x)}\ \_C1 = \left(\frac{\partial}{\partial x}y(x)\right) - y(x)$$

This is a linear differential equation that is easily solvable. The solution is

> **s919b := dsolve(deq919b, y(x));**

$$s919b := y(x) = \left(-\frac{1}{6}\_C1 + e^{(6x)}\,\_C2\right) e^{(-5x)}$$

Compare this solution to Maple's "one-step" solution:

> **s919c := dsolve(deq919, y(x));**

$$s919c := y(x) = e^{x}\,\_C1 + \_C2\, e^{(-5x)}$$

Once again it is not obvious that the two solutions are identical, but a little manipulation of *s*919*b* puts it in the form of *s*919*c*.

We are now moving away from a discussion of the standard techniques of solution for differential equations and on to the methods that make more complete use of Maple. It has been stressed that you need to have some familiarity with the classic methods of solution, if only to gain some exposure to a set of solvable differential equations. When you have gained that basic familiarity, you are in a position to take a critical look at any solution Maple proposes and determine for yourself whether it is believable or not.

Numerical Methods

It has been mentioned more than once that most differential equations, like most integrals, do not possess solutions in closed form, meaning that no solution in terms of a finite number of elementary functions exists for the "typical" differential equation. Only special ones have such closed-form solutions, and we have investigated a few of the types that have in the previous sections.

Since differential equations are solved to provide the answer to practical problems, it usually happens that initial conditions are known, or at least educated guesses can be made about possible values for initial conditions. With these as the starting place, the differential equation is used to compute a small continuation of the curve. Taking the end-point of the continuation as the new initial condition, another continuation is computed. This process is repeated enough times to plot the solution curve over the required interval.

We discussed in Example 9-18 the case where air resistance is proportional to an object's speed through the air. In aerodynamics, we learn that a better approximation to air resistance is that the drag, or retarding force, is proportional to the square of the speed. Let us return to the case of our skydiver, this time with a quadratic air resistance in the differential equation.

Example 9-20

Use the data of Example 9-18, but this time use as the differential equation

```
> deq920 := m*g - k*v(t)^2 = m*diff(v(t), t);
```

$$deq920 := mg - kv(t)^2 = m\left(\frac{\partial}{\partial t}v(t)\right)$$

This equation is solvable analytically, but we will try the numeric approach. The mass of the skydiver is known (50 kg), and the value for g is 9.8 m/s^2. We must determine k. Since k is related to the terminal velocity, dv/dt is 0 when the terminal velocity is reached; the equation for k is simply $0 = 9.8 \times 50 - 45^2\,k$; therefore, $k = 0.242$. The differential equation becomes, after substituting in the known values for the constants:

```
> deq920a := subs(m = 50, g = 9.8, k = 0.242, deq920);
```

$$deq920a := 490.0 - .242v(t)^2 = 50\left(\frac{\partial}{\partial t}v(t)\right)$$

The initial condition is still $v(0) = 0$. The numeric solution is found from the *dsolve* command, written this way:

```
> nsol920 := dsolve({deq920a, v(0) = 0}, v(t), type = numeric);
```

nsol920 := proc(rkf45_x) ... end

Maple returns a *procedure* as its solution. There are many things you can do with the procedure. The most useful for our purposes is the graphical. We graph the solution using *plots[odeplot]*. We know from Example 9-18 that a reasonable time for the plot is in the order of a few seconds or so (Figure 9.9).

```
> plots[odeplot](nsol920, [t, v(t)], 0 .. 10, labels=[`t`,`v(t)`], color = black);
```

The great power of Maple comes into play when you want to compare different types of air resistance. Once the formulas have been set up in Maple, it becomes a simple matter to reuse them. We will compare the two cases: air resistance is linear with velocity, and air resistance is quadratic with velocity as in *deq920*, and both having the same terminal velocity (Figure 9.10). The commands are:

> p1 := plots[odeplot](nsol920, [t, v(t)], 0 .. 10, labels=[`t`,`v(t)`], color = black): (Command for Figure 9.8)

> p2 := plot(44.99999999-44.99999999/exp(.2177777778*t), t = 0 .. 10, linestyle=2): (Solution to Example 9-18)

> plots[display]({p1,p2}); (Display the two curves on the same plot)

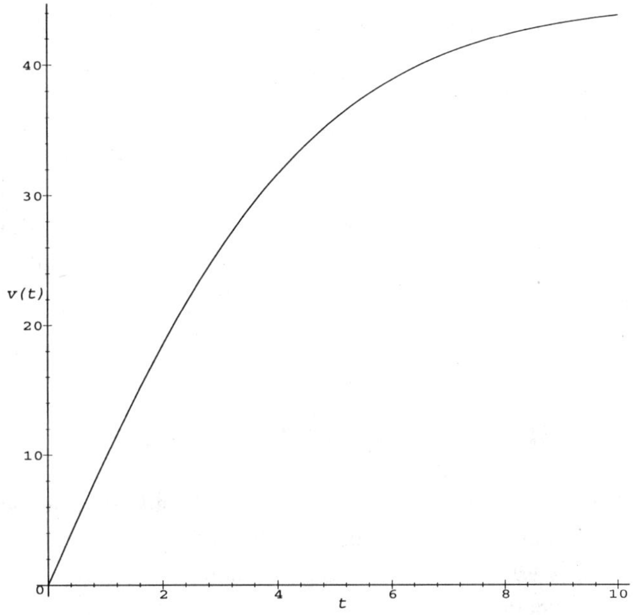

Figure 9.9 Velocity-Time Graph When Air Resistance Is Proportional to v^2

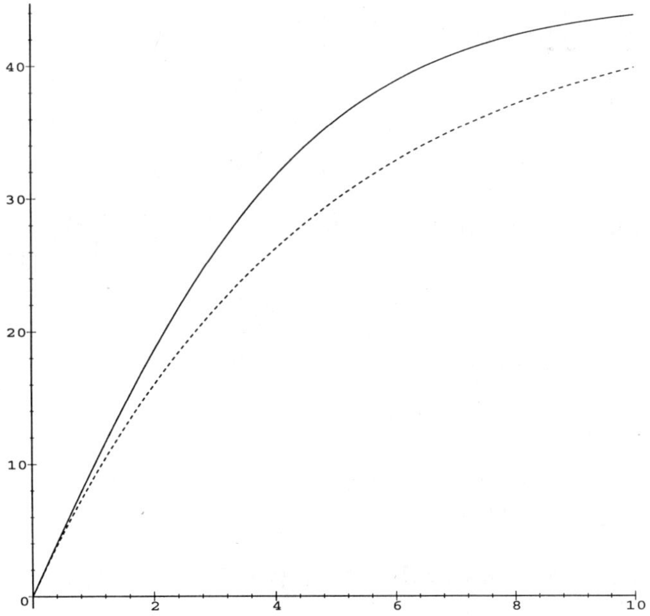

Figure 9.10 Comparison of Linear and Quadratic Air Resistance Terms

Note that we have used the plot option *linestyle* = 2 to distinguish one plot from another.

You can adjust other parameters in either of the differential equations and compare the new results to the old. The ability to change parameters, (or designs) and recalculate results quickly is one of the main advantages of computer-aided mathematics.

The Maple Lab for this chapter gives you some exercises in the solution of differential equations by numerical methods. They are perhaps the most important for applied work, where the ability to measure numerical results is so important.

Simultaneous Differential Equations

Maple can handle simultaneous differential equations. We give an example of a set of two simple differential equations. The equations say that the derivative of each is equal to the other, although in one case there is a minus sign, meaning that the derivative is actually the negative of the other function.

Example 9-21

Solve the system of differential equations

> sde921a :=diff(u(t), t) = - v(t);

$$sde921a := \frac{\partial}{\partial t}u(t) = -v(t)$$

> sde921b := diff(v(t), t) = u(t);

$$sde921b := \frac{\partial}{\partial t}v(t) = u(t)$$

Solution: Suppose that it is known that there is an initial condition

> ic921 := v(0) = 0, u(0) = 1;

Since all coefficients in the differential equations are numeric and there is an initial condition, a numeric solution is available:

> sol921 := dsolve({sde921a, sde921b, ic921}, {v(t), u(t)}, type = numeric);

sol921 := proc(*rkf45_x*) ... end

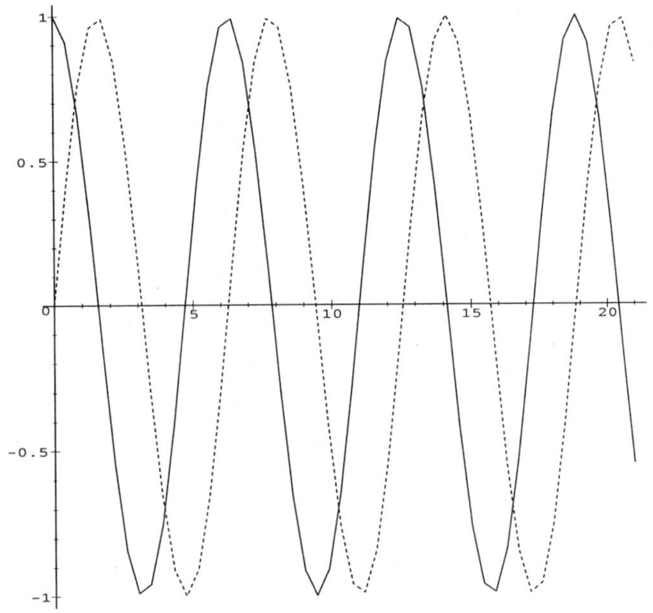

Figure 9.11 Numeric Solution to a Set of Simultaneous Differential Equations

Plotting the solutions gives us a clear idea of the nature of the solutions (Figure 9.11).

> p1 := plots[odeplot](sol921, [t, u(t)], 0 .. 21, color = black):

> p2 := plots[odeplot](sol921, [t, v(t)], 0 .. 21, color = black, linestyle=2):

> plots[display]({p1, p2});

We could increase the smoothness of the curves by using more points (by using the plot option *numpoints* = 200), but the graphs are pretty clearly those of the sine and cosine functions. This system of differential equations is simple enough for Maple to solve exactly:

> dsolve({sde921a, sde921b, ic921}, {v(t), u(t)});

$$\{ v(t = \sin(t), \ u(t) = \cos(t)\}$$

Your Turn. Solve the equations using the same initial condition as in Example 9-21.

> sde921yta :=diff(u(t), t) = v(t); sde921ytb := diff(v(t), t) = u(t);

$$sde921yta := \frac{\partial}{\partial t}u(t) = v(t) \ sde921ytb := \frac{\partial}{\partial t}v(t) = u(t)$$

Answer: ______________________________

The Laplace Transform Method

The power of a change of variable has been frequently emphasized throughout our discussions on techniques for solving algebraic and calculus problems. The Laplace transform is a *change of function*, so it is related to the concept of a change of variable, but it is a more elaborate method. It is the favorite of engineers, especially electrical and control systems engineers.

The Laplace transform of a function $f(t)$ is defined as

$$F(s) = \int_0^{\infty} f(t)e^{-st}dt \qquad (9\text{-}32)$$

The *inttrans* package in Maple automates the process of taking the integral in equation 9-32.

Example 9-22

Find the Laplace transform of

> ex922 := 3 + t + t^2 + sin(t) + exp(t);

$$ex922 := 3 + t + t^2 + \sin(t) + \mathbf{e}^t$$

Solution: The Laplace transform command requires that you list the independent variable in the expression, in this case t, and the target variable, which is by convention s, unless this variable is already being used for another purpose in a particular problem. The command is

> inttrans[laplace](ex922, t, s);

$$\frac{3}{s} + \frac{1}{s^2} + \frac{2}{s^3} + \frac{1}{s^2+1} + \frac{1}{s-1}$$

Notice that the terms, even those arising from the transform of the sine and exponential functions, are rational polynomials. Furthermore, the transform of a derivative of any order reduces to some multiple of s times the function whose derivative is being taken, and some constant terms related to initial conditions. For instance, the third derivative of $f(t)$ has Laplace transform

```
> inttrans[laplace](diff(f(t), t$3), t, s);
```

$$s(s(s\ \text{laplace}(f(t),\ t,\ s) - f(0)) - D(f)(0)) - (D^{(2)})(f)(0)$$

Upon examination of Maple's output above, you see that the derivatives have vanished and only terms in *s* and the Laplace transform remain. This allows us to replace a differential equation with an algebraic one.

Let a differential equation be given by

```
> deq922 := diff(y(t), t$2) - omega^2*y(t) = f(t);
```

$$deq922 := \left(\frac{\partial^2}{\partial t^2} y(t)\right) - \omega^2 y(t) = f(t)$$

Where *f*(*t*) is some (hopefully simple!) function. The Laplace transform of this differential equation is

```
> inttrans[laplace](deq922, t, s);
```

$$s(s\ \text{laplace}(y(t),\ t,\ s) - y(0)) - D(y)(0) - \omega^2\ \text{laplace}(y(t),\ t,\ s) = \text{laplace}(f(t),\ t,\ s)$$

We can isolate the Laplace transform term for the function *y*(*t*):

```
> student[isolate](", laplace(y(t), t, s));
```

$$\text{laplace}(y(t),\ t,\ s) = \frac{\text{laplace}(f(t),\ t,\ s) + D(y)(0) + s\,y(0)}{s^2 - \omega^2}$$

The differential equation is solved "in principle," meaning that if the terms on the right-hand side are all known, then *laplace*(*y*(*t*),*t*,*s*) is known. The only thing missing is a "dictionary," similar to the one we discussed in example 9-10 where we can look through the table of transforms and, if we are lucky, find the function producing that transform.

Example 9-22 has shown that it is possible to produce, at least in principle, a solution to a differential equation using Laplace transforms. We will not pursue the matter further in this book, as it is really an advanced subject, and there exist other methods, such as numerical integration, that are easier to apply to the types of problems we have encountered, or are likely to encounter, in basic calculus.

Your Turn. Find the Laplace transform for the differential equation in example 9-19, then isolate *L*(*s*).

Answer: ______________________________

Paper and Pencil Exercises

PP9–1

State the order and the degree of these differential equations.

(a) $\frac{\partial}{\partial x}y = 2y + 3x$

Answer: Order = _________, degree = _________

(b) $\left(\frac{\partial}{\partial x}y\right)^2 = 2y + 3x^2$

Answer: Order = _________, degree = _________

(c) $\left(\frac{\partial^2}{\partial x^2}y\right) - 2y = 3x^2$

Answer: Order = _________, degree = _________

(d) $\sqrt{\left(\frac{\partial^2}{\partial x^2}y\right) + \left(\frac{\partial}{\partial x}y\right)} - ay^2 = bx^3$

Answer: Order = _________, degree = _________

(e) $\left(\frac{\partial}{\partial x}y\right)^2 + 6x^2y\left(\frac{\partial}{\partial x}y\right) = x^3\left(\frac{\partial}{\partial x}y\right)$

Answer: Order = _________, degree = _________

(f) $\left(\frac{\partial^2}{\partial x^2}y\right) + 6x^2\left(\frac{\partial}{\partial x}y\right) + 5y = 0$

Answer: Order = _________, degree = _________

(g) $(x^2 + 2xy + y^2)dx^2 + 2xy\,dx\,dy - 4dy^2 = 0$

Answer: Order = _________, degree = _________

(h) $\sqrt{1 + \left(\frac{\partial}{\partial x}y\right)^2} = \frac{y}{x}$

Answer: Order = _________, degree = _________

(i) $\left(\frac{\partial^3}{\partial x^3}y\right) + qx^2\left(\frac{\partial^2}{\partial x^2}y\right) - r\left(\frac{\partial}{\partial x}y\right) + (3x^2 + 2)y = 4x$

Answer: Order = ________, degree = ________

(j) $\left(\frac{\partial}{\partial x}y\right) + 2\left(\frac{\partial}{\partial x}y\right)^2 + 3\left(\frac{\partial}{\partial x}y\right)^3 + 4\left(\frac{\partial}{\partial x}y\right)^4 = x^4$

Answer: Order = ________, degree = ________

(k) $\frac{1}{\frac{\partial}{\partial x}y} = \frac{1}{xy}$

Answer: Order = ________, degree = ________

PP9–2

State which of the equations in PP9–1 are linear differential equations.

Answer: ____________________

PP9–3

Use the method of separation of variables to solve these differential equations.

(a) $10x^2dx + 5xy\,dx = 0$ *Answer*: ____________

(b) $xy\,dy - 3(1 - x^2)dx = 0$ *Answer*: ____________

(c) $15y\,dx + 10x\,dy = xy\,dx$ *Answer*: ____________

(d) $\frac{ydy}{dx} = 3x + 2$ *Answer*: ____________

(e) $\frac{dy}{dx} + y = y^2$ *Answer*: ____________

Maple Lab

ML9–1 Radioactive Decay

(a) Solve the radioactive decay problem for radium, which has a decay rate of $k = -1.36\text{e}{-11}$ per second. Set up the differential equation and solve it, putting in the numerical value for k.

Answer: ____

(b) Find the number of seconds it takes for half of the atoms in a given sample of radium to decay.

Answer: ____

(c) Convert this number of seconds to years.

Answer: ____

Hint: The time it takes for half of the atoms in a sample of a radioactive substance to decay is called its *half life*. Let this time be represented by $t_{1/2}$. Then $N(t_{1/2}) = \frac{1}{2}N(0)$, where $N(0)$ is the number of atoms initially present. Then $\frac{1}{2} = e^{kt_{1/2}}$; therefore, $\ln(2) = kt_{1/2}$, $t_{1/2} = \ln(2)/k$.

(d) The half life of carbon-14 is 5,730 years. How much (as a percentage) of the original carbon-14 remains in the mummy of King Tutankhamen, who died around 1350 B.C.?

Answer: ____

ML9–2 The Normal Distribution

A differential equation of extreme importance in the theory of statistics is

$$\frac{dy}{dx} = -kxy$$

This is the differential equation for the normal probability distribution and is defined for all values of x.

(a) Have Maple find the general solution for $y(x)$ and write the solution here.

Answer: ____

(b) If you encounter difficulties finding the solution to the differential equation in (a), observe the message that Maple gives you. What further command is needed to enable Maple to find the solution?

Answer: ______________________________

(c) The initial condition is derived from the fact that $y(x)$ is a probability distribution. In this case, probability theory tells us that $\int_{-\infty}^{\infty} y(x)dx = 1$. Apply this condition to determine the value of the arbitrary constant in (a). What is the value of this constant in terms of k?

Answer: ______________________________

ML9–3 Mechanical Vibrations

In the study of mechanical vibrations, a system consisting of a mass hanging on the end of a spring is studied as an introduction to the study of vibrations. The downward force, in newtons (N), on the mass is $F_d = mg$, where m is the mass in kilograms (kg) and $g = 9.8$ meters per second squared (m/s^2). The upward force is $F_u = kx$, where x is the position of the mass in meters (m) and k is the *spring constant*, in kilograms per second squared (kg/s^2). The two forces balance when $mg = kx$. This is the equilibrium position for the mass, because when the forces are balanced, there is no tendency for the mass to move.

(a) What is the equilibrium position for the mass in terms of the other variables?

Answer: ______________________________

(b) If the mass is given a small displacement from its equilibrium position, there will be an unbalanced force and the mass will move. The differential equation for the position of the mass at any future time is found by applying Newton's second law of motion. The differential equation is, > *deq93 := m*g - k*x(t) = m*diff(x(t), t$2);*. Solve the differential equation and write the solution here.

Answer: ______________________________

(c) There are two constants of integration. To solve the problem completely we must supply information about the position and the speed of the mass at some time. Let these initial conditions be > *ic93 ::= x(0) = m*g/h, D(x)(0) = 0*. Recall that the latter equation is the way of telling Maple that $dx/dt = 0$ at $t = 0$. Solve the differential equation, including these conditions.

Answer: ______________________________

(d) Explain the solution in (b). Why is the solution not a function of time?

Answer: ______________________________

(e) Change one of the initial conditions to $x(0) = 2mg/h$ and solve the equation again.

Answer: ____

(f) Examine the coefficient of t in solution (e). Is the number inside the square root sign real or complex?

Answer: ____

(g) Maple strives for the most general solution based on the information it has been given. Since you know that m, k, and g are real positive numbers, give that information to Maple by issuing the commands, > *assume(m > 0); assume(k > 0); assume(g > 0);*. Simplify the solution in (e) after these assumptions have been made.

Answer: ____

(h) Describe solution (g). Do the oscillations continue indefinitely?

Answer: ____

(i) Add a damping term of the form, *-q*diff(x(t), t)* to the left-hand side of *deq*93. This damping term is a retarding force that is proportional to the velocity. Solve the differential equation using the same initial conditions as in (e).

Answer: ____

(j) So far you have been dealing with symbolic solutions. Find a numeric solution to the equation in (g) by substituting these values: m = 10 kg, g = 9.8 m/s^2, k = 0.1 kg/s^2, and q = 0.1 kg/s^2. Sketch the plot of the solution here.

Answer: ____

(k) Compare the frequency of oscillation in (e) and (j). To determine the frequency of oscillation for the solution (j), one way is to measure the time for a full oscillation on the graph. See if you can invent ways to make the measurement accurate.

Answer: ____

(l) Describe the general characteristic of a damped oscillation by observing the plot in (j). Make sure you have chosen an appropriate time interval for the plot so that the nature of the solution of clearly visible. What is a good choice for the time interval?

Answer: ____

Index